Harvesting the Wilderness

The TYOGA LUMBER COMPANY and ONOTA TOWNSHIP,

a Microcosm of the Upper Peninsula of Michigan

by:
John Parlin

with collaboration:
Bernice Anderson
Jim Carter
Echoe Deibert
Bruce Veneberg
Stacey Willey
Mike Zuidema

Harvesting the Wilderness

The TYOGA LUMBER COMPANY and ONOTA TOWNSHIP,

a Microcosm of the Upper Peninsula of Michigan

by:
John Parlin

with collaboration:
Bernice Anderson
Jim Carter
Echoe Deibert
Bruce Veneberg
Stacey Willey
Mike Zuidema

Edited by Jack O'Malley

Layout by Stacey Willey, Globe Printing
Cover Design by Stephanie Lake, Globe Printing

Soft Cover ISBN 978-0-9852108-1-6

Print Publishing and Coordination
Globe Printing, Ishpeming, Michigan
www.globeprinting.net

About the Cover

Drawing by Bill Botti, DNR (Ret), when interpretive signs were being developed for the Tyoga Historical Pathway in 1985

The authors have striven to use their resources to the best of their knowledge and never intentionally misused them. Opinions expressed or unless otherwise noted, are those of the authors. All profit from the sale of this book will be donated to the Alger County Historical Society.

In Recognition

This book is to honor

Mike Zuidema, DNR (Ret)
Without his vision and hands-on leadership, the narrative of the Tyoga Lumber Company would have become a distant memory. Thanks to Mike future generations can enjoy the area's unique history and biodiversity.

Charles L. Foreman – 1925-2009

C. Fred Rydholm – 1924-2009

TRIBUTE

The creation of this book was inspired by two dear friends and mentors, Charles L. Foreman and C. Fred Rydholm. Each in their own way rekindled my interest in history, encouraged my research, added to my source material and shared with me their love of the land and the people of what we consider to be God's country.

Chuck Foreman was born in Lansing, Michigan in 1925, served in the 87th Infantry division in World War II and was awarded the Bronze and Silver Stars. His wartime experience included the Battle of the Bulge and the liberation of Buchenwald. He returned home to graduate from Michigan State University and joined the Boy Scouts of America as a District Scout Executive. He came to Marquette in 1965 to serve as Executive of the Hiawatha Land Council of the Boy Scouts of America at the time when he began his long love affair with the Upper Peninsula. In the mid 1990's Chuck and I became acquainted through our mutual interest in the Tyoga Historical Pathway. With his long interest in the out-of-doors and his experience with the Boy Scouts, he knew a great deal about trail maintenance and the rich history of the area. He and his wife, Ginny, have a camp on nearby Laughing Fish Point and have been invaluable sources of and connections to the people and history of the area. On many occasions Chuck shared his war experiences with many local students and ultimately contributed to a book of his unit's personal wartime stories. Chuck was instrumental in the creation of Hiawatha Boy Scout Camp, the Marquette Maritime Museum, and the Senior Citizens Center in Marquette. His encouragement and mentoring kept me going and gave validity to my far-reaching quest for sources on this unique area of the Upper Peninsula.

Fred Rydholm was born in Marquette in 1924, graduated from Graveraet High School, obtained a BA degree from Albion College and returned to his beloved U.P. to teach high school science for thirty years. In addition he served on the city commission for 18 years and was mayor of the City of Marquette three times. Throughout his life, Fred was fascinated by the accounts, both oral and written, that told the history of the U. P. In 1989, after years of work, he published *Superior Heartland: A Backwoods History*, a 1600 page compilation of all the material that he had gathered since the age of twenty-one, much of it from his own experiences and from the people who had witnessed the events described. Over the years of our friendship, Fred told me many of the stories he knew, how he found them, and how he went about the process of weaving them together into a history. His last book, *Michigan Copper: The Untold Story*, builds the case for his thesis that groups from Europe and the Middle East from years prior to Columbus had removed large amounts of U.P. copper which was transported back to Europe and the Middle East.

His determination to preserve what he had collected inspired me, Bea Anderson and Mike Zuidema to embark on our own quest to describe the history of Onota Township and to assemble in one volume information drawn from a wide array of sources. I am forever indebted to Fred for his encouragement, his sharing of stories and his setting the precedent for this kind of history book of the region.

These exemplary citizens were dearly loved and will be hard to replace. They set the bar high.

~ John Parlin

Table of Contents

Tables and Graphs

PREFACE

by Jack O'Malley

With their castles and cathedrals, the English are proud of their culture and history. But coupled with this pride is a belief that America and Americans are lacking in history and culture, and moral standing. Historians, particularly those in the Ivy League colleges, have done much to dispel that notion by writing and publishing extensively about the New England states and their origins in the Massachusetts Bay Colony. The same spate of books and articles has not been found in the states of the northern tier especially the Upper Peninsula of Michigan. But that state of affairs is changing.

This volume is a vigorous attempt to address this issue and to articulate the very real and even profound history of this region and its people. It is meant to be a companion volume to the two works of the estimable C. Fred Rydholm: *Michigan Copper - The Untold Story* and *Superior Heartlands. A Backroads History.*

Like Rydholm, the authors of this work combine a textual history of the Upper Peninsula together with a rich array of sources from which their writing is derived. Unlike Rydholm, who concentrated on copper mining and history centered in Marquette County, this book emphasizes the logging industry and its impact upon the area of Alger County.

John Parlin, the principal author, along with Bernice Anderson and Mike Zudeima the one who conceived the need and the ambit of the book, has embraced the Upper Peninsula as his native land and it shows in the array of information he presents in the chapters he wrote.

Another writer, Mike Zudeima, detailed the logging industry and, by implication, the sophistication and complexity and mathematical precision required by that industry. A third writer is Bernice Perkkila Anderson who with remarkable prescience related the very fabric of the lives of her forbears.

Anderson, herself a Finn-American, remarked on the number of peoples attracted to northern Michigan in the glory days of copper and lumber: Germans, Italians, Scandinavians, English, French, Croatians, Slovenians, a smattering of blacks, and a scattering of Irish. She gave an account of the lives of these peoples, now brought together and how they subscribed to simple, humane, communal, old-fashioned values and a way of life that was world-wide and thousands of years old.

The intent of the authors is to provide another significant element in the history of the territory. They hope that some reader, having read these pages, will be inspired to write an economic history of the U.P. Perhaps another will pen a cultural history of this place that will mirror Anderson's rock-solid virtues and values in even greater detail. And perhaps a third will choose to write a political history that will take into account the political hegemony of downstate Lansing. And yes another might delve into whether the main line churches have had a significant impact on the spiritual lives and religious practices of the U.P.

More importantly, a poet should arise like Robert Service, the poet laureate of the Klondike gold rush or A.B. ("Banjo") Paterson, the bard of the Aborigines, or another who will tell the tale, in verse, of the copper rush.

This book is a necessary read for anyone who crosses the Mackinac Bridge in either direction because it lays down the history, culture, and moral strength of the people of the Upper Peninsula of Michigan.

INTRODUCTION

by Echoe Deibert

Lying beneath the waters of Lake Superior, beneath the vegetation of its southern shore, lying there for eons and eons is the evidence, the remnants of an ancient sea; its sands congealed to form the massive Brownstone of an ancient time. The rippled sands at Laughing Fish Point are now frozen into stone. The shoals stretch out a mile or more beneath waters only ten feet deep. Then suddenly the stony shelf drops off to greater depths where Whitefish swim this inland sea, and Lake Trout lie along the bottom's reefs.

This land, these waters drew people here for several thousand years. Copper-seekers passed this point; Anishinaabe came by canoe or overland to hunt the deer, to gather herbs for medicine and food; but most of all they came to net the whitefish, favorite of the People, to feast on them, to dry the rest by sun or smoke, then tote them off to their next camp.

In 1820 James Doty, a member of the Cass expedition, called this river emptying into Superior the Laughing Fish. Now it is called the Laughing Whitefish, this and the point of Laughing Fish Point, the swamp, and the lake upstream, the head waters of the highland springs arise in the heart of what is now Onota Township. The waters spewing up from underneath make all this possible: this story of the land, the people, the rivers and everything that happened here. This place where European settlers came by sailing ships; where the ship's crews helped them float their goods to shore in smaller boats. These people made crude shelters on the rock ledge for the night. Morning caught them heading south, seeking trails, lumber camps, towns grown up for eager immigrants wanting work. They had heard stories of camps and bosses looking for men to cut the trees, to build the kilns, to forge and sharpen tools, to drive the horses, to cook and bake and train cookees who would feed at least a hundred men four times a day.

Men came: some alone, some with families. Finns came to chop and saw the trees; Swedes had dug the iron ore at home, turned trees to charcoal, knew blacksmithing, and had made the finest Swedish Steel. And many came from other nations too. They all had dreams; these men looking for a place where they could build their homes. They left the "old country" for an honest job, a decent wage, the dream of work, of land, a family, a place to be a man again away from serfdom, sorrow, and scanty bread.

And so they came: Swedes to Rock River, Finns to Finn Hollow, and many other peoples scattered in their midst. They all set out to work and build their dreams; one of these dreams became the Tyoga Lumber Company.

AUTHOR'S NOTE

After spending some five years researching and writing this book, the best that can be said is that this represents an amateur's passionate exploration of a well-loved and historically rich community. In the early 1990's, Onota Township residents, Bea Anderson, Chuck Foreman, and I became aware of a common interest in the Tyoga Historical Pathway located on property owned by the State of Michigan and managed by the Gwinn office of the DNR. Mike Zuidema, DNR (Ret), had focused on this area because of its unique biodiversity and had developed a splendid interpretative trail to tell the story of the Tyoga Lumber Company which had been in existence from 1905 through 1907 with only a few traces remaining. While our initial intent was to write a short monograph on the Tyoga Lumber Company, we became intrigued by the wealth of information about Onota Township available at historical societies in Marquette and Munising as well as the Peter White Library and Olsen Library (NMU)in Marquette. This was capped off by the vast amount of information on the internet. The end result is a history of Onota Township in the northwest corner of Alger County from 1600 in the early 1900's with a focus on the timber industry and Tyoga Lumber Company. The appendix is relatively extensive for those who want to read further.

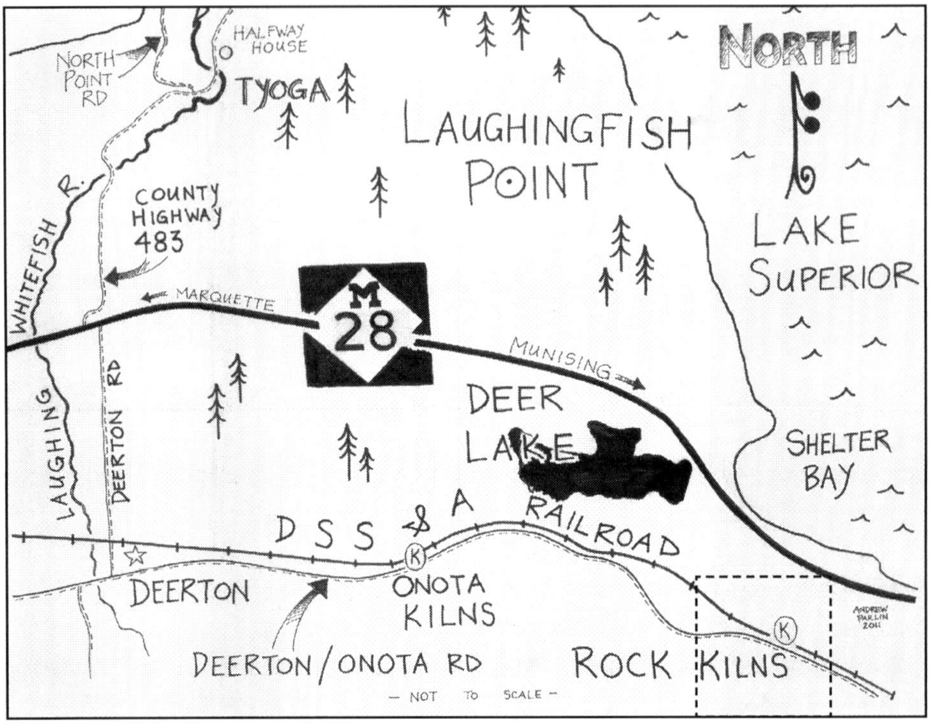

Courtesy Andrew Parlin 2009

Map of northern half of Onota Township.

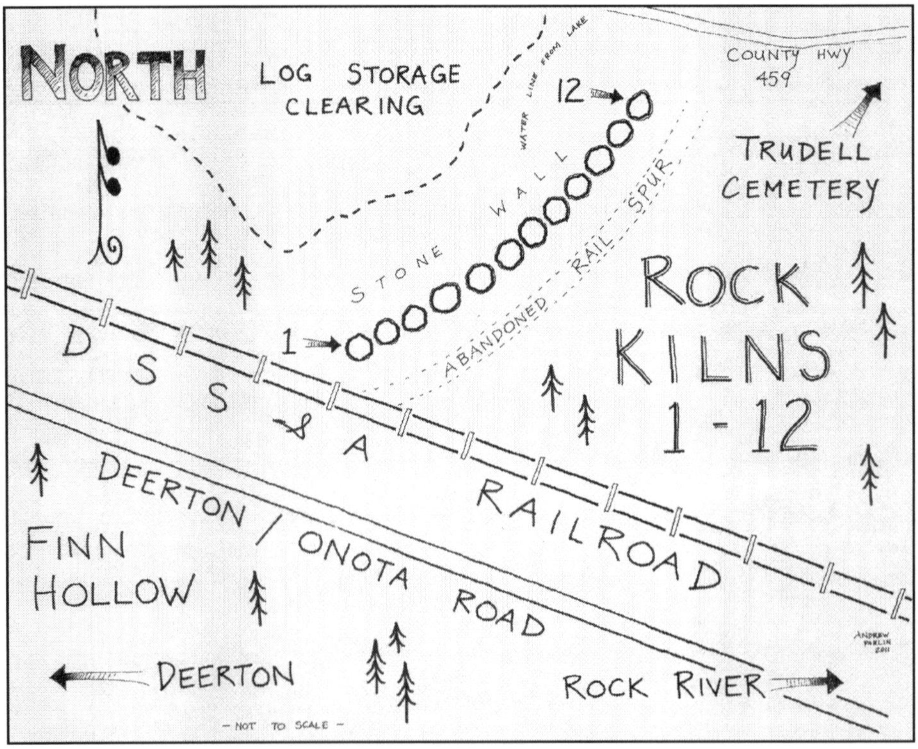

Courtesy Andrew Parlin 2009

A detailed map of Rock Kilns and Finn Hollow.

Time Lines

WORLD, NATION, STATE, AND MICHIGAN UPPER PENINSULA

3 Billion BC	Volcanic eruption pushes copper and iron to surface
10,000 BC	4th and last Ice Age – recession forms Great Lakes (see Appendix Ch. I, Item 1)
5000 BC	Warm cycle allows for wildlife propagation
3000 BC	Various peoples extract copper from open pit mines in Upper Peninsula during the Bronze Age.
900 AD	Ojibways migrate to Upper Peninsula from Canada
1619	Etienne Brüle – first recorded white man on Lake Superior
1660-1690	Jesuit Missionaries arrive in the Upper Peninsula
1670-1850's	Fur trading
1701 AD	Detroit settled
1783	British release Michigan Territory to the United States
1810	Gas light introduced
1820	Lewis Cass/Henry Schoolcraft (geologists) first expedition
1824	Hudson Bay Company established on Grand Island
1825	Erie Canal opens Lake Erie to Hudson River
1829	Welland Canal opens Lake Erie to Lake Ontario bypassing Niagara Falls
1835	American Fur Company opens trades with first sailing and steam ships on the Great Lakes
1836	Michigan gets U.P. and Isle Royale in exchange with Ohio for the Toledo territory
1836-1842	Ojibway Native Americans cede U.P. east of Chocolay River to U.S. government
1837	Michigan becomes 26th State
1840	Douglas Houghton (state geologist) lead an expedition to the Keweenaw and described the ancient copper mines
	Abraham Williams – first permanent white settler on Grand Island

1840's	Samuel Morse invents the telegraph
1841-1846	William Burt, inventor of Solar Compass, surveys U.P.
1844	Houghton and Burt discover iron ore in Negaunee

Courtesy Superior View Studios

Grand Island North Lighthouse
This building replaced the original 1855 lighthouse which was the highest in the world when built

1850's	Bessemer process replace the charcoal furnaces
	First transatlantic cable laid
1854	Treaty of La Pointe, Ojibway cede land east of the Mississippi River to U. S. Government
1855	Sault Ste. Marie Locks open
	Longfellow writes *Songs of Hiawatha*
	Northern Lighthouse on Grand Island, World's highest
1857	First Railroad in U.P., Marquette to Ishpeming
1861-1865	Civil War – 90,000 Michigan troops
1867-1877	Munising (aka Schoolcraft) Charcoal Furnace, iron smelting for Civil War
1868	Michigan becomes nation's largest producer of lumber
1868	City of Marquette fire consumes most of the downtown resulting in an ordinance requiring stone public buildings

1871-1899	Stagecoach Road for mail delivery ("Old" State Road – aka Rte. 94) – Marquette to Munising via halfway house
1876	Alexander Graham Bell discovers the telephone
1879	Thomas Edison develops the light bulb and phonograph
1880's	Pine logging on Laughing Whitefish River begins
1881	Detroit, Mackinac and Marquette railroad completed from St. Ignace to Marquette Glenwood (Onota) Charcoal Kilns – closed 1910 First telephone in Marquette County
1882	First car/railroad ferry across straights
1885	Alger County split from Schoolcraft County Onota Township established Michigan Tech (Michigan College of Mines) founded
1886	D.M.&M. becomes DSSA (Detroit, South Shore, and Atlantic) Railroad
1887	Grand Hotel on Mackinac Island opens
1889	Phone cable laid across straights of Mackinac
1890	National Football League established
1890	George Shiras III from the state of Pennsylvania takes first night photo of wildlife in U.P.
1890's	Wilhelm Roentgen creates x-rays
1890's	J. C. Olmsted plans Presque Isle Park in Marquette for Peter White
1894	Ore boat trimmers strike in Marquette
1895-1922	Munising Leather Co., subsidiary of Cleveland Cliffs Company (CCI)
1899	Northern Michigan University, originally the Marquette Normal School, founded.
1902	Munising Paper Co., subsidiary of Cleveland Cliffs Deerton School opens
1903	Wright Brothers first flight
1905	Teddy Roosevelt sworn in for 2nd term Einstein's Theory of Relativity published
1906	CCI builds model company town of Gwinn, Michigan
1907-1956	Atlas Plywood Co., CCI subsidiary in Munising
1908-1955	Munising Woodenware Co., CCI subsidiary
1908	Ford Model T
1909	Superior National Forest – becomes Hiawatha National Forest in 1931

1914-1918	WWI
1920	M-94, road from Marquette to Munising completed
1922	Last log drive on Laughing Whitefish River – CCI
1927	M-28, clay road Marquette to Munising built north of and parallel to M94
1937	M-28, paved road Marquette to Munising completed
1957	*Mackinac Bridge completed
1959	St. Lawrence Seaway completed

Courtesy J. M. Longyear Library

Laughing Whitefish logging crew, circa 1900
C.D. Johnson with saw (third from right) and
Charles Larson, son of John, fifth from right

** Uses of Mackinac: straights, bridge, island, county, fort and boat race*
Uses of Mackinaw: city, flat bottom boat, heavy cloth, woolen

CHARCOAL KILNS AND TYOGA LUMBER COMPANY

1860	Halfway house built near mouth of *Laughing Whitefish River for sled dog teams carrying mail from Marquette to Sault Ste. Marie.
1867	Charles Schaffer comes to Munising to haul pig iron and charcoal.
1869	George Wagner builds a saw mill and a quarry at Rockport. (Laughing Fish Point).
1870-77	Bay Furnace was operating in present day Christmas then known as Onota.
1870-71	Dock and sandstone quarry in operation at Rockport employed 40 men. A second quarry opens one mile inland connected by narrow plank tramway.
1873	Present road to Tyoga campground completed. Halfway House just north of Tyoga on Laughing Whitefish River was used for mail stagecoach from Marquette to Munising.
1876	Charles Johnston arrives via Canada from Sweden. Three brothers follow soon after. After eight years of working on building a railroad Detroit Mackinac and Marquette (DM&M) Johnston runs logging camps to make ties for the railroad and timbers for mines, concentrated on his property on Rock River.
1877	Bay Furnace (Onota) burns and Onota name transferred to Glenwood between Deerton and Rock River
1879	Union Fuel Co. began building kilns along DM&M
1880	John Burt erects eight kilns at Onota (Glenwood). Lindquist family arrived from Sweden.
1884	Charles Schaffer, aka the "Charcoal King", purchased timberlands and kilns of Union Co. including Whitefish, Rock River and Onota. Chan Brown arrived at Rock River via rowboat from Marquette. He manages the store and post office at Rock River.
1885	Alger County split off from Schoolcraft County
1888	A new boarding house 30 x 60' was constructed at Rock River housing 60 men. Fourteen new cottages at Rock River village were built.
1889	Tram road was completed from mill to Rock River siding (Rock River village)

* *Laughing Whitefish: from Baraga Ojibwa Dictionary (1853)*
"Bapiwin" Laughing and "Atikameg" Whitefish

1892	Lath mill addition planned for the saw mill. New depot was built at Rock River village. Fire at the Johnston settlement destroys the outbuildings and most of the season's cut wood.
1893	Charles Schaffer builds a saw mill at Onota (Glenwood).
1898	Last inhabitants of Rock River leave due to the successful introduction of the Bessemer process (1855) rendering the charcoal-fired iron furnaces obsolete by 1900.
1905	Tyoga Lumber Company was incorporated and settlement is built.
1906	A post office is established at Tyoga which closed July 31, 1907.
1907	Tyoga Lumber Company was sold to Cleveland-Cliffs Iron Co. (CCI) – Deed recorded July 12, 1907. Tyoga ceases to exist as a company.

1869	Bay Furnace Company, 6 miles west of Munising
1875	Renamed Onota, burned in 1877
1836	Renamed Christmas, Michigan
1881	Onota name transferred to current location, 3.5 miles west of Rock Kilns. Onota, "quiet waters" was named by Nah-ben-ash of the Grand Island band of Chippewa

Chapter I

NATIVE AMERICANS of the Upper Peninsula of Michigan

Introduction by Echoe Deibert

This land is why they came. The person who first entered these waters, these woodlands, these meadows ripe with living things; that person came to satisfy a need, sometimes if only to satisfy his desire to see what he had never seen before. But most came to work, to be able to provide what was necessary for their families.

Thousands of years ago the people who came were wanderers. They followed the waterways, seeking game and fish, berries and sap. They were fishers, hunters, and gatherers by turn of season. They too came for the sake of family. There were always lone individuals who struck out on their own, but they were not common. There was benefit in numbers in the event of accidents, sickness, or even in wisdom shared. There was concern for new-borns and for the elderly, therefore they came as family. We have scant evidence of their having been here.

Then came the people who call themselves the Anishinaabe, the Native Americans of Algonquian descent. There is always a reason why one leaves their native lands, the Anishinaabe came for many reasons, but the primary one was food. Their prophets had told them of the white men coming from across the sea. The land along the northeast coast of this country was already crowded with many people all speaking dialects of a similar tongue. The people who hunt, fish and gather what the land provides need lots of area in which to find food, and find materials for dwellings and clothes. The Anishinaabe learned from their wise men that they were to seek a land where "the food grows on water." They were also told that if they did not leave their homeland in the northeast they would die. When the Europeans came, diseases came too and the Native Americans who chose to stay in the lands along the eastern shore soon died. The native peoples had little or no resistance to diseases that had never been known on this continent before. Many clans and tribes who did not leave died by the hundreds, and the thousands, until finally there were tens of thousands dead and dying; many of their names are no longer remembered by anyone, anywhere; and they did not all die of disease. Conquest is such messy business.

~~~~~

Our narrative begins with a speculation by some anthropologists that the U.P. was inhabited by aboriginal peoples probably as far back as 6000 B.C. Numerous remnants, including mounds and artifacts of the Adena people, also known as middle woodland natives or mound builders, indicate their presence in northern Michigan between 100 B.C. and 800 A.D. Because many of the original people of the U.P. were hunter/gatherers and therefore nomadic, there are relatively few artifacts that could be found at their temporary camp sites which would enable archaeologists to document their cultures. Spear points dated 7,000 BC have been found in the bed of Deer Lake in Marquette County.

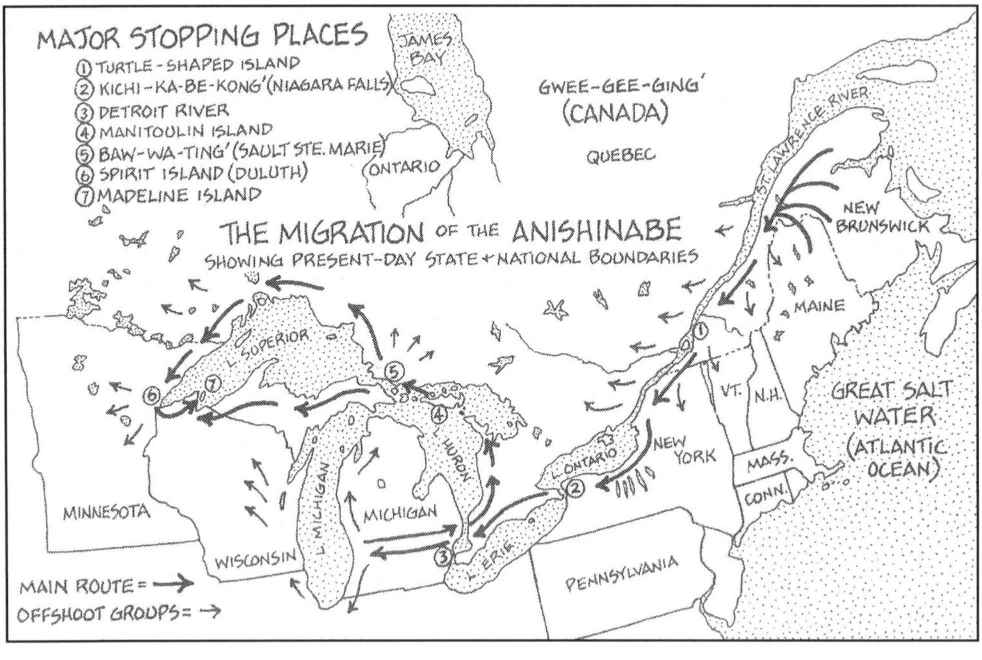

Migration of the Anishinaabe (Ojibwe) from Maritime Provinces to Great Lakes.
The Mishomis Book: The Voice of the Ojibway by Edward Benton-Banai

There are no written records of the migration of Native Americans to this area of Upper Michigan. However there are Native American records kept on birch bark scrolls which the Native peoples have saved for hundreds of years and which their elders can interpret. Native Americans have systematically replaced deteriorating records with new scrolls. Birch scrolls have been used for over 400 years.
~~~~~

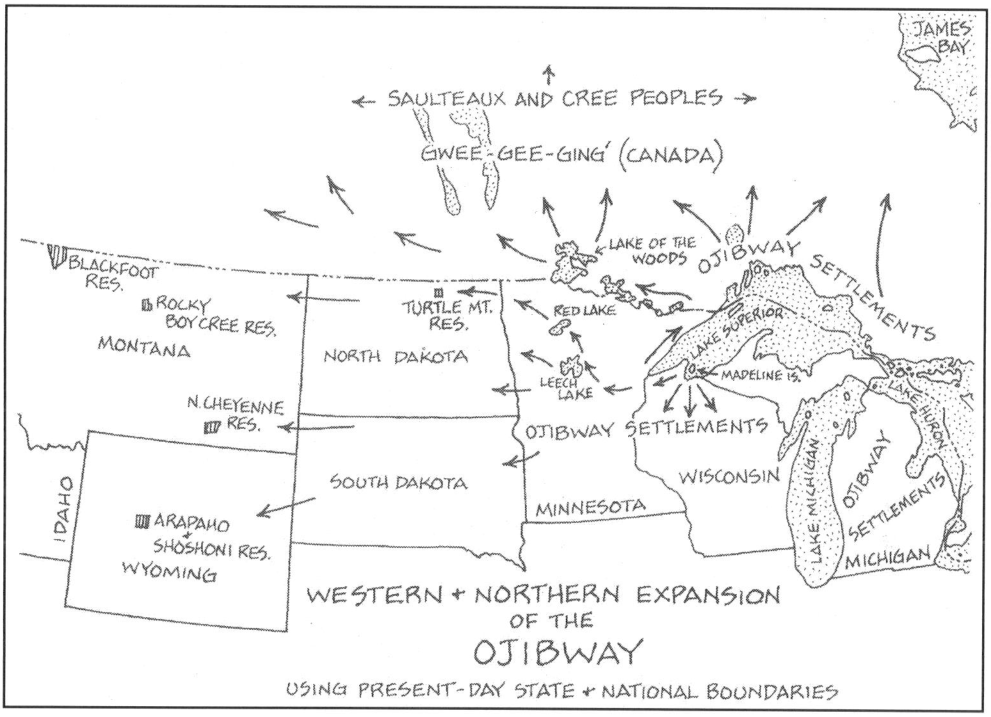

Migration of Ojibwe from Great Lakes.
The Mishomis Book: The Voice of the Ojibway by Edward Benton-Banai

The Ojibwe who do the recording of their history and songs and who are the keepers of these records are members of the Grand Medicine Society, or the Midewiwin. When these scrolls are brought out for ceremonial use, they are called mide-wiigwaas which refers to the "birch-bark scrolls of the medicine men."

The Ojibwe people and their closest kin, the Odawa and Potowatami, know they came from near the Salt Sea (Atlantic Ocean). They know they moved in stages following the spirit of the Megis shell, the sacred shell through which the Creator blew his breath. They stopped at various locations until the people settled where there is "the food that grows on water." This is in reference to the wild rice found in shallow lakes, primarily in Northern Minnesota.(see Appendix Chapter II, Item 10)

Drawings from *The Voice of Ojibway* illustrate the migration beginning in the Maritime Provinces of Canada and spreading to the west, north, and south. (see Appendix Chapter II, Item 9)

By the 1600's, when European explorers began to arrive in the area, the Ojibwe had occupied most of the Upper Peninsula for hundreds of years. The Dakota Sioux were their most dreaded enemies. They came from the lands further to the west which eventually became known as Wisconsin and Minnesota.

The word Ojibway is the preferred name of the people also known as the "Chippewa." This name is thought to mean the "puckering" of the moccasin where it is sewn together from the toe toward the ankle or of a prisoner's face while they are being burned at the stake. Chippewa is a corruption of "Otchipwe." The "t" sound was eventually dropped. The "b" sound within a word required less energy than the more plosive "p" sound. A living language always changes. Native Americans of the Upper Peninsula refer to themselves as Ojibwe or Anishinaabe (First People). The Ojibwe are one of the many tribes who speak variations of the Algonquin dialect. Others are the Ojibwe, Ottowa and Potowatami Tribes. These people are collectively called "The People of the Three Fires."

The Ojibwe in Alger County were sometimes referred to as the Noque, because they wintered near Little Bay de Noc on the Stonington Peninsula of the north shore of Lake Michigan. The Noque Indians were a sub-group of the Ojibwe who lived primarily in the vicinity of Green Bay but also traveled further north seasonally.

In the mid 1600's, the Jesuits came to Lake Superior with the goal of converting the Native Americans to Catholicism. At that time Native Americans major interaction with the Europeans was through the fur trade, first with the French and then with the English. By 1718, many Native Americans had acquired guns which they used primarily for hunting. Despite the numerical advantage of Roman Catholic priests in the region, Rev. John Pitezel and the Methodist missionaries succeeded in converting the Grand Island Band of Ojibwe to the Protestant religion in the 1800's.

Some of the Ojibwe in the northern half of the Upper Peninsula served as guides and manpower to assist Frenchmen who were called "Voyageurs." They were the people who gathered furs from the Indians and transported them to collection centers primarily Sault Ste Marie for the Upper Great Lakes.

Courtesy Superior View Studios

Hudson Bay Company Voyager Canoe, John Jacob Astor's American Fur Company was the dominant player in the Great Lakes after 1700

The coming of Europeans to the new world eventually forced the Native Americans out of their home territories and onto lands designated as Reserves. These were lands the white settlers did not want. As an example, many Native Americans who are of the Grand Island Ojibwe band, live east of Munising on relatively poor land that has been known as Indian Town for over a hundred years.

Although most of the Native Americans from the Alger County area wintered at Bay De Noc (the Noque Indians) because of its milder weather, some Native peoples wintered near the Carp River south of the present city of Marquette and at Sault Ste. Marie. It was not until native peoples of Upper Michigan gave up their hunter/gatherer existence that they and their families stayed in one place or on a reserve. This was encouraged by the European settlers and potential employers. Farming was successful primarily on the northern shore of Lake Michigan where weather and soil were more favorable than the northern U.P. In Alger County this trend toward farming came later with the sale of clear cut land.

W. B. Hinsdale in one of his four volumes of *Archeological Atlas of Michigan* (1931) indicates locations of Native American trails, settlements, and grave sites. It is of interest that he makes no mention of Native Americans in Onota Township, which was the location of the Tyoga Lumber Company in the early 1900s. The Tyoga area was used for hunting, fishing and gathering but there is no evidence of use for semi-permanent settlements, rituals, or burial. The Rock River area was a place favored by these nomadic people as a source for birch bark. As the Upper Peninsula became more settled and connected with the rest of the country, the Ojibwe turned to such occupations as mail carriers, makers of maple syrup for bartering, and as commercial fishermen.

An enlightening discussion of the attitude of the French and later the Americans towards the Native Americans from the early 1700's to the early 1800's prior to the founding of the United States is included in the appendix (Chapter I item 7 *Native American Land Sessions in the United States* by Charles C. Royse, published by the Bureau of American Ethnology in 1899.)

The Native Americans, especially the Ojibwe, often became loggers, or began supplying camps with deer meat. One family had two brothers drown while breaking up the log jams after the spring break-up. Another family had a member build two-thirds of the houses in Manistique, MI in the late 1800's and early 1900's. The hunter-gatherer style of living had waned. They began to learn new jobs such as potato farming, raising live stock, and carpentry. Some still trapped animals for their furs but the animal populations were greatly reduced. Almost all of them still picked blueberries and made maple sugar. Some began to do this commercially. All of them fished for the whitefish, especially in October when they would catch large numbers. They would dry them, smoke them, store them for winter. When this was done, their future seemed more secure.

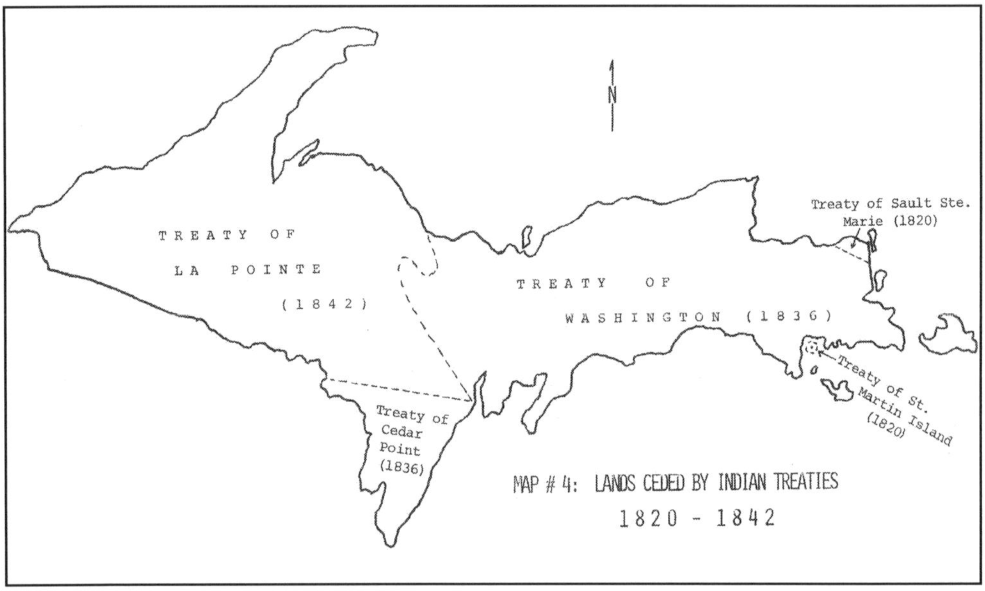

Land ceded by Indian treaties to Federal government 1820 to 1842.

Treaties, their objectives and effects

Between 1808 and 1855, there were at least twenty treaties between the United States Government and the Ojibwe. These are typical examples of unfulfilled and one-sided treaty agreements the Federal government made with the Native Americans. The underlying policy was to push them ever westward leaving only a few small parcels for reservations which were primarily on undesirable land.

Recently some payments from the Federal Government have been forthcoming at a fraction of the current values as they do not include a current inflation factor . The Treaty of 1820 at Sault Ste. Marie and the Treaty of 1836 at Washington had the greatest impact on the U.P. Ojibwe. (Appendix Chapter I Item 1 – Treaties of 1826, 1836, 1842, and 1855)

Courtesy Superior Alger County Historical Society

Above - *Native Americans near Bay Furnace, circa 1870.* **Below,** *Fishing in the St. Mary's Rapids. This area downstream from the Sault Ste. Marie locks was a prime fishing area for Native Americans, guaranteed for life in Treaty of 1820 and then claimed by the Federal government in 1855*

Treaty of 1820 – Lewis Cass at Sault Ste. Marie:

The Chippeway tribe of Indians cede to the United States the following tract of land: Beginning at the Big Rock, in the river St. Mary's, on the boundary line between the United States and the British Province of Upper Canada; and, running thence, down the said river, with the middle thereof, to the Little Rapid; and from those points, running back from the said river, so as to include sixteen square miles of land.

The United States will secure to the Indians a *perpetual right* of fishing at the falls of St. Mary's, and also a place of encampment upon the tract hereby ceded, convenient to the fishing ground, which place shall not interfere with the defences of any military work which may be erected, nor with any private rights.

> This "perpetual rite" of fishing at the falls of St. Mary's was subsequently revoked in the treaty of 1855 at Sault Ste. Marie.

The treaties of 1826, 1836, 1842 and 1855 along with maps are included in the appendix of Chapter II, Item 1.

There is a glimmer of optimism for the economic opportunities for Native Americans with the coming of the Native American casinos in the 20th century. In 1986, the Pequot tribal nation opened the Foxwoods Casino in Connecticut. Currently this is the largest casino in the world with a complex that covers five million square feet hosting more than 40,000 visitors per day. As of this writing, there are now nine casinos in the Upper Peninsula that are beginning to show a significant impact on social services, healthcare, and education not only for their people but for others in the community. The new Tribal Center in Munising provides social and medical services with both modern and "traditional" medicine. The Native Americans are largely responsible for the resuscitation of the American Soo following the closure of Kincheloe Air Force Base in 1977. Whether casino revenue becomes another white man's curse will be dependent upon the wisdom of the Native American leadership in the U.P.

This chapter of history has been distressing to research and write. Treatment of the Native Americans is as devastating as the other major atrocities of our time, slavery in the United States and the holocaust in Europe. As a former physician, reading these treaties reminds me of a diabetic patient who first loses a toenail, then a toe, and then the entire leg, and is then put on dialysis and dies. White settlers brought more than treaties, they also came with alcohol, guns, and devastating diseases including measles, small pox, typhoid, and diphtheria resulting in a Native American mortality rate of 60-70%. Tuberculosis was felt to have been present in North America prior to the coming of the Europeans. Native American were not savages. They have very close communities with morals that are absolute.

After the treaty of 1842, this once proud race was no longer able to live independently off the land and became dependent on Federal government handouts while living in deplorable conditions. It was not until the late 20th century that the Native Americans were able to win legal settlements requiring the Federal Government to fulfill their treaty obligations.

Additional material related to Native Americans is included in Appendix Chapter I, Items 2-10.

Chapter II

EXPLORATION AND SETTLEMENT OF THE SOUTH SHORE OF LAKE SUPERIOR

by John Parlin

Pre-Columbian Expeditions to Canada and the Great Lakes

There is a substantial credible, body of evidence that supports the theory that white men came to Canada and the Great Lakes, probably in several groups, as far back as 3000 B.C. Much of the evidence gathered in support of this thesis is the result of a lifetime of research by Fred Rydholm of Marquette. His thesis is fully elaborated in his book *Michigan Copper, The Untold Story.* He makes a strong case for his hypothesis that people from Mesopotamia came across the North Atlantic, up the St. Lawrence to Lake Superior and Isle Royale where they extracted a huge amount of float copper. Because Lake Superior was higher at that time, water flowed south from the west end of Lake Superior to the Mississippi. Ships laden with copper went down the Mississippi to the Gulf of Mexico, around the Florida Peninsula, and followed the Gulf Stream north easterly back to Europe.

Although there is no written record of these journeys, many stone structures along the described route have been identified as ancient dolmens. These memorial stones or way-marks were used as navigational aids. Dolmens have been found in the Huron Mountains north of Marquette, and on the Keweenaw Peninsula. The continuing accumulation of evidence in regard to how the copper was shipped may well prove this theory. Fred Rydholm is in agreement with the Reverend John H. Pitezel's conclusion in *Missionary Life* in 1857 that "we must leave the reader to his own speculations, respecting the people who, long ago, delved into the mining of Lake Superior for hidden treasure."

In his 1865 book, *Discovery of America*, J. P. Kohl alludes to the possibility that Normans (Norse, North men, and Irish) came to the east coast of Canada prior to Columbus. His theory was that their route was from Europe to Iceland, to Greenland, and then to Newfoundland and Labrador. "If at any point in prehistorical time America received inhabitants from Europe, this would seem to have been the most probable route. Here lies that 'Helluland' (Stoneland – probably Baffin Island) and that 'Markland' (Woodland – probably Labrador) which I have already spoken of as fragments of America, and which, long before Columbus, had been reached and had been named by the Normans of old. This narrow part of the ocean, too, it is, which, in our days has been found to be the most suitable for connecting the two great divisions of the Earth with an electric wire." The modern era of transatlantic cable became a reality in 1866.

More current archeology has settlements on the east coast providing further evidence of the presence of people from northern Europe prior to Columbus.

Kohl concludes his description of this era by saying "it is to be regretted that the earliest history of the French voyages across the ocean is so very obscure. We have no authentic information of the way in which they developed their marine nor how they learnt the secrets of the Spaniards. The maps which they made of their discoveries are lost and the names of their bold heroes, the contemporaries of DaGama and Magellan, are nowhere mentioned."

The Canadian author, Farley Mowat, has written an elaborate description in *The Farfarers* of the European movement of the Norman and Viking eras to North America which echoes Kohl. He maintains that the period of Viking expeditions to eastern Canada extended from about 870 to 1300. In approximately 980 A.D. Vikings led by Eric The Red moved from Iceland to Greenland. He and his followers developed a colony on Greenland of up to 5000 people who farmed and harvested walrus ivory, whale, and seal blubber. In the late 1400's Greenland's climate once again became very hostile because of the "little" Ice Age. Leif Ericson, in addition to Helluland and Markland, describes Vinland as "the land of meadows" which is felt to be an area in Newfoundland dating to approximately 1000 A.D. With the current retreat of the ice covering Greenland, we may well see renewed human activity on Greenland.

Early French Explorers, The Jesuits, and Voyageurs

While the Atlantic coast of North America was settled primarily by the English and Dutch, Mexico and South America were settled by the Spanish. The French took the lead role in exploring Lower Canada, the St. Lawrence, and the Great Lakes. Beginning in 1534, Jacques Cartier made the first of a total of three trips to the area which he named New France. In 1535, he founded the settlement of Mont-Royale, the present site of Montreal.

In Sheldon's Early *History of Michigan*, Jacques Cartier was told by the natives that "there was far away to the westward a country called Saguenay (Saginaw), where there were great stores of gold and copper; and that there were three great lakes and a sea of fresh water so large that no man had ever found the end" [of Lake Superior].

Charles Whittlesey in his *Ancient Mining on the Shores of Lake Superior* states that copper locations were a "profound secret from all Strangers." Native Americans were warned they would die if they revealed the secret. The Jesuit Fathers found silence to be the typical response of the Chippewas.

As with all early transatlantic explorers, the ultimate goal was finding a route to China. On Samuel Champlain's return to France from his expeditions in the early 1600's, he expressed a "likelihood of a discovery of a passageway to China beyond the Lake of the Hurons".

Falling short of their goal, the French explorers turned their attention to the goods they could obtain in St. Lawrence and the Great Lakes.

> Cartier showed the Indians a silver whistle of his steersman, or a golden chain presented to him by a king, and asked them if they had ever seen such metals in their country – if upon this they said they had seen things shining, it was clear that the land was full of gold and silver mines. If they shook their heads on being shown such articles that was not taken as proof that no mines existed, but merely to show that the wicked or jealous Native Americans wished to keep these treasures secret.

Gabriel Sagard, in his 1836 *History of Canada*, states, "The interpreter Brüle (1592-1633) assured us that beyond the Freshwater Sea, Lake Huron, there was another very large lake which empties into it by a waterfall, which has been called "Saut de Gaston," of a width of almost two leagues [6 miles]; which lake and the Freshwater Sea have almost thirty days journey by canoe in length, according to the account of the savages, but, according to the interpreter of Brule's account, they are four-hundred leagues in length."

As judged by Jesuit standards Brüle was not highly regarded by the Jesuits because of some of his less savory personal habits. "Alone in the wilderness, not another European, man or woman, within hundreds of miles of him, it is perhaps not to be wondered at that he adopted the Indian way of life and became as promiscuous in his habits as the red men."

In the *Immortal Scoundrel* by Cranston, Etienne Brüle is depicted as Champlain's agent and is given credit for the discovery of Lake Superior as shown on Champlain's map of 1632 which included all of the Great Lakes.

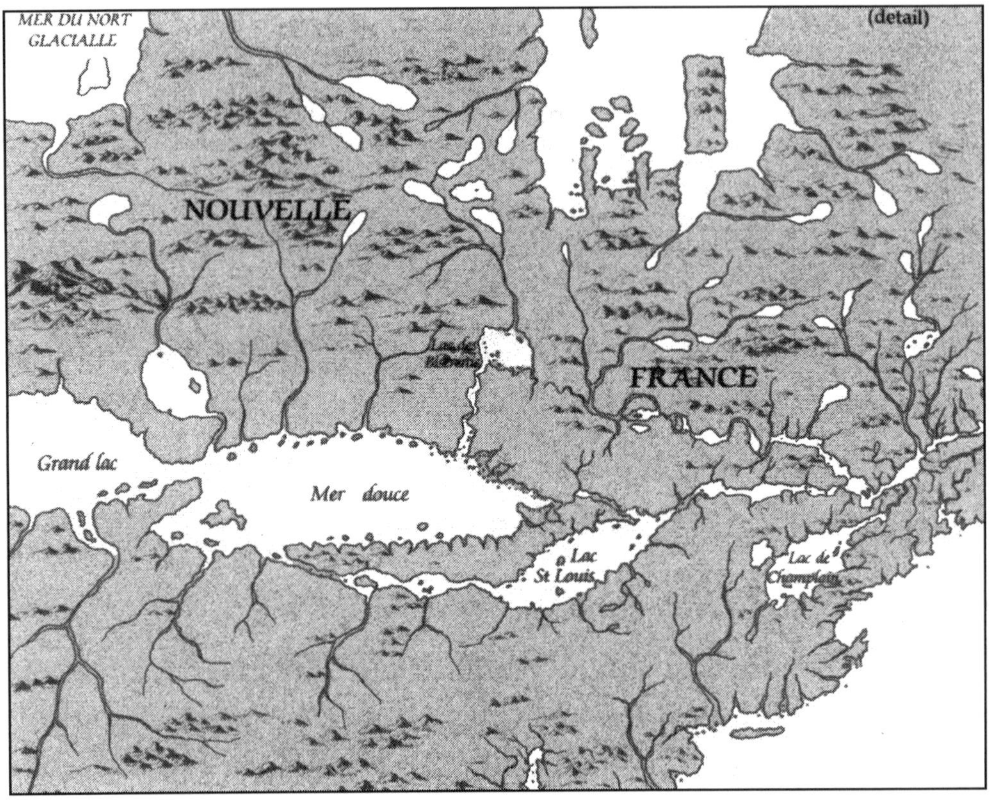

Courtesy Library, Archives Canada/NMC

Champlain Map 1632, Lake Superior labeled as Grand Lac

Samuel Champlain (1580-1635), the founder of Quebec City in 1608 and representative of The Colonization Company of France, became known as "The Father of New France." Champlain had come to realize the Natives held vast information and knowledge about the land which the French both lacked and desired. In an attempt to gain this information, Champlain and the French began to increase their interactions with the native population through promised assistance and trade. The French even went so far as to assist their allies, the Hurons and Algonquins*, in raiding and attacking the Iroquois. Champlain was able to further his connection with the Natives by exchanging youths between the Native Americans and the French groups. This allowed each side to learn from the other and help expand their trust. After waiting years in vain for the Natives' promised help, Champlain and his men attempted to explore Canada's interior on their own. Realizing that Native help was necessary for exploration, he continued to increase his ties with the indigenous people in order to gain the needed Native assistance. He used religion and assimilation as means to bring the Huron Natives into the fold of the French. Champlain and the Augustinian Recollects (later replaced by the Jesuits), then attempted to "civilize" and

* Chippewa or Ojibwe are a subset of Algonquins.

Christianize the Natives. The French even went as far as to marry French men and Native women in order to spread Christianity.

The next 150 years (1650-1800) were an active time for Jesuits who formed settlements and worked to convert native people to Christianity. During the same period in time, the French Voyageurs continued to develop an active and lucrative fur trade licensed by the French merchants in Montreal.

J. G. Kohl in *Discovery of America* states "We are greatly indebted to the Jesuit missionaries for further disclosure of the districts around the St. Lawrence. They were, for the most part, brave and well-informed men, who, now traveling on foot, now in barked canoes of the country extended their excursions far to the north. Almost all of the subsequent discoveries of importance were under their direction and from their careful reports and knowledge of a large portion of North America has been derived.

And, in especial, they were the first to explore and make known that wonderful basin, that lake system of the west of Canada, in all its extent and capabilities."

In 1668 at the time of the founding of Sault Ste. Marie, the Jesuit Father Jacques Marquette (1637-1675) remarked "here, where the evergreen of the unbroken forest was contrasted with the matchless beauty of the foam-crested waters, and handiwork of Deity was everywhere strikingly manifest, the weary Voyageurs found a settlement of 2000 of nature's own children, the hospitable Chippewas. "The chiefs of the tribe gave the travel-worn missionaries a cordial reception, Sheldon continues in *Early History of Michigan.*

> We welcome you as brothers and we will profit by your words,' was their assurance, when the object of the long wearisome journey had been explained. The Chippewas informed their guests that, beyond the foaming rapids and the clear, placid St. Mary's River above, was a lake, called by them *Gitchi Gomee* (Great Water), larger than either Lake Huron or Lake Michigan, which was then named Lac des Illinois by the French. Beyond the western limits of *Gitchi Gomee**, the country was said to be destitute of trees, while countless herds of deer and buffalo roamed over the vast prairies. But that favored portion of the Earth was inhabited by the Sioux (a.k.a. Dakotas), a warlike band, between whom and the Chippewas have long existed a deadly hatred. Thus were those choice hunting-grounds, to the Chippewas, like paradise to our first parents, protected from intrusion by the weapons of destruction.

Missionary outreach was intensified by Jesuits René Mesnard (1605-1661) beginning in 1660 and Claude Allouez (1622-1689) in 1666. Soon after his

---

* Gitchi Gomee. "great water" is the Ojibwe name for Lake Superior Gitchi Gummi is a later spelling.

arrival at Sault Ste. Marie, Mesnard went up the river in a birch canoe and coasted along the southern shore of *Gitchi Gomee*, reaching the head of the Keweenaw Bay, to which he gave the name St. Theresa. The day of his arrival being the anniversary of that patron saint. Sheldon adds that on a later trip to Western Lake Superior, while his Native American guide was carrying the canoe for a portage to Lake Superior, Father Mesnard "wandered into the woods, and was never again heard of."

Sheldon relates that Claude Allouez traveled with an entourage of 400 Native Americans. By this time the French no longer referred to the indigenous people as natives or Native Americans (aborigines), but "savage companions," an unfortunate moniker to describe a wild and uncivilized human being. His exploration extended along the south shore of Lake Superior between the Sault Ste. Marie and Fond du Lac, the area between Duluth and Grand Marais, Minnesota. Father Allouez, along with Father Marquette, is credited with creating the first map of Lake Superior (1668-72) titled *Lac Tracey ov Svperievr.* Upon return to Sault St. Marie, Allouez announced that

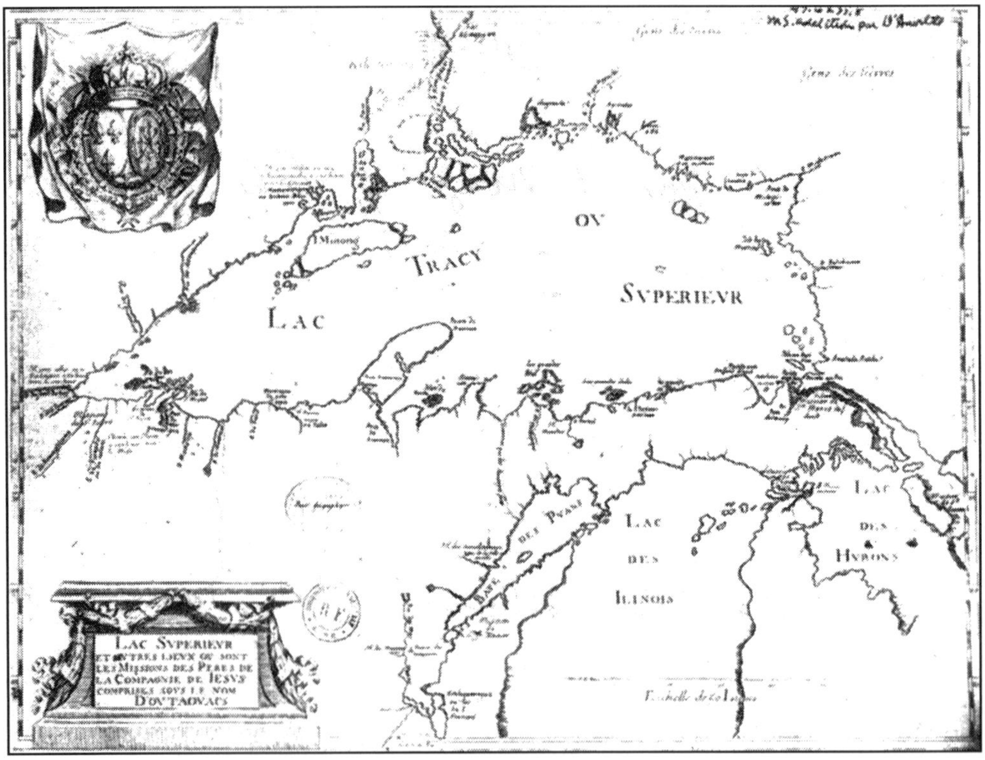

Courtesy Illinois State Museum – 2000

Jesuit Map of Lake Superior – 1671
Note this map includes Au Train River and Lake.

"this lake, shall henceforth bear the name of M. de Tracey, in token of the obligations of the people of this region owed to him as viceroy of France."

When one considers that these early French explorers were not engineers and that to

> note the geographical features of the country formed no part of their requirements, this map is regarded as a remarkable production. The whole coast, 1600 miles in extent, as well as the islands, were explored...the savages respect this lake as a divinity, and offer sacrifices to it because of its size, for it is 200 leagues long and 80 broad [600 x 240 miles] and also in consequence of it furnishing them with fish upon which all the natives live, when hunting is scarce in these quarters.

That the discovery and mining of copper in those regions is not a recent event is supported by Sheldon's comment in 1856 that

> it frequently happens that pieces of copper are found, weighing from 10 to 20 pounds. I've seen several such pieces in the hands of the savages; and, since they are very superstitious, they regard them as divinities or as presents given to them to promote their happiness, by the gods who dwell beneath the water. For this reason, they preserve these pieces of copper, wrapped up in their most precious articles. In some families, they have been kept for more than 50 years; in others, they have descended from time out of mind, being cherished as domestic gods.

Transport of goods and people by water was a necessity prior to the development of the railroad and roads because of the daunting terrain of the interior. With the coming of the fur trade and the need to transport large quantities of pelts, the Voyageurs developed a 32 foot expanded version of the Native American canoe capable of carrying 12 men and cargo.

By the late 1700's, birch bark canoes had evolved into the French bateau, a flat-bottomed wooden boat measuring up to 58 feet, capable of transporting 10 tons. The English counterpart was the Mackinaw which became the common fishing boat of the 1800's. The Mackinaw was also of wood and flat-bottomed, measuring 18 to 24 feet. One or two schooner-rigged sails were frequently added. With the opening of the Soo Locks in 1855, these smaller craft were eclipsed by larger sailing and steam vessels.

Courtesy Canadian Heritage Gallery 1999

Fur trader Bateau 1897 a flat bottomed boat up to 58 feet that could carry up to 10 tons and sometimes included a small sail

Alger County
The "Modern Era" Begins - 1800 – 1900

By the late 1700's, the American Fur Co. had established a trading post at Bay Furmada on the mainland south of Grand Island. By the last quarter of the 18th century, writing in the *Rites of Conquest* Ireland observed that "in rum they [the British] had found the perfect trade good [with Native Americans]: cheap, addictive, and immediately consumed." By 1800, the British had taken over the fur trade. By 1835 both sailing and steam-driven vessels were utilized by the fur trade on the Great Lakes.

The Hudson Bay Company had its first outpost on Grand Island in 1824. Abraham Williams lived year-round on Grand Island making him the first permanent white settler in the area.

Michigan became a state in 1837. Schoolcraft County followed in 1843. In 1885, Schoolcraft County was divided. The east half kept the Schoolcraft name. The western half became Alger County.

There were a number of expeditions along the south shore of Lake Superior. Douglas Houghton, who had been appointed Michigan State Geologist in 1838, lead an expedition in 1840. This description of the copper he found on the Keweenaw Peninsula precipitated the cooper rush that was to last for the next 60 years. The famous "Copper" expedition in 1840 was also chronicled by geologist, surveyor and explorer Bela Hubbard. He included drawings of the shoreline between Tahquamenon River and the Madeline Islands. (see page 142) Charles W. Penny also kept a valuable diary of the exploration.

Another exploration Houghton lead in 1845 ended with his drowning in a storm near Eagle River harbor which is on the southwest side of the tip of the Keweenaw Peninsula.

Following his tenure of territorial governor of Michigan from 1805 to 1831, Lewis Cass went on to become Secretary of War and head of the Indian Department. By 1836, Henry Schoolcraft had become Michigan superintendent of Indian Affairs and was instrumental in a number of the treaties with Native Americans in the 19th century. (see chapter II)

William Austin Burt (US District surveyor) completed his survey of the Lower Peninsula before 1841. In that year he began his survey of the Upper Peninsula eventually including the copper area of the Keweenaw. In 1844, iron ore was discovered in Negaunee thus beginning a long history of mining which continues to the present day.

The first written description of the mouth of the Laughing Whitefish River in 1855 by Schoolcraft relates that "a curious flux and reflux is maintained. From there is a line of sandstone coast 'til reaching the bay of the Chocolate River." A subsequent excursion from Grand Island to Marquette is described in *Harper's New Monthly Magazine* in May, 1867.

> The wind played a fast-and-loose game with our sails, and teased us along a mile or so an hour, just enabling us to go ashore at Laughing-Fish Creek and cook supper, and build a lean-to of poles and boughs, with a bed of pine and hemlock, when the sun set with a glorious display of colors. The color of the water of the lake is a clear blue, and where the sunlight illuminates it with the slanting morning ray near the cliff, or overshadowing forest, it becomes a cool bright green, nearly like lemon yellow shaded with a pale blue. The night was bitter cold for July, and we built a rousing fire and fared sumptuously on roasted fish, fried fish, good bread, and coffee, with huge lumps of maple sugar and good sweet milk.
>
> There joined us for the evening two men [Mallette brothers?] who lived near, on a point of rocks, in a snug house with garden and stables for cows and pigs and fowls, and who seemed delighted to find several pairs of fresh ears into which to pour their sidesplitting jokes and adventures of summer and winter; giving us in such brilliant brief extracts a complete history of as much of their lives as they cared to have us know. Lemm said next day that they were a curious set. The old one came from the East; had plenty of money and everything, seldom went to

> town or any where, fished a little, hunted less, but was always bragging about his rifle, 'when he knew so and so', and the like. The reader must imagine Lemm's conclusion to be a very modest reference to his own trusty gun.

Much of the iron used to forge guns and weaponry for the Civil War was smelted from iron ore using charcoal as fuel. Because of an abundance of iron ore and timber, the smelting of iron became a major industry in the U.P. in the second half of the 19th century. Schoolcraft Furnace in Munising was one of the earliest of the many furnaces created to meet this demand. Charcoal production consumed 450,000 acres of Upper Peninsula timber between 1860 and 1900. By comparison Hiawatha Forest (East and West) has approximately one million acres. The 1870's in Alger County witnessed the coming of the timber industry, railroads, roads and the export of long square timbers to the East coast and Europe for ship masts.

By 1900, Cleveland Cliff's Iron became the dominant land holder and employer for mining and timber industries primarily in the central U.P.. Offshoots of the CCI forest division include Munising Leather, which used hemlock bark in its tanning process, Munising Paper which is currently Neenah Paper, Munising Woodenware and Atlas Plywood. By the end of the 19th century, Alger County had been converted from ancient forest land to an area with active timber and related industries. With the disappearance of the forests, farming and other occupations became much more prevalent after the turn of the 20th century. The traders corridor opened by the French missionaries led ultimately to the discovery of vast stretches of timber and mineral resources. Despite the seemingly endless quantities of undeveloped land, it took less than 100 years to transform a wilderness into a resource for human consumption.

Chapter III

This Land of Tyoga

Introduction by Echoe Deibert

By the late 1800s the White Pine was pretty well gone. Then they began the cutting of the hardwood trees. This was a major business as iron ore mining began to develop. In order to make pig iron from the rock-bearing ore, they needed charcoal because it burned at an intense heat to melt the rock and release the iron. They made charcoal from the hardwood; then they had to transport it to the blast furnaces. Soon there were dray lines, and then there were railroads. There were kilns set up near the hardwood sites. Change came galloping relentlessly to meet the demands for more trees, more charcoal, more iron shipped to the steel mills to the south and east.

As adaptable as the Native Americans were, so were the Europeans. They were willing to learn new jobs. The Europeans always had the advantage because of prejudices against the Native Americans which were held by many people, especially the employers. One immigrant man was different. He hired both Native and African Americans to work in the woods for him. For this he received a visit from the Ku Klux Klan during the 1920s. Fortunately they were full of bluff.

In the place we call Tyoga* at the turn of the last century there was a village, a railroad spur, and a lumber company. Most of it is lost in time with few traces left of its existence. When one lumber company moves on to harvest other forests, it doesn't move very far; often just a mile or two away. Then business begins again. South of Tyoga is an area that was logged next. You'll read about it as you read your way from page to page.

The dream of Tyoga is still alive. Now we visit it and follow a well-marked trail that "shows and tells" its history. Many of descendants of the people who spent time there in its hey-day still live in the area. Most are hardy, friendly, hard-working, willing to spin stories and spill them into listening ears. In this book you will learn about some of them, and the way they lived. It will be worth your while. You may even recognize them should you dig into your family's past. Where did you come from long ago?

We all came from somewhere.

---

*Tyoga is said to be a Native American word meaning "meeting of two rivers" or "at the forks." Tyoga is not to be confused with Tioga, Pai, Tioga River west of Marquette, or Tioga Tavern, a local watering hole on M-28 just west of Tyoga.

THE TIMBER INDUSTRY AND THE FOREST OF TYOGA

by Bruce Veneberg, Michigan DNR (Ret.)

Europeans came to the land of Tyoga for many reasons: one family said a relative of theirs in Finland had gone down to the ship to see his uncle off for America and just got on board. Another man said his grandfather came because life was so hard in Sweden. Like most countries who had a king over them, all wild game was called the king's property. Another said there was a compulsory draft and every man had to serve in the military at the whim of the ruling regent. Many came to find work so they could have food enough to feed a family. By the late 1800's America had become their dream.

Iron ore was discovered in Michigan's Upper Peninsula on September 19, 1844. Beginning in 1841 William Burt's survey made it possible for land to be purchased at government land offices. The boom to harvest timber and mine iron and copper was ready to begin.

Courtesy Superior View Studios

Square-cut white pine as large as 4 feet x 4 feet and 120 feet long. Many of these were used for ship masts in both United States and Europe.

Courtesy Superior View Studios

Long drivers with peaveys moving spruce for pulpwood downstream to a mill. The peavey is a cant hook with a painted end. The pike pole was also used by drivers.

Michigan became a state in 1837. This was followed by the Homestead Act of 1862. Some wealthy men gave money to others who were to go out and file a claim on a section of land. A section was 160 acres. The filing fee cost $26.00. That man then was required to build a cabin with at least two windows in it, and to live there for no less than five years. (The lumber rights went to the man who loaned him the money.)

The lumber barons bought up all the land they could. The way of life shifted for everyone, Native Americans as well as Europeans, as progress, purchase and providence began to unfold. Logging camps began to pop up all across the U.P. The cutting of the big pines began in earnest. Saw mills grew up overnight. Some small towns had as many as five or six mills within the city limits. Everything changed for everyone.

Pennsylvania lumbermen created the Tyoga Lumber Company in Alger County in the year 1905. They had sized up the lands on Lake Superior's south shore and figured the hardwoods growing there would be a wise investment. They chose a location in Alger County near the mouth of the Laughing Whitefish

River. The river mouth and nearby Duluth, South Shore and Atlantic Railroad would provide ready access by both ship and rail to markets around all of the Great Lakes and beyond, even as far away as Europe.

The Pennsylvanians were not the first lumbermen on the scene along Superior's shore. The first to arrive were the adventuresome entrepreneurs best exemplified by Robert Dollar. Dollar started with nothing, worked diligently, and built a logging firm that included a mill on the Tahquamenon River which became known as Dollarville in Luce County just west of Newberry. His earliest timbering jobs were just west of Grand Marais and also at Au Train, both of which are on Lake Superior's southern shore where water transport was easy. Prime red and white pines were sought out. The subsequent logging operations consisted of, at least in part, hand-squared timbers which were shipped overseas to hungry markets in England, especially valuable for ship's masts.

The next big wave of logging along Lake Superior had to await the arrival of logging railroads. The first railroad was the Burt and Gamble built east of Grand Marais in Alger County around 1881. Grand Marais became a boomtown of 3,000 by 1900, decreasing by 90% in the year 2000. Other parts of the Upper Peninsula, drained by streams flowing into Lake Michigan, saw intense activity in earlier years because these rivers had more volume and could be used for floating logs to sawmills along the northern shore of Lake Michigan, especially in the Manistique area. The streams draining into Lake Superior were mostly smaller and shallower, making log drives impracticable. The Tahquamenon, Dead, and Ontonagon Rivers were the major exceptions as they were large enough to support repeated log drives.

The Laughing Whitefish River area did not contain sufficient red and white pine to warrant investments in river or railroad logging. The exploitation of the stands of virgin hardwood and hemlock had to wait its time to play its role in the economic development of the Upper Peninsula. In the year 1905, that time had arrived.

The forest was impressive in its own way. The land was covered with a mix of species, but was dominated by sugar maple (*Acer saccharum*). Native Americans recognized the maple's value in another way. Each family group had locations where they went every year to boil down the sap into maple sugar. This was a major item in their yearly quest to meet their needs nutritionally and as a flavoring for foods and tea. Their favorite locations were often adjacent to lakes or streams. These water ways were their main means of travel to and from their traditional sites. In this way future loggers followed suit but their use was always to float their logs down river.

The other deciduous trees in Tyoga's hardwood stands were red maple, yellow birch, elm, basswood, and cherry. White ash was present in the uplands and black ash on the wetter areas. Ironwood, also known as hop hornbeam, *Ostrya virginiana,* with the brown scaly bark was capable of attaining saw-log size and could be found scattered about, although it never was used for sawed lumber. Ironwood was very strong and was often shaped into handles for tools, and also for wagon and sleigh parts. The beech found there marked its westernmost boundary in the Upper Peninsula.

Hardwood stands also contained significant amounts of hemlock. Hemlock lumber is similar in nature to red pine. It is strong and very hard when dry. Unfortunately, it had to compete with the more commercially available red and white pine. Pine lumber was much easier to mill and shape and thus was always favored. Hemlock also had the frequent trait of having an internal defect called ring shake where the wood separated along annual growth rings. Hemlock with ring shake simply fell apart or split when sawed into boards. As a result, the hemlock stands were usually bypassed by early loggers.

Courtesy Superior View Studios

Hemlock bark taken by sled for use in the tannery in Munising.

Hemlock was a significant component of the hardwood stand at Tyoga although hemlock was rarely used for lumber. There were limited markets for the hemlock bark which was used for tanning leather. The trees were cut down in the spring during the time the tree was growing which made the bark loose and more easily removed. The bark was peeled off with a spud, a slightly curved steel chisel-like tool with a wooden handle. It was then cut into 4-foot lengths, and placed in neat piles usually about four feet high. The bark was then hauled to the nearest railroad by horse and wagon in the summer or by sleigh in the winter. The hemlock tree itself was usually left in the woods to rot. In 1902 the Munising Paper Mill was organized to make paper out of hemlock, just one more example of the timber industry striving to eliminate waste. Hemlock added some value to the Tyoga Lumber Company's coffers but by itself would not have been a profitable venture.

In addition to hemlock, hardwood stands also contained a minor number of the much favored and sought after white pine. The white pine growing in hardwood stands were usually the biggest and the best in quality and often were rated as "Cork Pine". They were regarded as the best of the best. Cork Pine floated higher in the water than did smaller diameter white pine that grew on the sandy soils out on the plains between Munising and Grand Marais. The Cork Pine had a higher ratio of lighter heart wood in the log, which along with the thick cork-like bark, was the reason it floated higher in the water. Early lumbermen sought out this Cork Pine and even ventured into hardwood stands where it was economically feasible to do so. Although the Pine Era had ended by 1905, there were still some Cork Pine left to cut.

White spruce was occasionally found lightly scattered about in hardwood stands. They were tall, straight trees, easily sawn into lumber. Log decks under close scrutiny as shown in old logging photos gives witness to the presence of white spruce that were cut during the Pine Era.

Many generations of hardwood and hemlock occupied the Tyoga timberlands. Each generation nurtured the next by enriching the once thin glacial soils. The maple trees that the Pennsylvanians saw in 1905 were primarily 2-3 feet in diameter. Yellow birch, elm, hemlock, and white pine often grew larger than they do today, simply because they hadn't been cut over previously. Some pine had diameters of 4 to 4½ feet which was not uncommon. Timber crews measured the diameter of a tree at a level of 4½ feet above the ground.

Bob Graves, retired logger of Shingleton, MI, relates that he cut a sugar maple in the early 1950's that was 4 feet in diameter that was located about 10 miles northeast of Munising. The log had 1,000 board feet in it and weighed nearly 4 ton. It was delivered to the Munising Woodenware plant for manufacturing into bowls. It took the Woodenware's D9 bulldozer, Caterpillar's largest machine at that time, and another machine from the adjacent Atlas Plywood Company to unload the huge log and move it into the mill. The Woodenware superintendent had complained about small logs being delivered but was silenced after moving this monster.

"Cork" pine, cutting the giant white pines up to 200 feet and 300 years old, began with an ax under cut and completed with a crosscut saw.

Hardwood trees were tall and if straight enough would yield 2-1/2 to 3 sawlogs. Each sawlog in those days was considered to be 16 feet long. Each hemlock tree contained fewer logs because it retained more top limbs. The white pine that survived until 1905 was taller than all the other trees in the stand. Commercial heights of 4 logs were not at all unusual.

The south shore of Lake Superior is subject to winds sweeping in from the northwest off the plains of Canada. An "Alberta clipper" can easily blow down decaying old-growth timber. Before the arrival of the lumberman's ax, wind was the primary source of hardwood regeneration. A large tree or group of trees would be blown down and its prolific neighbors would quickly re-seed the opening which had been created. Violent winds could blow down entire stands of hardwoods. Many acres were blown down in such a storm in the Chapel Lake area of what is now the Pictured Rock National Park of central Alger County. Richard Kloptic, a retired lumberjack, had lived out his last years in the woods not far from there, he knew what the winds could be like in a big blow. He had seen the results. Huge trees looked like match sticks thrown about over large tracts of land.

No matter what type of storm, the evidence of previous blown-down trees can be found in most hardwood stands by the uneven soil mounds found scattered about on the forest floor. The mounds were created when the blown down tree and its roots rotted away adding to the soil that had already been pulled up. A cavity was created when the roots were jerked loose. The lumberjacks called them "cradle knolls." He could lay down in them and use part of the bank as a pillow. Its sloping sides fitted nicely along the curve of his back. This was only a tongue-in-cheek reference because no respectable lumberjack took time to sleep in the woods.

The natural forces that had created the forests of Tyoga perfected their choices over the millennia. The forests were powerless in the face of the lumberjack's relentless ax and their two-man cross-cut saws. The future of the Tyoga Lumber Company looked bright.

Logging At Tyoga

The first step in the logging process began before the lands were purchased. Landlookers, men skilled at evaluating timber volume, were hired to size up the amount and quality of the timber. They also mapped the area and noted the logging problems that needed to be considered. There were a considerable number of these men to be found wherever logging was active in the Lake States. John M. Longyear and "Moccasin Bill" Oliver were local landlookers who could have been involved in the timber cruising phase of Tyoga. Other of Tyoga's landlookers were more than likely nameless men who plied their trade and left no record of their contribution.

Courtesy J. M. Longyear Library

Courtesy Mary Braamse Edgar, grand-stepdaughter of Bill Driver

Left, *John Monroe Longyear (1850-1922), photo circa 1870's, a landlooker, primarily in western half of U.P. and Keweenaw.* **Right,** *William "Moccasin Bill" Oliver, Jr. (1852-1932) A timber cruiser in the central U.P and New York chronicled his work in a diary with daily entries between 1874 and 1931.*

The landlooker's work demanded an understanding of surveying practices. The Upper Peninsula had been previously surveyed by U.S. government contract surveyors. Tyoga forest lands were first recorded in August of 1841 by William A. Burt who surveyed the township or the outer boundary of a 36 square mile block. The interior lines of this township were done by Alfred Millerd in June of 1852 and checked by William Burt in 1854. Section lines were blazed around each square mile. Posts and scribed, blazed trees were located at each of the four corners of the section and at points one half mile in between. The future landlookers knew how to find these corners, even though the work had been done 50 to 60 years previously. The landlookers would also have known that excellent hardwood timber existed on Section 35 and 36 of the Twp 48N/Range 22W based on notes recorded on the original Burt land survey – "Very good quality of land and supports a heavy growth of sugar maple and hemlock." This timber must have been exceptional as similar remarks were rarely noted by the surveyors.

Once the selected corners were located the landlookers could find any 40 acre parcel in a section, which is a one mile square area. They kept track of distance by counting paces or using a 66 foot surveyor's chain. A tally of the number of paces was kept on a green hardwood twig by making a series of bends; one bend per 100 paces for example. The bends could be counted and they would then know how far they were from their starting point. While using a compass to determine direction of the pacing, they counted trees in a strip of land usually 33 feet wide by 1320 feet long, equaling two strips per 40 acre parcel. The species, diameter, and number of logs per tree were recorded in a tally book. The recorded data could then be extrapolated into the timber volume per each 40 acre parcel.

The Tyoga lands totaled 7000 acres. A good landlooker could cruise 160 acres per day under ideal conditions; only about half that amount could be covered in the winter when snow and snowshoes slowed their pace. At that rate, it would have taken forty-four days to examine these lands in summer and eighty-eight days in winter. Usually two men worked on a project; one ran the compass and paced while his partner did the cruising. Larger groups of four and six men were common on bigger jobs such as the one at Tyoga.

Courtesy Superior View Studios

Hungry loggers – big meals

Courtesy Superior View Studios

Bunkhouse for smoke, conversation, and music from the old country

Courtesy Superior View Studios

Dining room ready for breakfast – another day begins

Landlookers worked all twelve months of the year. Spring and early summer were the months most dreaded because of the unrelenting attacks of insects. The men camped out on the project area, preferring the autumn season, as it was cooler and there were fewer insects. Also, when the leaves fell, visibility was excellent in the woods. Winter work was difficult, partially because the daylight hours were shorter. Then there was the additional weight of heavier winter clothing. This, including the snowshoes, made it harder to move about and each took its toll by wearing down the landlookers physically.

Courtesy Superior View Studios

Crosscut Saw, typically for 4 to 8 feet in length

A vivid description of the life of a landlooker can be found in the diary of "Moccasin Bill" Oliver who made daily entries from 1874-1931. The original makers of snowshoes were Native Americans who lived where snow was common during the winters. The landlookers were appreciative of their ingenuity and adopted their ways. The snowshoe preferred in the Upper Peninsula was the tennis racket style with the tail, to keep the snowshoe going in a straight line as you take the next step. The toe could be rounded or pointed, seldom up-turned. The wood used was white ash and the lacing was rawhide. Some preferred moose, others caribou, some used deer hide. In Michigan they came to be known as the "Michigan Snowshoe".

The reports given by the landlookers must have been favorable as the decision to purchase the Tyoga timberlands was made. The deed for the properties was duly recorded in the Alger County Courthouse on March 23, 1905. The logging buildings at Tyoga were constructed, food for hungry loggers was stockpiled, logging tools and horses were purchased, and logging trails were marked by the camp foreman as assisted by the landlookers. The landlookers' job was then over, and they left to cruise more timber stands elsewhere.

The process of logging began. The logging road network had been identified. Typically the roads were brushed out and leveled by an advance crew of lumberjacks in the fall of the year. Timber cutting would begin as soon as the full crew was assembled. Typical logging camps were anywhere from 20-100 or more men. The men were recruited from the local population and supplemented by newly arrived immigrants. Finnish lumberjacks were the most common in Alger County in the early 1900's. Swedes and French Canadians were abundant as well.

The Alger County lumberjack population also contained a sizeable number of Croatians and Slovenians (formerly named Yugoslavians). These two different nationalities had left their homelands in south central Europe where many of them had worked as lumberjacks or miners in their native land. They were well suited for the work and conditions found in the Upper Peninsula.

Women could be found in these early camps working as cooks or laundresses. Some of the lumberjacks were married and lived in separate quarters along with their wives and children. The Tyoga settlement was a small village, so women and children were common, much more so than in the white pine logging days prior to 1900.

Courtesy Superior View Studios

Typical U.P load on sled pulled by horses on iced road

Lumberjacks were strong young men who expended lots of energy sawing and chopping. The saw they used was a two-man crosscut saw, usually 6 foot long with a wooden handle attached at each end. In those days kerosene was used to wash the pitch off the saw. The felling saw had fine teeth; the crosscut saw teeth were coarse. The process became easier in the 1880's with the addition of "raker teeth" designed to clear sawdust and free the kerf (the width of the cut).

Loggers tried to choose partners wisely, as the work was meant to be a 50-50 split. Once a good partner was found, they tried to stick together. Some of them formed partnerships which lasted for many years. Pairing tended to be by nationality because of the common language bond. This made for easier communication, especially in dangerous situations. Logging could be tricky and difficult in the old days when so much of the work was done by hand.

Teamsters knew how to handle horses, and many men specialized in this trade. Many of them were French Canadians and Swedes. Horses could be frustrating to work with. Many men had an impressive repertoire of expletives at their command which could be released at a moment's notice. A few men could communicate to their teams by pursing their lips and whistling. Often these people loved their animals, cared for them, and trained them to respond to their commands. This wasn't possible in some camps because of the large numbers of men and horses, the latter being owned by the company.

Hardwood stands were usually brushy, much more so than the open, park-like white pine stands of the previous era which tended to be open underneath. With the extended branches of the pine causing deep shadows, very little day light actually reached the forest floor.

When approaching a hardwood tree for cutting a certain amount of ax work was necessary to first remove smaller trees or saplings out of the way of the falling tree and from around its base where the sawyers had to stand. When the tree was analyzed for felling direction, a directional notch was made consisting of a shallow cut made with the crosscut saw. This was at a 45 degree angle cut made with a single bit axe in the direction of the fall. The single bit was more versatile as the squared off back side could be used to pound a steel wedge into the back-cut. This was a cut made to begin the falling action. When it was deep enough into the tree the crosscut saw was used to complete the cut. A wedge would be driven in to help keep the blade free from pinching as it continued into the heart of the tree and beyond.

Bob Graves from Shingleton indicated that double bits were also used along with a special wedge with a wooden handle attached to the steel. The side of the double bit ax could then be used for pounding without distorting the thin steel around the eye of the ax which would result in a dangerously loose handle.

A double bit ax had an advantage in that one blade could be used to crop the soil covered roots while the other was kept extra sharp and reserved for cutting limbs or undercuts. This was work for strong, skilled men.

As the winter wore on, snow depths levels, usually leveled off at 2-1/2 feet. After a big storm, the depths could get to 4 feet before compacting back down under the snow's own weight. Snow was both a curse and a blessing. The snow made it difficult for the lumberjack to escape from the area when the tree was falling; even under the best of conditions it was difficult to estimate where a hardwood tree would fall. Mistakes were sometimes made which resulted in injury or death, and hard hats had yet to become standard equipment for loggers.

Al Kaiser from Manistique related his experience of cutting hardwood logs in the 1930's in Alger County east of Melstrand. One winter he was working with his sixteen year-old brother cutting down a tree; when it began to fall, it suddenly shifted direction. His brother could not escape because of the deep snow. He was crushed to death.

Despite the hardships and hazards, winter was the favored time for the hardwood logging in the early 1900's. Slippery snow made it a lot easier for the horses to skid a heavy log or load of logs to the nearest landing. The haul roads from the landing to the sawmill at Tyoga were iced so that heavily loaded sleighs could move along easily. Hardwood is heavy. A 24-inch diameter log that is 16 feet long weighs almost 3,000 pounds. A number of logs were loaded on a sleigh. The load was kept fairly small; much smaller than the huge white pine era loads so commonly pictured.

Courtesy Superior View Studios

Water Wagon
Roads were iced in the evening to make it easier to haul loads the next day.

Courtesy Superior View Studios

World's Fair Load of Logs (36,055 board feet) near Ontonogan River 1893

Courtesy J. M. Longyear Library

Largest known load hauled, 1898, Shingleton, Michigan

Courtesy Superior View Studios

Deckers

Larger loads were a hazard to create. Lumberjacks had to use a cross-haul system consisting of a pair of log ramps over which the log was pulled up by a chain hooked to a pair of horses. A man with a cant hook had to risk life and limb to straighten out any log that went up wrong.

Courtesy Superior View Studios

Cross-haul technique, used for loading wagons or railcars

Smaller loads were also easier to haul up and down small hills that could not be avoided. A pair of horses hauled the smaller loads, but more horses were used depending on the distances and the size of the hills. “Big wheels” were used for hauling over terrain that was unsuitable for regular hauling.

Courtesy Superior View Studios

Big Wheel, an alternative method for moving large logs on dry ground

Courtesy John Parlin

Big Wheel and Sleigh Exhibit House near log slide, part of Pictured Rocks National Lakeshore, U.S. National Park Service, standing here John Parlin is 6'3"

Courtesy Superior View Studios

Skidding Hardwood, a tremendous strain on horses on dry ground.

Courtesy Superior View Studios

Blunt Tipped Cant hooks at ready for stacking log decks

Logs hauled to the site of the mill were then stacked by deckers by using a combination of ramps, chains, and manpower equipped with cant hooks. The dimensions and grade of the timber was then recorded by the Lumber Scaler. The scaler gin pole made loading easier. The work of loading large stacks of timber on sleds that needed to be moved to a location; or when it was necessary to load the trains with logs, timber or cut lumber the gin pole was the answer. One of these contraptions

Courtesy Superior View Studios

Lumber Scaler, he measured dimensions and grade of lumber

was first used in this area by Charlie Johnston of Rock River. The loggers had never seen one before, so they thought Charlie had invented it. It was actually a form of derrick used to lift objects into place, such as ship's masts. Now it was the logger's highly appreciated device.

Courtesy Superior View Studios

Horse-powered gin pole to load railroad car

Once the logs were in the mill yard they had to be cleaned of debris. First they were put in the mill pond and cleaned; and then lifted up to the saw on the second floor by use of a jack ladder or a gin pole. Tyoga probably used both techniques.

Courtesy Superior View Studios

Gin pole used to stack deck near mill

Courtesy Superior View Studios

Jack Ladder, the logs were pushed into mill pond for cleaning and raised to saw on second level of mill

Winter-cut hardwoods were usually frozen and contained very little sap. This allowed the logs to be stored for a longer period of time before they would stain. Staining is a process which begins in all logs after they are cut and this marks the start of the decaying process. Hardwood logs cut in the spring and early summer stain in a few weeks. Winter-cut hardwood logs usually remain stain-free until the weather warms up in late May or June. Once a log is stained, its value for lumber is gone. Stained logs can be used for railroad ties, but their value is minimal. A company like Tyoga could not market stained logs so they preferred to cut in the winter.

1906 began a time of transition for the lumberman. Logging opportunities had changed. The white pine era and its massive volumes of logs that could be floated to the mill were gone. The hardwood lumberman had to deal with the difficulties of cutting and moving the heavy hardwood timber. Hardwood logs are about 30% heavier than pine and do not float dependably. Some lumbermen tried floating logs to the mill anyway. Today the effort to salvage sunken hardwood logs from Lake Superior's bays is evidence hardwoods did not float well. Some of this wood is well preserved and can now be salvaged and cut into boards.

Logging methods and equipment had changed little since 1850. The needed bulldozers, chainsaws, and logging trucks were not invented yet. It was too difficult and expensive to produce a hardwood saw-log in 1906. Hardwood logs had another disadvantage over pine in that the market for hardwood lumber was much more limited since it was used primarily for furniture and flooring. In each case the quality of lumber had to be high, as the product produced had to be pleasing to the eye. Hardwood lumber had to be sold to a limited number of furniture and flooring mills. In addition, the southeastern United States was entering into the lumber market and could produce high quality hardwood lumber at very competitive rates.

Pine could be sawed into boards and distributed to retail lumberyards all over the country. Pine markets were strong and provided the materials needed for housing a rapidly growing U.S. population. As immigrants were pouring into the treeless plains looking for free or cheap land, they needed large quantities of lumber for their homes and farm buildings. American cities in the East were growing quickly as well. There is a direct correlation between the settlement of the Great Plains and the demand for pine lumber. As shown in the accompanying graph shows the value and relative production of Michigan white pine versus hardwood between 1860 and 1930. The number of logging camps, sawmills, and planing mills in 1900 according to the amount of revenue per each state affected are listed.

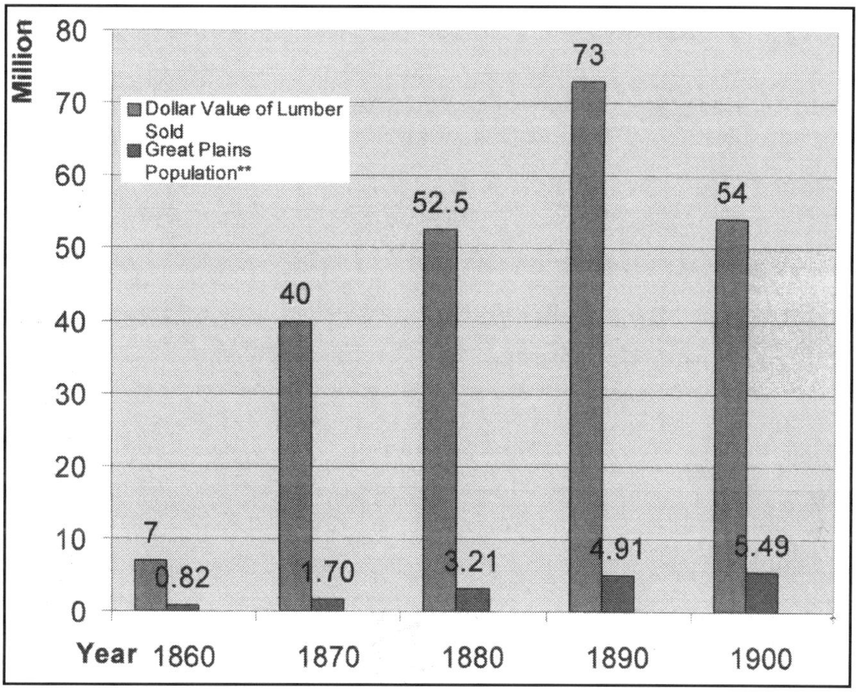

Graph by Bruce Venniberg

Dollar value of Michigan's lumber and Great Plains population, 1860-1900

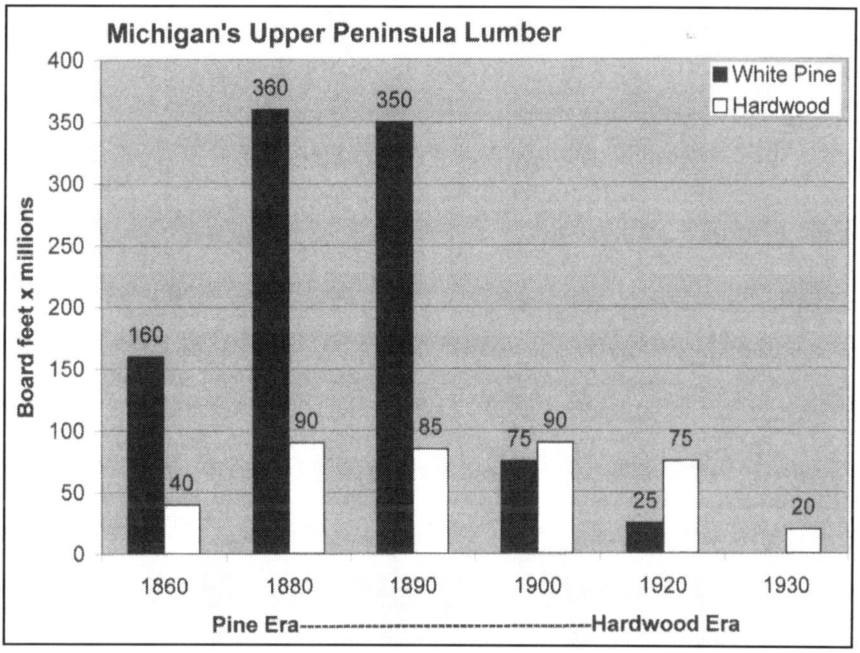

Graph by John Parlin

Upper Michigan's production of white pine and hardwood in boardfeet, 1860-1930

All of these factors added up to put the Tyoga Lumber Company at economic risk. The great profits realized by the pine-era lumbermen simply were not there. As a result, the economics of the timber business in the Upper Peninsula played a significant role in the early demise of the Tyoga Lumber Company.

TIMBER INDUSTRY PRODUCTS - 1900
Millions of Dollars (Net Profit)

	Logging Camps	Saw Mills	Planing Mills
Michigan	20	43	18
Wisconsin	18	40	24
Pensylvania	14	28	21
Minnesota	16	25	16
Indiana	4	19	7
Washington	12	19	7
Arkansas	7	18	9
Ohio	4	18	13
New York	4	11	43

Table by John Parlin

Today's Tyoga Forest

Forest cutting practices in Tyoga's time were based solely on economic choice. As a result, the largest and best quality trees were cut and removed from the forest. Fortunately, a hardwood stand contains a complex mix of sizes and species. Harvesting the large trees did not result in a stand being clear-cut in contrast to the devastated appearance of the Kingston plains southwest of Grand Marais following the clear-cut during the boom years of 1880-1910 which were followed by devastating fires. Slashings and trimmed branches were left where they fell. Fires followed. Pine burns with great intensity, hot enough to destroy all the humus along with its nutritive value. What had been sustaining soil was now gone. Only the sand was left. Gradually a few plants took hold, and then years later smaller shrubs moved in. In the meantime travelers through this area lost hope that there would ever be a forest here again. The process of re-creating humus requires over a hundred years for certain plants to begin to grow again. (It takes a lot to feed a pine.)

As each small plant dies, it returns its nutrients to the soil and gradually more plants are able to come in and find the trace beginnings of humus. The plant's seeds may be borne by animals, especially birds whose droppings add to the nutritive value. Some are carried in on fur or feathers. Some are borne by winds.

When the soil is enriched through this long process, eventually trees like poplars and birch gain a foot hold as they recycle nutrients with their falling leaves and after thirty or forty years after their demise their trunks rot. The process keeps grinding along. No one living at the time of the fires in the late 1800s and early 1900s will live long enough to see a forest of pine begin here again. Not even their great-grand-children will see them standing tall.

But gradually the soil will become enriched again. Then the pine will begin to take hold, their seeds having been borne in by nature's ways. Three to four generations later people may see the beginning of the large trees, especially the pine, once again growing on the sandy plains. Note the appearance of the Kingston plains in 2009.

Courtesy John Parlin

Kingston Plains,2009 the remains of white pine clear cut were followed by fires southwest of Grand Marais in later 19th century.

The clear-cutting of hardwoods was in vogue at the time when charcoal or wood chemical markets were active during the latter half of the 19th century. Such was not the case at Tyoga. The uncut smaller hardwoods had the ability to continue growing, to reproduce, to establish and maintain an ongoing forest community which can still be seen today.

The ownership of the Tyoga lands was passed on to the Cleveland Cliffs Iron Company. Cliffs had professional foresters on staff and managed their lands for sustained long term yield with a management philosophy that is practiced today by the Forestland Group, the current owner of Cleveland Cliff's forest division. These lands have been properly cut over numerous times. The hardwood forest still thrives there. It even contains a good amount of hemlock as it once did. Despite the intensity of logging over 100 years, 90% of the forests in the U.P. remained intact in 1941.

By contrast, modern timber harvesting has become highly mechanized with 348 million cubic feet of timber harvested in 1992, almost twice the amount for 1965. In Michigan, 54% of production was in the Upper Peninsula, and 41% in the northern Lower Peninsula. Michigan produces over 2/3rds of its total timber volume as pulpwood. In the southern Lower Peninsula, however, saw log volume produced is twice that of pulpwood, mostly in high value species such as red and white oak. In the U.P., "mechanical felling devices", such as tree shears and feller-bunchers have replaced chainsaws; industry-wide, a mechanized feller can be used 50% of the time on tracks that are not too steep and on timber that is not too large. These machines move at a terrifying rate, reducing a tree to a relatively neat stack of timber in 30 seconds.

Jack Pine "crop" renewal is said to outpace usage in the year 2008; hardwoods have generally not been a part of modern reforestation. As a rule, mechanical felling is better than manual felling – in its niche. Keeping the specialized tool in its niche is what logging planning is all about knowing well ahead of the scheduled harvest what logging system is needed. For the sake of the U.P. timberland, we hope the harvesters use this high-powered equipment appropriately.

Timber production fluctuates with the economy. WWI and the automotive industry created a great demand for timber. By 1921, a post-war slump began that temporarily slowed cutting to a bare minimum. In the years following, conditions improved somewhat. During the Depression, employment continued, although at rock bottom wages. By 1930, Alabama had seven times the number of mills as Michigan with 92% of Michigan's forest gone by 1929 – the U.P. fared better than this overall number for the state. The Depression in the 1930's was the catalyst for the pulpwood era and the coming of the "rubber tire years" with the hauling of jack pine initially by gasoline-powered trucks that were replaced by diesel power. (General Motors developed its first diesel in 1934.) In the beginning of the 21st century, 250 truckloads of pulpwood at 20 cords/tandem truck are needed daily to feed the Mead Paper Company in Escanaba.

After 1900 came the "Great Migration" to the U.P. with men attracted by opportunities in both the farming and mining industries. Cutover land was billed as prime for farming, and tractors had replaced horses. In 1937, Andrew Stihl invented the chainsaw with that name brand continuing its dominance in the chainsaw market today. The large McCullough saws were the preferred brand through the 1960's with the two-man 25 pound McCullough 325 being the saw of choice.

In 1872, Michigan mills sawed lumber that would load a train of cars 2,470 miles in length – enough to build one million frame houses. The state marketed $1,000,000,000 in total sales of forest products, greater than any other state or county of like area.

After the 1890's, the "lumber barons" moved westward in their quest for virgin timber. The charcoal-fired iron furnaces were no longer cost effective and closed by 1900. With the development of the Bessemer process, air being forcefully blown into molten iron in 1855 and the opening of the locks at Sault Ste Marie in that same year, iron ore from the Marquette range was shipped to furnaces on Lake Michigan and Lake Huron.

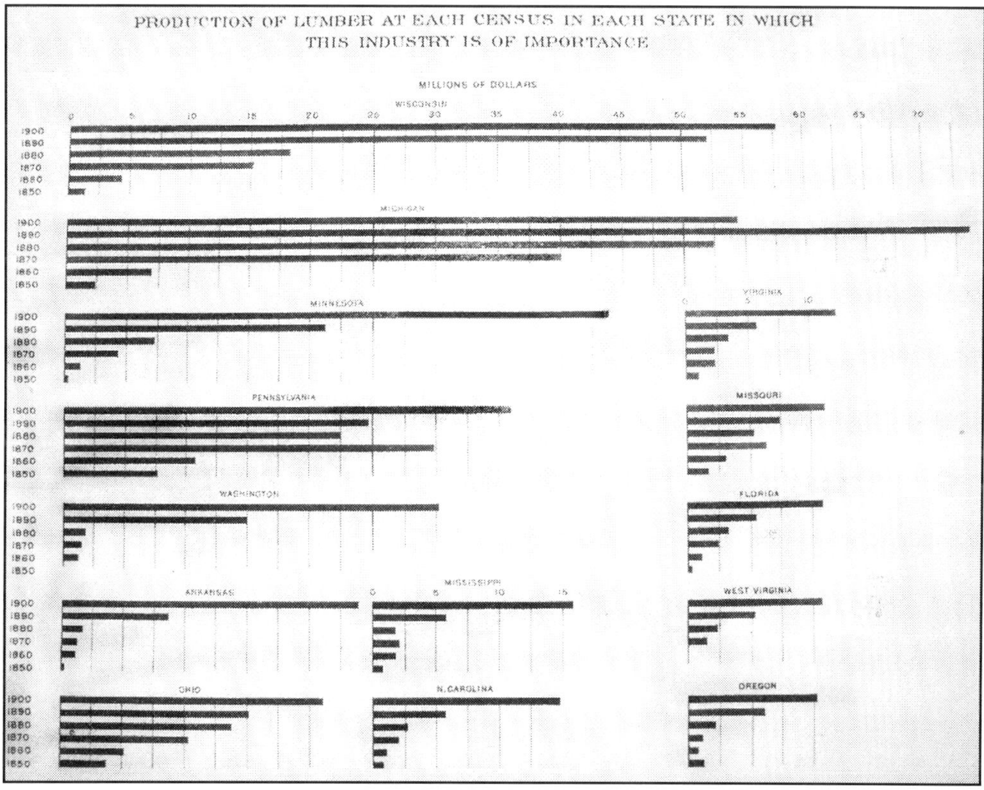

1910 U.S. Census

Production of lumber in millions of dollars by state, 1850-1900

By 1920, Henry Ford had amassed 400,000 acres in the Western U.P. to supply the 100 board feet of hardwood used in each Model T Ford after its invention in 1906 – the beginning of industrial forestry in the U.P. During this era, hardwood tended to be selectively cut as clear cutting would stimulate regrowth of aspen.

If the 1906 Pennsylvanians were able to see their forest today, they would probably think of it as worthless. The large trees they valued are all gone. The sawlogs now growing there are much smaller but highly valued in today's market place. In fact, they are worth many, many times more than they were in Tyoga's heyday. The trees too small for sawlogs are marketed as pulpwood for the larger paper mills in the Upper Peninsula. Beyond the borders of the Tyoga Historical Pathway, lumber and paper products from a portion of the 7100 acres owned by Tyoga Lumber Company are now sold in the worldwide marketplace. The Pennsylvanians had the right idea. They simply did not have the right time.

Appendix Chapter III contains current statistics on timberland throughout the state of Michigan and in the U.P.

Chapter IV

TYOGA LUMBER COMPANY 1905-1907

by Mike Zuidema and John Parlin

The Upper Peninsula is full of long forgotten villages, logging camps and homesteads which flourished around the turn of the Twentieth century. Scores of towns and camps were created as the lumbermen began to harvest the virgin forests of this land of Hiawatha. Most of these settlements thrived, waned and then disappeared as the great forests were cut and the mills moved on to other locations. Mother Nature then proceeded to fill in the openings, which were home to so many of our ancestors, with brush and trees. Today, only the observant eye can detect evidence of the activity that once took place, due to the growth of trees and brush.

As with many lumber operations at the turn of the 20th century, the Tyoga Lumber Company was a relatively small operation and was short-lived. The company was located in the northeast quarter of the southwest quarter of Section 35, Township 48N and range 22W, north of M-28 and two-thirds of the way to Laughing Fish Point on North Point Road. Leonard Harrison of Pennsylvania began accumulating property in Section 35 beginning in 1902 by purchasing 1200 acres from Daniel Powell in Marquette. On March 7, 1905, a company was formed by a group of investors from Western Pennsylvania led by Mr. Harrison. Three days later it was sold to a second group for $41,000. The parent company in Pennsylvania was named the Tioga Lumber Company. To avoid confusion, the subsidiary in Alger County, Michigan, was named the Tyoga Lumber Company with L. L. Bailey as the president (Alger Co., liber 18, page 20 Tyoga Lumber Company to L. L. Bailey, recorded 3-23-05). At the time of the Tyoga purchase, three to five million board feet of pine and hardwood had already been cut from the 7,000 acres of virgin timber.

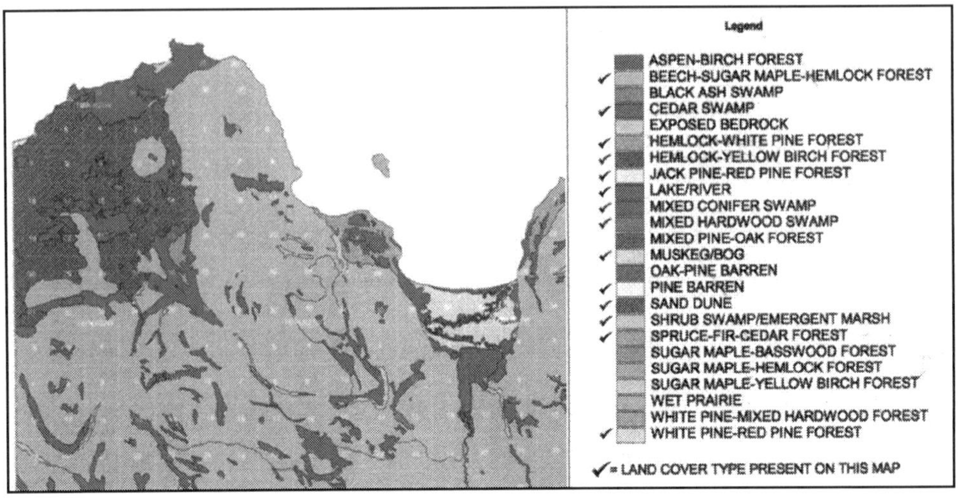

Courtesy MSU Board of Trustees

Vegetation of Alger County, 1880 The Tyoga region was composed of cedar swamp, hemlock-yellow birch forest, and beech-sugar maple forest.

Courtesy MSU Board of Trustees

Land cover change, 1800-1978
Most of Onota Township remains northern hardwood.

Comparison of the maps of vegetation in Alger County from 1800 to 1978 depicts the changes brought by the lumber industry. The 1800 combination of cedar, pine, hemlock, and yellow birch in the lower terrain and hardwood higher up would be a fair description of the regenerated timber in the 200 acre Tyoga Historical Pathway in 2000, 100 years after logging by the Tyoga Lumber Company and its successors.

The first press article of the coming Tyoga Lumber Company was in the *Munising News* on February 24, 1905:

> WILL HELP ONOTA. Location of Mill North of Deerton. Last week Pennsylvania parties bought about 3,000 acres of land in the vicinity of Deerton, in Onota township, Alger county.
>
> The new concern will be known as the Tyoga Lumber Co.

> and 'tis said they will install on the Laughing Whitefish this season at least a 50,000 board foot band [saw] mill.
>
> The timber which the eastern concern has bought with the land is mostly hemlock, it being estimated there is at least 30,000,000 feet of that wood on the tract in question. There is also said to be about 1,000,000 feet of pine on the tract.
>
> The lands lie close together about three miles north of Deerton.
>
> The D.S.S. & A. will run a spur from the mill which is believed to have at least a six years' cut of timber on site.
>
> It is further understood that the Tyoga Lumber Co is in the market for other lands adjoining theirs (with a total of 7,000 acres).
>
> The mill is to be modern in its equipment. Its location in Onota township means not only a considerable addition in population but also, perhaps, the establishing of a good-sized mill town in the western part of the county.

The Daily Mining Journal of Marquette expanded on the topic in an article on April 20, 1905:

> Within the next two days the Tyoga Lbr. Co., a Pennsylvania corporation that recently acquired holdings of Marquette and Alger county timber lands, will begin operations in the vicinity of Deerton. The company lands, which lay within four miles of the main line of the DSS&A, will be reached by means of a branch line, which will be built beginning at a point one and a quarter miles east of Deerton and extending north three miles to the site where the sawmill will be erected.
>
> The present plans of the company are for the construction of a sawmill with a daily capacity of 150,000 feet*. It will be a single band mill and will be fitted throughout with new machinery and labor saving devices in every department. In addition to the manufacture of lumber, the company will utilize their refuse, so far as is possible in the manufacture of lath. A lath mill with a capacity of 30,000 feet daily will be built along with the sawmill plant, and a 20,000 capacity shingle mill will also be constructed. All the hemlock and pine slabs will be worked up in the lath mill and the hardwood slabs will probably be shipped to this city [Marquette] and sold for fuel.

**Assuming planks were 12" x 16', the mill could produce 35,000 board feet of hardwood or 50,000 board feet of pine, hemlock, or cedar every 10 hour shift which equates to 7 and 10 miles respectively if laid end-to-end.*

> The company owns over 3,000 acres of land in an almost solid body in Marquette and Alger counties, and their holdings will probably be augmented within a year by an additional 4,000 acres that are now held under option. The character of lumber in the vicinity of Deerton is good and the lands are what is known in the lexicon of the woods as "heavily timbered." Figuring on an estimate of the timber supply and the mill capacity, it will take the company over 10 years to cut out their stock.
>
> Frank Hammond, general manager of the company, who will have charge of all the woods operations and the construction of the plant, expects to take a force of men to Deerton tomorrow, to begin work.
>
> The old Onota clubhouse, the "Halfway House", will be transformed into a lumber camp where 40 men will be housed while the sawmill is being built. Mr. Hammond says that by June 170 men will be employed and that the timbers used in building the mill will be cut at once. The sawmill will be completed by September 1 and 20 cottages for use of the employees will be built as soon as the lumber can be sawed out. In the woods a considerable quantity of hemlock and cedar will be cut this summer, in order to stock the mill until about January 1, 1906 when hardwood will be sawed.
>
> Although the office of the concern will be at Deerton, the company will also maintain headquarters at Marquette, and will probably handle the bulk of its business through offices in this city.
>
> The grading for the railway branch will be done by the company, and the surveyors of the DSS&A will probably lay out the route as soon as the engineer returns from Houghton.

It is estimated that from 1905 to 1907, 150 people were employed to run this newly founded lumber company; 40 men to operate the mill while the remainder of the crew cut timber and moved it from the woods to the mill. There is very little left today at the site to indicate the size of the operation for those two years. Only the stone base for the 9-foot band saw and foundation for the steam-generating plant remain. A drawing by Mike Zuidema, DNR (ret.),shows the stone base as it would have looked when built in 1905. Today the bolts for the base of the 130 horsepower Atlas steam engine are still in excellent condition.

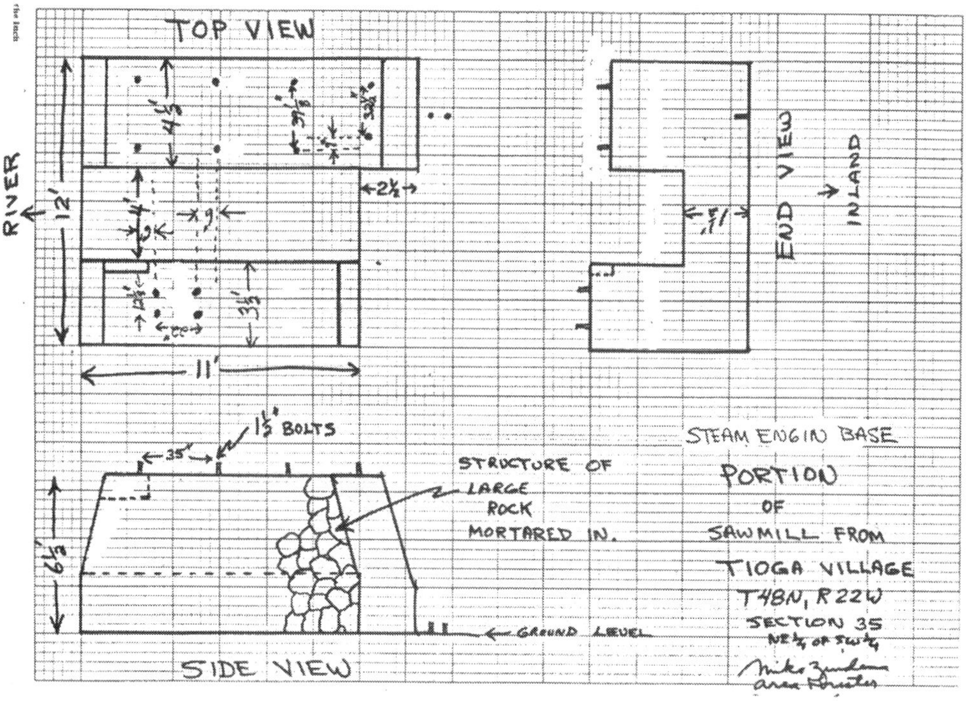

Courtesy Mike Zuidema, DNR (Ret.)

Tyoga Lumber Company steam engine base

In response to a query by Mike Zuidema, the curator at the Henry Ford Museum replied that

> there is virtually no doubt in my mind that your [Tyoga steam] engine is the same as the one at Nahma. Most side-crank engines were available either right or left hand; a few makers designed their machines so that they could be set up either way with no modifications. Of course, if it were a center-crank machine, the wheel would simply be placed on one side or the other. I'll take a stab at the horsepower, although this is only an educated guess. Assuming that the engine ran at 100 r.p.m. and that the effective cylinder pressure was 50 p.s.i., then the theoretical horsepower would be about 170. Subtract about 25 percent for mechanical inefficiency and you get a figure of about 130 h.p. which is probably pretty close. Of course, if the effective pressure was higher (quite possible) or the r.p.m. was greater, then output would be correspondingly higher. 130 does seem quite reasonable though.

Tyoga steam engine base 2007, the photo including Tina Hall and John Parlin was taken facing west. Laughing Whitefish River is just on the other side of foundation – note bolts on top of right side of base.

The mill machinery, including a 20" planer, was considered a valuable asset by C.C.I. with their purchase of Tyoga in 1907.

Some of the waste from the mill consisting of slabs and sawdust was used for lath with the remainder used to fuel the steam-operated band sawmill. An article in the Munising News of July 20, 1906, reports that Game Warden Eddycited the lumber company for illegally dumping sawdust into the stream.

> Deputy Game Warden Eddy, of the Soo, was in Alger and Marquette counties recently. He reports wholesale slaughter of deer. At the plant of the Tyoga Lumber Company, near Dearton, Mr. Eddy found the state law violated by the company, which was depositing its sawdust and other refuse in the stream on which it is located. The management agreed to burn the refuse in the future, so no arrests were made. The state game warden's department proposes to see that the mill company observes the law.

The concern for water quality in the U.P. was an enforcement issue long before the coming of the DNR-DEQ.

Courtesy Jim Johnson, Marquette

Photo 1 - Tyoga Lumber Co. mill, 1906 - East side of river – note railroad east of mill and river (photo taken facing west).

Courtesy Jim Johnson, Marquette

Photo 2 - Bunkhouse/Dining Hall ("dingle style"), 1906
East of river – note mill steam-power stack to the north (photo taken facing north).

Courtesy Jim Johnson, Marquette

Photo 3 - Schoolhouse/Store/Post Office/Bunkhouse, 1906 - Looking south from roof of mill. Note large stack of timber west of the river and close proximity of the railroad to east side of river (photo taken facing south). The Laughing Whitefish River is frozen.

Courtesy Jim Johnson, Marquette

Photo 4 - Schoolhouse/Store/Post Office/Bunkhouse, 1906 (photo taken facing north)

Courtesy Jim Johnson, Marquette

Photo 5 - The Superintendent's House, including covered porch, west side of river and opposite the south end of the mill (photo taken facing southwest).

We are indebted to Jim Johnson of Marquette for purchasing original photos of the Tyoga settlement of 1906 at an auction some ten years ago. These depict a complex consisting of the mill, boarding house, post office, dining area, schoolhouse, store, and saloon on the east side of the Laughing Whitefish River and boarding houses on the west side. The house on the west side of the river with a porch and a cat in the yard belonged to the superintendent (photo 5). Although saloons were not usually a part of lumber camps because of the hazards of mixing alcohol with logging, the sale of alcohol was permitted as Tyoga was considered a settlement. The Bunkhouse/Store (photo 3) was taken standing on the corrugated roof of the mill facing south. In photos 1 and 2, note the smokestack for the steam boiler adjacent to the mill. Also notice the narrow gauge railroad track extending to its termination at the mill. Follow this rail line south along the east side of the river noting the close proximity of the rail to the riverbed and the huge pile of wood stacked on the west side of the river (photo 3). The rectangular log building was probably a dingle-style cabin with bunks in one building and dining in the other, the two parts connected by a roofed breezeway (photos 2-4).

Courtesy Jim Johnson, Marquette

Photo 6 - Housing, 1906, West side of river (photo taken from roof of mill facing southwest). Note bridge over river.

On April 22, 1905, *The Daily Mining Journal* printed an announcement stating that

> The Tyoga company will build the grade as soon as men can be secured to do the work, and the scrapers and all tools necessary for the railroad construction being on the ground…

Additional articles concerning the camp and the railroad construction appeared on April 25, May 15, and June 24, 1905, in the *American Lumberman*:

> **Calumet, Apr. 25** – Another large industry has been started in Marquette county. The Tyoga Lbr. Co., a Pennsylvania corporation which recently acquired large holdings in Marquette and Alger counties will begin operations this week in the vicinity of Deerton. The company will build a single band saw mill with a daily capacity of 50,000 feet. A lath mill with a daily capacity of 30,000 and a 20,000 capacity shingle mill will also be erected. All the hemlock and pine slabs will be worked up in the lath mill and hardwood slabs sold for fuel. The company owns a solid tract of 3,000 acres of heavily timbered land and has an option on 4,000 acres more, which will be taken up within a year, insuring ten years work at least. Frank Hammond, general manager, says he will have the mill completed by September 1.

Twenty Houses will be built for the use of employees as soon as the timber can be sawed. The DSS&A will build a four mile branch to reach the mill. The survey of the branch line of the DSS&A at Deerton [sic. Onota] from the main line to the site of the Tyoga Lbr. Company's new town, was completed last week. The Tyoga company will build the grade as soon as men can be secured to do the work, and the scrapers and all tools necessary for the railroad construction being on the ground. The branch will be three miles in length and will leave the main line at a point half a mile east of Deerton. Several carloads of horses and lumbering equipment arrived last week at Deerton from Pennsylvania, having been in use in the camps of the Tioga company in that state.

Calumet, May 15 – Surveys for the branch line of the DSS&A at Deerton [sic. Onota], from the main line to the site of the Tyoga Lbr. Co.'s new town, have been completed. The Tyoga Lbr. Co., will build the grade as soon as men can be secured, all the tools being on the ground. The branch will be three miles long. Several carlods of horses and lumbering equipment have arrived at Deerton from Pennsylvania where the Tyoga [Tioga] Co. operated camps last season.

Calumet, June 24 – The Tyoga Lbr. Co., has a force of 40 men at work in the woods clearing a site preparatory to the erection of its saw mill plant, three miles north of Deerton [sic. Onota] and the company expects to have its plant in operation October 1. Clearing the right of way and grading for three miles of railroad to be built from the main line of the DSS&A is in progress. It will be finished by July 1.

In *The Daily Mining Journal* of July 14, 1905:

The Tyoga Lumber company has 80 men at work at the scene of its lumbering operations near Deerton. A crew of ten carpenters arrived here yesterday on their way to Deerton to engage in building the sawmill plant which will be completed about September 1. The machinery for the mill is ready to be shipped from Menominee as soon as the railway spur from the main line of the DSS&A is completed to the Tyoga company's town site. The work of grading the track has been in

> progress for some weeks, but the scarcity of men has interfered with operations and it is now thought that the branch will be completed until the middle of September. A number of men can find work in the woods or on the railway extension by applying to the companies.

In September of 1905, *The Daily Mining Journal* published a schedule of excursion rates for fishermen and campers on the DSS&A (see Appendix Chap. VII).

On February 16, 1906 *The Munising Rebuplican* notes that things were taking shape by October when

> three carloads of machinery arrived last week and the mill building is nearly completed. The company is engaged in logging its first cut of timber which will consist of hemlock and white pine. The company had estimated that the lands would keep the operation going for six years. It was announced that company would cut 5,000,000 board feet with over 2,000,000 board feet of saw logs currently on the rollways at the mill with sleighs actively hauling in wood from the camps. With the 1905 cut of 1.5 million board feet divided between pine and hardwood and 2,000,000 board feet of hemlock, the operation would yield a total of 5,000,000 feet for the year 1905.
>
> The mill is ready to begin operation as soon as the belting arrives. The machinery has been tested and the fires kept alive under the boilers for the past two weeks.

The February 16 article continues;

> to reduce horsepower for moving cut timber to the mill, the company expects to install a railroad during the coming summer and after this year will do away with team hauling from the woods. A complete railway logging equipment will be ordered this summer, and tracks laid from mills to the lumber camps.

While most of the 7,000 acres were within three miles from the mill, some were as far as six miles away. As the crow flies, the distance from the trailhead of Tyoga Historical Pathway to Deerton is 3 miles, to Shot Point and Shelter Bay 4 miles, and to the south end of Laughing Whitefish Lake (Peter White Camp) and Rock Kilns 5 miles.

Although the year of 1906 was one of intense timbering at Tyoga, no articles are available for that year. Following the severe winter of 1907 the company was sold to CCI. In the *Munising News* of February 8, 1907:

> Camp laborers are said to be very scarce this winter despite the fact that high wages are paid. Snow is getting so deep in the woods as to materially interfere with lumbering. If much more of the "beautiful" [snow] falls 'tis altogether likely several camps in the vicinity of Munising will be obliged to suspend operations for the winter.

According to Mike Zuidema,

> Felling crews were finding it extremely difficult to walk from tree to tree and the horses weren't much better skidding; the winter was only at the half-way mark

The final mention of Tyoga in the Perron Archives at the Peter White Library was an announcement in the *Daily Mining Journal* of July 15, 1907.

> Moving Tyoga Mill. The CCI Co. is shipping the mill erected in Onota township by the Tyoga Lumber Co. a few [2] years ago, to Munising. The CCI Co., last spring, purchased the plant, the company equipment of various kinds at its timberlands, getting them, it is said, at a bargain. The mill is a single band one, its machinery being practically new and in excellent condition. It is now being shipped to Munising where it is expected it will be erected and operated, though probably not this season. It's operation by the Tyoga company was not a success and the promoters of that company, Wellsboro, Pa., men, pulled out of the enterprise losers, though it is said that they might have made money if they had stuck to their guns.

C.C.I. records from the Northern Michigan University archives entitled *Tyoga Logging Report* detail the Tyoga operations between the date of purchase by C.C.I. on March 18, 1907, and Tyoga's closure the following September (Chapter VIII, Item 1). A millwright and carpenter were added in August. Closing down a business can be a difficult time given the absence of a cook and the presence of the summer's winged-pests.

Although there is very little written material, we have been fortunate to have interviews by Fay Swanburg, with primary sources Fay edited of the *Alger County Historical Society Centennial Edition 1885 through 1985*. The interviews are the basis for some comments about individuals involved with logging in Onota Township and its Laughing Whitefish River, both before and after the Tyoga Lumber Company:

- In 1870, the Irishman Curt Mullaly was one of the early settlers to arrive at Laughing Fish Point. He subsequently opened a lumber camp on the Laughing Whitefish River just south of M28. His house in Deerton was the site of the first school.
- George Wagner was the first individual connected to logging on the Laughing Whitefish in the early 1870's. George had been in negotiations with the state of Michigan for building the stagecoach line between Marquette and Munising via the halfway house with negotiations broken off by Wagner in a letter to Governor Bagley. (Appendix Chapter VII, Item 3) Instead, Wagner turned his attention to building a Jacobsville sandstone quarry on the west side of Laughing Fish Point as well as a saw mill, which was very close to the site of the halfway house, approximately one-half mile north of the Tyoga mill site.
- George Burtis had a steam-powered mill at Tyoga that was purchased by C.C.I. and moved to Munising in 1895.
- Eli Johnson operated a shingle mill on the Laughing Whitefish River powered by a waterwheel.
- Charlie Johnston, the proponent of the gin pole, was logging between the 1870's and 1898, cutting white pine in the Rock River basin He utilized three dams to move the timber to his mill. He was also a competitor to the adjacent charcoal industry of Charles Schaffer, "The Charcoal King". Charcoal kilns required large amounts of hardwood. This may well be the same C.D. Johnson in a photo of a Laughing Whitefish logging crew circa 1900. He is holding the saw, third from the right.
- Dan McEachern brought a crew of 33 men, 4 women, a baby and a dog, primarily from the Blaney Park (at the junction of US 2 and 77) that worked six days per week at Tyoga. (Dan McEachern's Crew photo page 244) His crew cut pine approximately two miles from the mill. The second crew was a "company crew" that cut hemlock and hardwood with the two groups having separate horse barns. Many of those men that built and logged at Tyoga came from Pennsylvania. There were also a few Native Americans from the reservation at L'anse (base of Keweenaw Peninsula) as well as "freelance" Native Americans providing deer meat, probably from the Grand Island band. At that time, L. L. Bailey was the president of Tyoga Lumber Company.
- The men working in the mill made $30 a month with room and board or $1.75 per day without room and board. Fifty men lived in the boarding house. Dances were held at the general store on Saturday evening.
- Prior to 1900, Robert Dollar had a logging operation upstream from Tyoga. He is remembered for the Dollar Steamship Company which

built the *SS President Hoover* (a 22,000 ton 665 ft. ocean liner) for $8 million in 1832. The depression was a bad time to launch a luxury liner.

- With the closure of the Tyoga Lumber Company, George Saul had an operation ½ mile from Laughing Whitefish River between 1906 and 1909 which debarked and floated hemlock. In 1908, CCI had a spring drive of hemlock from Deerton. At that time, there was a dam at the Halfway House and a second dam just south of M28, now the location of a gravel pit.
- The largest logging operation on the Laughing Whitefish River was that of Peter White Lumber Company on land around Laughing Whitefish Lake, seven miles south of Deerton, between 1910 and 1915. At that time, sawdust from his mill was used to "pave" roads prior to the use of gravel. Peter White was also the first owner of the franchise to run a mail service between Marquette and Munising via the Halfway House in 1871. This was a part of Peter White mail route between Green Bay, Wisconsin and the Soo.
- Art Skinner, a black drayman, provided hauling services for Tyoga Lumber Company and had a logging operation of his own into the 1920's.
- Leo Johnston had a logging operation on Sand River between 1927 and 1929.
- In an extensive interview by Fay Swanburg with Charles Risku (born in 1901), he tells of his father, Isaac, who worked at the Tyoga in the early 1900's, arriving in Alger County in 1900. Isaac cut virgin timber in what was considered wilderness, skidding logs with horse. There was no decent road between the lumber company and Deerton so they would travel along the river's edge. Charles Risku recalled that a geared-down water-powered saw was initially used at the mill. This may have been the Wagner Mill. In the 20th century, he relates that logging operations hired prisoners at 60 cents per day. In the 1950's, there were many Native Americans from reservations in L'anse, the Sault, and Escanaba working for jobbers in the pulpwood business. Isaac Risku built the original Tioga Tavern in Deerton which was moved to its current location on M28 by Jack Patton in the 1950's.
- Walter Alexander at age 95 was interviewed by Fay Swanburg in 1979. He described life at the Tyoga Lumber Company. "Board included bunk of straw and blanket. Put your own straw in bed from horse barn. Top bunks were real hot – couldn't sleep. Chore boy kept fire going in "sleeping camp" (chore boy was a man about 30 years old), also brought

> in wood, water for washing – helped the cook – kept busy cutting fire wood. Used potbelly stove. Good fishing in LWF river. Caught brook trout. Good deer hunting – used 30-30. Had to eat venison at neighbors house as it was against the law to eat at camp. [Other sources said Native Americans provided deer for Tyoga.] Log drives to down LWF belonged to different companies. A lot of peeled hemlock. Rafted to Munising. Logs cut upstream. All open around camp from cutting. Could see a "long ways." He mentions the very beautiful Mrs. Chamberlain who was very popular at dances and caused fights. Her husband, Ross Chamberlain, and Dudley Thompson also lived there "with families". He describes a band saw mill with the Halfway House. It was used as a hunting lodge following the Tyoga Lumber Company. A deer license for five animals cost 50 cents. A number of Winchester rifles were available with the most popular being the .44-40 and the 30-30.

In the 1920's, so many logs were floated to a dam at the mouth of the Laughing Whitefish River that it created a two mile backup. The last official log drive on the Laughing Whitefish was in 1922 by C.C.I. Logs floated down streams were frequently "branded" on the ends so they could be sorted out at the terminus. We do not know if this was the case on the Laughing Whitefish as there was no requirement to report brands in Alger County. It consisted of one-half million board feet of timber with Art Skinner in charge. The last drive on the Sand River was in 1926.

The existence of the Tyoga Lumber Company between 1905 and 1907 was both brief and intense with significant preparation prior to operation and a period of activity to dismantle the operation and move it to Munising after the purchase by CCI. The lumber company went from a thriving operation to a ghost town in two years, partially defeated by the weather and partially "frozen out" by CCI who owned surrounding lands.

Chapter V

PEOPLES AND OCCUPATIONS OF ALGER COUNTY

1850 – 1910
by Bea Anderson

The need for vast quantities of timber to operate the charcoal industry had a major impact on the forests of the area prior to 1900. Most of the forest lands acquired by the Tyoga Lumber Company were forests that had not been cut over. The timber harvest could be used for charcoal production as well as timber for mines, railroads, and construction.

Rock River Settlement

Laughing Fish Point just east of the mouth of the Laughing Whitefish River is a site where many immigrant families landed in the Upper Peninsula after arriving from Europe. One of these families was that of Gus Lindquist from Sweden who arrived in June of 1880. They spent the night camping on the flat rock and in the coming days the family stayed at the Halfway House about a mile south of Lake Superior. Meanwhile Gus walked south and east to the Rock Kilns to find work to house and feed his family. Although his English was minimal, his iron working skills spoke for themselves and he was promptly hired as a blacksmith. "I watched them until I saw the drill at just the right heat to work with. Then I jumped behind the anvil and grabbed a hammer and went to work on the drill."

Gus then built a house on the north side of the D.M.&M. tracks near the house of the head charcoal burner Albert Trombly. They were the only two houses in that area of the kilns. His family story by his son, Fred, is of great historical value, as well as delightful reading. – *They Came to Laughing Whitefish.*

Charlie Johnston of Varmland, Sweden, was born in mining country. In 1874 he arrived in Alger County via Canada. He was followed by his three brothers in 1876. At first he worked building the DM&M Railroad near Seney, Michigan, and encouraged many of his countrymen to come to work on the railroad. Some followed him to Rock River, giving him an experienced and skilled work force. Many were skilled before arrival and had been recruited by various employers to come to America. It was joked about that his section of the railroad was built on the snoose, a smokeless Swedish tobacco similar to Copenhagen placed inside the lip.

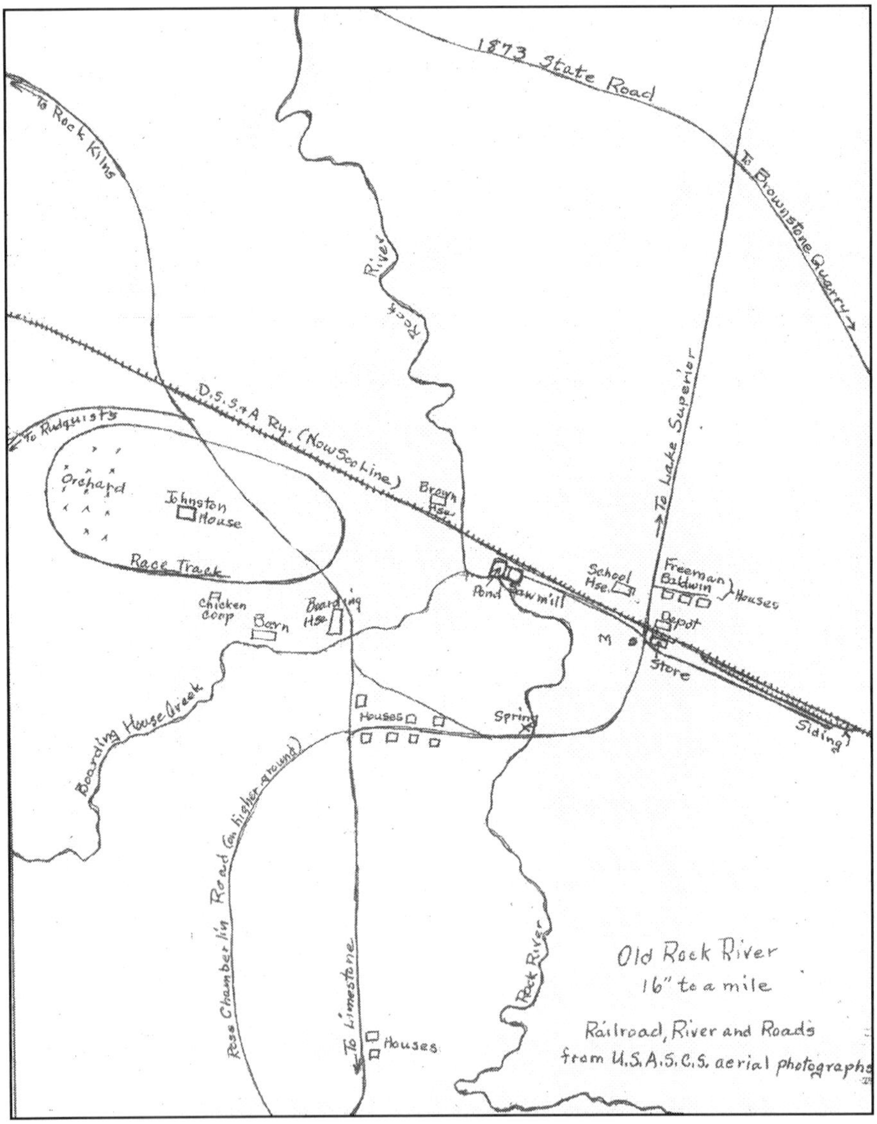

Courtesy Bernice Anderson

Johnston settlement at Rock River, 1880-1896. The Schaffer settlement at Rock River Kilns was north of DSS&A Railroad and one mile to the west.

In 1884 Charlie began to concentrate his efforts on the Rock River settlement. In 1885 the Johnston timber mill got out three million feet of pine and one million feet of hemlock. By 1887 there was a new schoolhouse, DSS&A railroad depot, and store plus a post office in Rock River. He also built a huge family home on the high ground west of the road from Rock River to Chatham. It had five Swedish fire places that ran from floor to ceiling which were very efficient in heating the house using hemlock bark for fuel. Charles had some saddle horses and a one-quarter mile race track that circled around the house.

In May, 1887, he built a sawmill and in June, 1887 his store burned. A new store was built along with 14 cottages. He also had a boarding house 100 feet long and two stories high with dining and kitchen areas on the first level and the men sleeping on the back side of the upper level. He also had a large orchard on the east side of his home.

By 1889, seventy-five men were employed year-round in the mill. The town was booming. The old Rock Kiln River map shows the relationships between Johnston and Schaffer settlements, the Rock Kilns*, Finn Hollow, and Tyoga. The Rock River DSS&A train depot is shown in photo below.

Courtesy Linda Stiuvitz

DSS&A Railroad depot on northeast corner of intersection of Rock River-Chatham Road and DSS&A. The School house was on the northwest corner.

**Rock River Kilns, the original name when built in 1882, has been shortened to Rock Kilns.*

Johnston had a two-story horse barn 150 feet long. It held 50 teams of horses. The second story was for hay and was also a good insulation. It was told how he pampered his saddle horses. He would have them pastured for the summer at Bay Furnace (later called Onota and now Christmas). They were not to be ridden to the furnace site but were walked the 10 miles by some of his crew.

Johnston and Charles Schaffer each had his own store in Rock River and Onota respectively. Both had a similar policy of granting credit to the men who worked for them. Everyone else paid cash. Johnston was always looking for business opportunities. In 1889 he organized the Rock River Brownstone Company as brownstone (sandstone) was in great demand for public building.

A few years later on August 26, 1892, a big fire consumed nearly all of the wood the Johnston crew had cut that season.

Almost all of the forests for miles around, including south to Silver Lake, north of present Highway 94 and west of Chatham, were extensively cut for local charcoal kilns and saw mills. Some whole logs were shipped to England for ship building. If one stood in the clearing for log storage behind the Rock Kilns, he could probably have seen Lake Superior one mile to the north. In 1898, with the pine gone and Johnston's home burned, Klondike fever took the Johnston family to Alaska.

During the years 1875-1900, many of the Rock River workers at the kilns and timber mills had purchased homesteads on cut-over Rock River timberlands south of the area, mainly in the area of Chatham and Limestone. Those who weren't needed at the kilns and Rock River mill spent the summer working at their respective homesteads. They would then return again when winter conditions were good for hauling logs on the frozen roads. The Rock Kilns ran all year long because they had such a huge supply of logs on the elevated clearing to the west above the kilns. The "Whitefish Trail" going south from Rock River to Chatham was the route the people used. It was a little west of the present Rock River-Chatham Road built in 1926. The Onota-Au Train Road was built in the 1920's. Much of the fill for this road was taken from remnants of the kilns and also soil from Finn Hollow across the railroad grade south of the Rock Kilns.

HALFWAY HOUSES

In 2011, it is hard to realize that in 1870 there were few roads, no telephones, or our present day technology. Most travel was by foot, horse or stagecoach. No motels, rooming houses, etc. to stop at when the weary traveler was affected by weather and road conditions. All over the Upper Peninsula and probably much of the country, there were way stations or halfway houses, usually called that because the distance between two areas of some population required more than one day travel. During the winter in the 1870's Upper Peninsula mail carriers traveled by foot, sometimes with snowshoes or standing on the runners of a dog sled. In the summer months they brought the mail by boat.

The Halfway House of this narrative was established in the late 1860's to provide winter lodging for the Sault Ste. Marie mail carrier along the early land route between Marquette and the Sault Ste. Marie. The Halfway house was 2½ miles north of M-28 and just north of the second bridge on present North Point Road to Laughing Fish Point. Today there is a historical marker on the west side of the road indicating the old site.

The building was a 10 x 12 foot log cabin, maintained by the government. It could accommodate 6 men and their sled dogs on the two day trip from Marquette to Bay Furnace and was open to anyone who traveled that way and needed shelter. Some sources say this construction was by the government and occurred following completion of the Old State Road about 1873. Another source says George Wagner, one of the founders of the Bay Furnace Co., built a saw mill near the halfway house by the Laughing Whitefish River in 1869; he may have built the halfway house as well.

A local resident, Russell Voce, said his grandfather's first job after arriving here from England was carrying mail from Marquette to Munising in 1874 with a St. Bernard dog. As carrier for about 30 pounds of mail, the dog had a buckskin double bag that flapped over him with a strap around his chest. The journey took four days each way. He stopped at the halfway house where the government sent men to cut firewood as there was no caretaker.

With construction of the DM&M railroad between Marquette and St. Ignace in 1881, travel along the road almost stopped, ending the need for overnight lodging. Over the years the building was expanded and improved and served a number of people for various purposes. The hewn log structure was eventually given a second floor and was graced with a huge fireplace and a mahogany staircase. Both private and public sportsmen used it as a lodge and a shelter for fishermen during the annual Lake Superior fall fishing.

In 1905 the lodge was included in a package sold to Tyoga Lumber Company. Forty men were housed here while the mill and company housing were being built. Another sportsman's club followed, using it as a men's club. In other years it was used for river drives, usually men employed by CCI as well as by trappers, hunters and fishermen. The mammoth logs eventually deteriorated and part of the site was used for road fill.

"Moccasin" Bill Oliver was a well known landlooker of this time. He kept a daily journal between 1874 and 1931 and traveled all over the U.P. of Michigan and many other states during his long career of landlooking. Two entries regarding the halfway houses were included:

Nov. 12, 1877 – I started for Munising on foot. Got as far as the halfway house.

Nov. 30, 1877 – Started from the halfway house about 7:20 a.m. Reached Marquette at 1 p.m. Rode about 2 miles. [walked the remaining of approximately 20 miles]. My bill at the halfway house was $1.50. Weather cold and stormy. NW winds."

Another halfway house of the early 1890's in the area was one mile east of Rock River village. During the construction of the Butler and Brownstone Quarrys, a halfway house was built by Cyrille Doucette Jr. along the railroad between Rock River and Au Train. It was utilized by travelers and temporary residents. The store at the Rock River village owned by Charlie Johnson supplied the needs of the location.

Owners, Occupants, and Proprietors of the Laughing Whitefish Halfway House:

1869	Constructed by George Wagner (?)
1870	Charles Sawyer (logger)
1880-1886	Mr. and Mrs. Fritz , Proprietors
1892	Sportsman Club – Peter White, A. Joplin, Charles Schaffer
After 1900	Intermittent use by CCI log division.
	Men's club for drivers, trappers, hunters, and fishermen
1905	Mr. and Mrs. Eugene Carp – Proprietors
	Temporary housing for 40 Tyoga Lumber Co. employees
1906-1909	George Shaw, hemlock logging
1907	Sold to CCI as part of Tyoga Lumber Company land

Courtesy J. M. Longyear Library

Bay Furnace, 1870-1877, Molten iron fills the piglets.

Courtesy J. M. Longyear Library

Schoolcraft Furnace a.k.a. Munising Furnace, circa 1900

Whitefish, Onota, and Rock River Charcoal Kilns 1879 -1905

Charles Schaffer of Grand Rapids left home at the early age of 16 and proved to be a self-made man. He arrived in Munising in 1867 to help his brother fulfill a contract with Schoolcraft Furnace. By 1880, John Burt, son of surveyor William Burt, erected eight kilns at Glenwood (now Onota). In 1882, Charles Schaffer, leased the financially troubled Union Fuel Company from Marquette. He continued to acquire large land holdings and Iron Furnace's stock. Schaffer's charcoal-fired iron furnace at Bay Furnace had burned down in 1877. The beautifully restored furnace by Hiawatha National Forest is shown in the 2009 photo below. In 1879, Schaffer was employed by Hiram Burt (son of William Burt) to oversee the building of charcoal kilns at Whitefish, Onota, and Rock River.

The first workers to erect kilns were French Canadians, soon followed by Swedes who were trained iron workers. Then the Finns arrived and people from the British Isles. Land sold for 25 cents an acre. Wages ranged from 10 cents an hour to 30 cents for masons.

The Rock kilns were completed in 1880. In no time, Schaffer became known as the "Charcoal King". Another expansion occurred in 1890 when J. W. Belkamp (Schaffer's brother-in-law) bought half interest in the Union Fuel Company. At the time, nine kilns at Rock River were producing twenty carloads per month; with the addition of three kilns, output reached 45 carloads. In 1890, Schaffer built several sawmills in Onota along with many log homes to house his employees. He built a three mile narrow-gauge railroad from Onota Hill south toward Silver Lake to harvest saw logs of hemlock and pine. A stone roundhouse at Onota serviced the donkey engine known as "Little Tommy".

Courtesy Ben Blackmer – 2009

Bay Furnace, 1870-1877, restored by Hiawatha Division of National Forest Service

Courtesy John Parlin

The "Round house" remains on a hill above Onota Kilns 2007

Other construction at Onota included a store, freight depot, telegraph office, company office, and a second story living quarters. Schaffer then added an icehouse, chicken coops, pigpens, slaughter house, barns, and blacksmith shop. In 1881, Schaffer built a large frame house for his family which they referred to as "The Big House".

Courtesy John Parlin

THe foundation of a Schaffer barn and stable at Onota 2007

Courtesy Alger County Historical Society

Munising Beehive kilns – 1870 Stereopticon view
These kilns were similar in appearance to the Rock Kilns.

The first set of kilns at Onota were completed 3.5 miles west of Rock Kilns at Glenwood (Onota) where the company had its headquarters. Four charcoal kilns at Whitefish three miles west of Onota were completed by 1881. Schaffer commissioned John LeBray to operate the Whitefish kilns as well as run a boarding house on the north side of the tracks. The four kilns on the south of the tracks held 34 cords. The burn was regulated by removing or adding two rows of brick around the base of the kilns. Schaffer refused to pay LaBray in cash. LaBray was paid in notes which were only good at the company store in Onota.

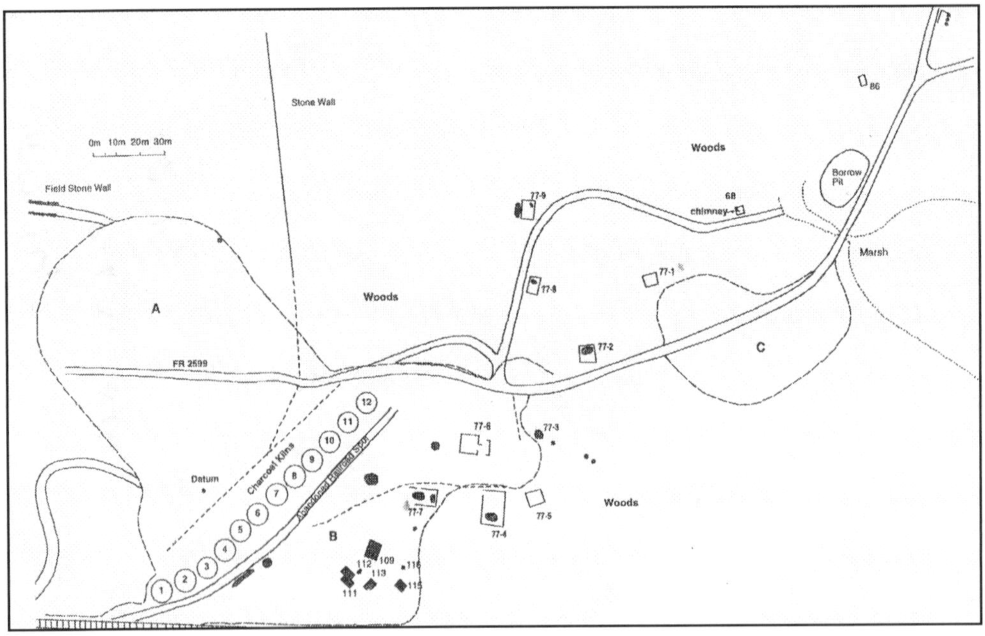

Dan O'Rourke

Rock River Kilns settlement

Courtesy Alger County Historical Society

Charles "Charcoal King" Schaffer, circa 1920 Charcoal production had ceased by the turn of the 20th century.

Following the fire of 1868 in Marquette, an ordinance was passed requiring that all public buildings be constructed of stone. Stone quarries were developed to provide sandstone for building for construction and limestone for charcoal kilns. Tub-type and beehive designs were preceded by circular above or below ground temporary kilns. Records indicate that Charles Schaffer supervised the construction of kilns at Rock River. Tub kilns (#2, 3, 4, 5, 6, 7, 8, and 10 numbered from south to north from the DSS&A railroad grade) were completed in 1880; beehive kilns (#1, 11, and 12) were constructed in 1890. Kiln 9 was also a beehive-type but date of construction is unknown. A stereopticon photo shows the Munising kilns in 1870, identical in appearance to Schaffer's beehive kilns at Rock River, Onota and Whitefish. Sequential photos show Charles Schaffer in 1920, five people in 1940.

Courtesy Mary Braamse Edgar, grand-stepdaughter of Bill Oliver.

Rock Kilns 1930 - On right is Ida Oliver Moore, only daughter of William "Moccasin Bill" Oliver. Man in photo is Wayne Moore.

Courtesy J. M. Longyear Library

Remains of beehive kiln at Onota, 1966. By 2010 only 50% of the kiln remained.

Kilns were built with local limestone. Beehive kilns were built using the corbel arch configuration with two-foot thick walls. The tub type with vertical stone walls required metal bands around the circumference to support a brick roof.

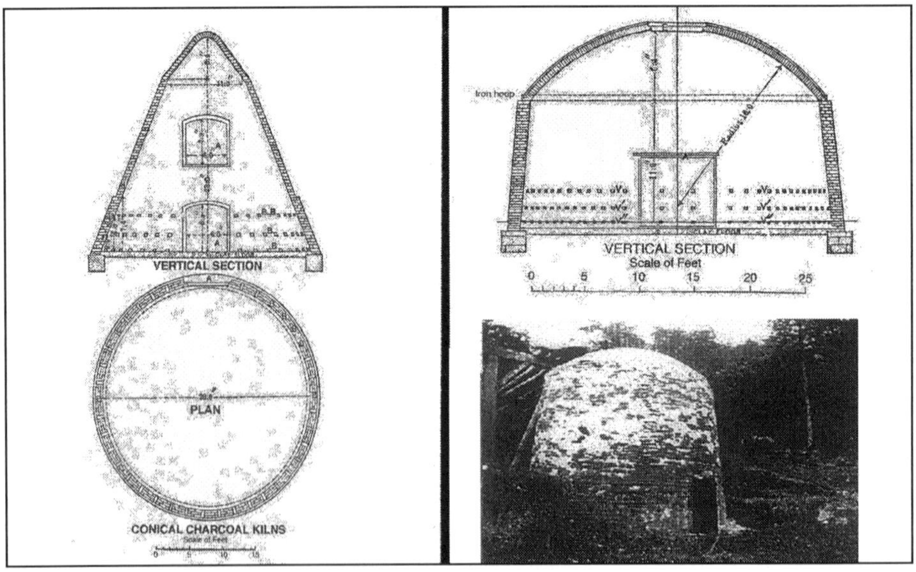

Courtesy Dan O'Rourke 1997

A beehive kiln (left) in contrast to the diagram and photo 8 a tube-type charcoal kiln (right).

Each beehive kiln required 35 cords of stone, 40 barrels of lime, and 40 loads of sand and held between 35 and 50 cords of wood. Wood was piled as high as 20 feet in the clearing behind the kilns for easy top-loading with the

Courtesy Michigan Tech Archives and Copper County Historical Collections

Top-Loading of beehive-type kilns. Rock Kiln apparatus had similar appearance loading from the open area behind Rock Kilns.

Courtesy Ben Blackmer

Remains of Rock Kiln #12 in 2009.
The original mortar is no match for a century of U.P. weather.

bottom removal of the finished product. Operations continued throughout the year. The four-foot cord wood was packed to eliminate air space and burning was controlled with brick vents at the base. A week later charcoal was raked out through the bottom openings, stubborn brands (coals continuing to burn) doused with water and scooped up with a large dustpan-like scuttle basket. The charcoal was packed in boxcars on the railroad siding for shipment to Marquette.

TUB vs BEEHIVE-TYPE KILNS

	TUB	BEEHIVE
	#2-8, 10	**#1,9,11,12**
Diameter	28-30’	26-30’
Height	26-28’	20-28’
Burn time	10-12 days	6-8 days
Cooling Time	5-6 days	3-5 days
Capacity	40-45 cords	35 cords
Production	1800 bushels	1400 bushels
Acres Hardwood	**1-1/2**	1-1/3

Courtesy Dan O'Rourke

Rock River archeological studies in 1977 and 1997 (Appendix Chapter VI, Items 1 and 2) describe a stone wall over 830 feet in length that runs north from a point above Kiln 12. It goes straight for 540 feet then veers west and it continues 270 feet to a bluff overlooking a small lake. The wall was about 3 feet high and rough laid. It is a single line of stones for the first 300 feet north of the southern most point at which point it becomes a double row of stones and a depression in the center; this depression at one time held a water pipe.

Courtesy Gregg Bruff, Munising

Recreation of Rock Kilns at work.
Drawing by Gregg Bruff in 1910 is displayed at Rock Kiln now owned by Hiawatha National Forest USDA Forest Service.

Local sources claim that water for the kilns was kept from freezing by charcoal on the wall. A pump house was built for the main source of water north of the kilns and later was replaced by a windmill in 1889. A gasoline pump was used to bring water from the lake to the kilns.

Wood for the kilns was brought either by wagon, sleigh, or flatcars backed in along the spur of the DSS&A which terminated at Kiln 12. To transport the charcoal to the furnaces, 4-foot sided half-gondola cars were used. The track was a standard gauge. (*See Appendix Chapter V, Item 3)* At that time, these would have been steam engines with oil head lamps. The drawing by artist Gregg Bruff for Hiawatha National Forest recreates the kilns at work.
Many of the workers at the Schaffer kilns were Finns with some of them also working for C. Johnston at his saw mill or logging for the kilns. The combined Finnish community numbered 200 people. Some families built homes in the open hollow south of the kilns on the other side of the DSS&A track. About twenty families or more lived there, hence the name Finn Hollow. The Finns came here after the Canadians and Swedes. All had various reasons for leaving their home country and were encouraged by family or friends who were here in America.

Courtesy John Parlin

The first floor walls of Jon Maki's hay barn built in 1915 at the northwest corner of Whitefish Road and DSS&A grade 2007.
Note blackened (creosote) stones taken from Whitefish kilns.

Some of the people arrived by stagecoach along the trail from Munising to Marquette. Others came by rail or by boat, dropped off at the dock for Laughing Fish point. They then walked one mile south to the halfway house on the Laughing Whitefish River. From there they proceeded by whatever means possible to get to the Onota or the Rock Kilns with many stories told of incidents occurring during that era.

In the late 1880's a huge amount of snow fell in the U.P. for a depth of 8 feet in the woods. With the temperature in May of 100°F, the snow melted in three days and all of Finn Hollow was flooded. Household goods, etc., had to be hauled to high ground to dry but it was said all was back to normal in three weeks. Not many years ago hunters south of the area found remnants of stone dams built across small ravines to prevent flooding of buildings; many of the stones were removed and several large borrow pits were created to build the present Onota-Au Train Road.

There were 23 coal-fired iron furnaces in the U.P. at their peak in the late 1800's. With the end of the charcoal iron industry in 1900, a visitor would not have to look far to find blackened stones from the kilns used for buildings. These stones were used for the first floor wall of a 3-story hay barn built by John Maki in 1915 on the south side of the junction of Whitefish Road and DSS grade. Other examples of remaining foundations using stones taken from kilns are found at Onota.

Courtesy John Parlin

Gravemarkers for three Trudell sisters who died from diphtheria in 1885 in the Rock Kilns settlement.
Gravemarkers by Marvin June, U.S. Forest Service (Ret.) and Andrew Parlin, 2010

Fires were a special hazard in those days with all of the wood products around and with the drought; just a few sparks could create a fire. In 1866, fences were erected along the DM&N railroad along the clearing of Rock Kilns and Whitefish Kilns. In May of 1887, fires raged in many areas including the

Anna River and Rock River; Marquette was enveloped with smoke. All night foghorns were blaring to assist the navigation of steamers and other water craft.

In 1893 Charles Schaffer built a new saw mill at Onota to bring some additional revenue from timber holdings in addition to from the production of charcoal. By 1896, the hardwood was depleted and the manufacture of charcoal was ending. Many families moved to newly opening farmlands in Limestone and Onota Townships. By that time charcoal-fired furnaces were being replaced by the Bessemer process with ore shipped to furnaces below on the lower Great Lakes.

A small cemetery, outlines of a few foundations, and ruins of the kilns are all that remain of the Rock River Kiln settlement. The 15-foot square cemetery has grave markers for the three Trudell girls who died of diphtheria in 1885. These are currently maintained by a descendent, Marv June. He discovered the graves while working for Hiawatha National Forest Service.

By 1898 the last inhabitants of Rock Kilns had left because a deep depression in the iron business beginning in 1892 that coincided with a nationwide depression. All of the above occurred prior to the building of Tyoga in 1905.

OTHER OCCUPATIONS OF EARLY SETTLERS

Ice Harvest

Courtesy Superior View Studios

Cutting ice required a saw, peavey, tongs, and crosscountry skis.

In the late 1800's there was no refrigeration. Ice harvesting was a necessary function although not too profitable. Harvesting ice in the winter provided the only source for cooling until the turn of the 20th century. In 1874 Mr. Rease established an ice business in Marquette. It began as a modest business but expanded year by year. By the late 1880's it averaged $7,000 to $10,000 per year. The local need for ice increased so much that two new ice houses were built. In addition to providing ice of the best quality to private homes, hotels, saloons, etc., Mr. Rease provided ice for the large steamers and other vessels that entered Marquette port. Most of this ice was harvested from the Dead River and Lake Michigamme.

Charles Schaffer harvested ice for his needs from Deer Lake. Ice was also harvested from Au Train Lake. Many small resorts had small houses for ice storage, built in a shady area. Sawdust was liberally used between the ground and the ice and on top to help insulate and prevent early melting. Some local people in later years did the same for their personal needs. All this effort would become history with the invention of the refrigerator.

Quarries

Courtesy Superior View Studios

Jacobsville sandstone quarry

In 1870, George Craig and Henry Smith of Marquette owned property on the west side of Laughing Fish Point and hired the civil engineer, George Cummings, to evaluate the stone of this site. In 1871, Craig and George Wagner teamed up and signed on investors from Chicago. Their quarry was located 50 feet east of the now submerged remains of the quarry dock. Marks from blasting can be seen on the sheer wall on the west side of Laughing Fish Point. A second quarry opened further to the east and was connected to the dock by means of a plank tramway with horse-drawn cars transporting the stone down a gentle incline to the landing. The quarry opening was stimulated by the new law requiring stone public buildings in Marquette but closed after one year because of high cost to ship even though the stone was of excellent quality.

In 1899, Brownstone Quarry was opened by the Rock River Brownstone Company. The first large sale was for the construction of the jail in Au Train. Many public buildings in Marquette were constructed with stone from this quarry. Quarries opened 1/4th mile from DSS&A Railroad and a spur was laid from there known as Ward's Siding. Another boarding house, also called the

Halfway House was built by Cyrille Doucette Jr. along the railroad between Rock River and Au Train was a boon for travelers and temporary residents. Apparently there were a number of buildings built on a hill overlooking Lake Superior and the quarries. The Johnston store at Rock River supplied the needs of this area. Close to the Brownstone Quarry settlement was a 1,000 foot dock reaching out into Lake Superior. This made it possible to ship stone by water or rail.

In 1891, Butler Brownstone Company opened a quarry northwest of the Brownstone Restaurant, between Rock River and Au Train. Incorporators were Thomas Butler and some Rock River Brownstone stockholders. A sidetrack was extended to the site and more docks were built. There is a little quarry at Onota, just south of the present Onota-Rock River Road.

In 1892, Nathan Kaufman built a large fishing and hunting lodge on the north shore of Silver Lake. This was located north of Rumley and a little southeast of Peter White Camp. "Moccasin Bill" Oliver mentions getting a flat stone eight foot long for the huge fireplace from one of the quarries. Because he mentions Charles Schaffer, the stone probably came from Schaffer's property at Onota where the quarry was located. This quarry was in operation until 1900. In later years, many people collected stone from this quarry for flagstone walks, etc. (see Appendix Chap. VI, Item 3)

Harvesting Ferns

As the forests were cut and the canopy disappeared, ferns sprouted in abundance. Native Americans and early settlers harvested some for personal use and sale locally. Soon it became a business opportunity for those living in the area for sale to a wider area.

In 1921, Ed Phillipson moved up North from Lower Michigan to harvest ferns. He set up a large tent for 25 workers by the railroad at Deerton for living quarters, while many picked ferns on their own property. Women bunched them, 25 to a bundle. This was done in their own homes and sometimes in the sauna. Pay was 50 cents per 1,000 for picking, 25 cents per 1,000 for bunching. A good picker could harvest 10,000 ferns in a day. In the big tent they were packed in crates and shipped to various greenhouses. A similar operation occurred out on "the plains" west of the present Forest Highway 13 (south of Munising). This place was known as Dunklee's. It was in operation somewhat longer than the one at Deerton and most workers commuted on a daily basis.

Blueberry Picking

Courtesy Superior View Studios

Before roads were built, the DSS&A provided transportation for these urban blueberry pickers with day excursions from Marquette.

With the cutting of the forests and forest fires, there was a tremendous production of blueberries in the U.P. The Chippewa Native Americans gathered them for their diet and then with the coming of the white man they sold them to the various settlements. It wasn't long before whites replaced the Native Americans among the bushes in the Marquette and Alger County area. In 1877 buyers paid $2.50 per bushel. With the opening of the DM&M Railroad in 1881, it was easier for pickers to get to the best picking east of Marquette. Special "blueberry" trains ran daily from Marquette at 7 a.m. and returned in the evening. Cost was 20 cents per trip. (see Appendix Chap. VII, Item 3) Many took the whole family and during a good day got their winter's supply. Some also went picking for needed income. In some areas in the U.P. entire families spent up to two weeks in tents picking berries to be sold. Agents would go from tent to tent to buy the berries. Most of those involved said it was an enjoyable experience because of the freedom. For the youngsters there were always small lakes to swim in after the picking was done.

Jo Marsh was born in Virginia and came to Munising in 1906. In a few years he went into commercial fishing and also used his boat to take area blueberry pickers to Miners and Beaver beaches along Pictured Rocks. He hauled crates of berries back for shipments to buyers in the big cities. If storms arose and lasted too long so that he could not sail to pick up the berry picker and their crates, he took a horse and buggy over rugged, twisting trails to bring food to the pickers who lived in tents on the plains.

Fishing

Courtesy Superior View Studios

The catch of E. T. Shoblanske - Munising, Michigan October 1910
Lake Superior continues to be a fine fishery for lake trout and whitefish.

Fish were a valuable addition to the menu of many of the settlers. As a rule, venison, duck, partridge, rabbits and geese were among the options of meat on the table. Almost everyone did some fishing on the streams or lakes based on the equipment they had. Remember in those days there was no radio for S.O.S., no radar telling you where the fish were or how deep, etc. Many a kid fished with a sapling pole cut by the stream and a few feet of line and simple hook. However, the fish were plentiful and fairly easy to catch, particularly the grayling. They were fished into extinction in most of the continental United States. There were few commercial fishermen at that time. A few were based at the mouth of the Laughing Whitefish River. Ed and Louis Malette were French fish peddlers. They purchased land on the western side of the river's mouth and did their fishing from that site. Jack Makela had a fishing site in the same area and he fished and trapped until becoming a state trapper. Au Train fishermen used the first camp on Laughing Whitefish Point now owned by Don Paajanen. Many young boys living in the area scooped fish as they lay like cordwood when the water behind the dam on the Laughing Whitefish was low. The fish were huge. It was said that three fish could fill a wheelbarrow.

Shelter Bay (previously known as Sucker Bay) also had a small commercial fishery with some charter boat operators and cabins on the Bay. Native Americans called the area of the mouth of the Rock River “zhooshqua nabe konk sibi” Zhooshk means “slippery rock”. Anyone trying to walk on the stones there along Shelter Bay will agree 100%. The Doucette boys had guided fishing parties for years. Several fishermen had cabins and fishing vessels in the area, known as squatters on this CCI property. In 1926, CCI began leasing land to the people, some of the parcels near Laughing Fish Point. The first lease was to the Doucette boys. Among others were Jack Ronta, Isaac Maki, Andrew and John Parkkila, Waino Russo, Sander Mattson, Arvid Puro, Herman Kujala, Andy Kainilainen, and Toivo Partonen. Some of these men did only commercial fishing. It was a hard life. The nets were made of a thread that did not dry fast and was not as durable as the nets of later days. They had big reels beside their cabins where they placed the lengthy nets to dry and to spot tears in the net which required mending when the nets were dry. As a child, I remember helping my uncles with this task; we used a tool called a shuttle. Rock River enters Lake Superior about one mile east of Shelter Bay. Spring fishing for rainbow (steelhead) was a much anticipated event in the area.

Courtesy Superior View Studios

Thill's fish store in the near background was the outlet for commercial fishermen in the Marquette area beginning in the early 20th century.

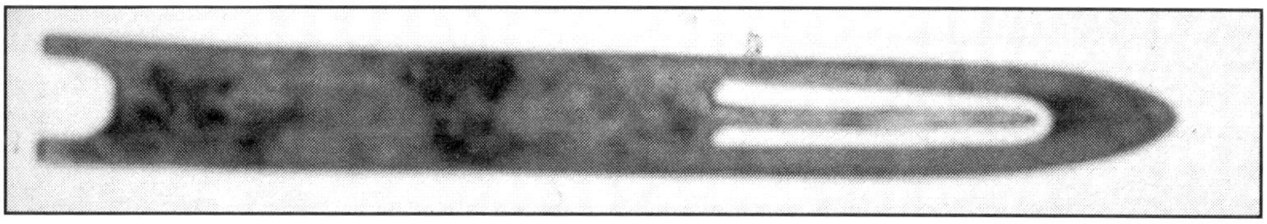

Courtesy Echoe Deibert

Shuttle for net mending found on shore of Munising Bay

Farming

Most of the new settlers did some farming in the homeland. As they arrived they bought homesteads here with small gardens. Others farmed on a large scale as that was their primary occupation. Farmland replaced cut-over timberland beginning in the 1860's. Improvements in farming equipment greatly improved productivity as horses were replaced by tractors. Tractor kits could be purchased from Montgomery Ward but you had to supply the engine. The tractor made all the difference in rasing crops in this hostile environment.

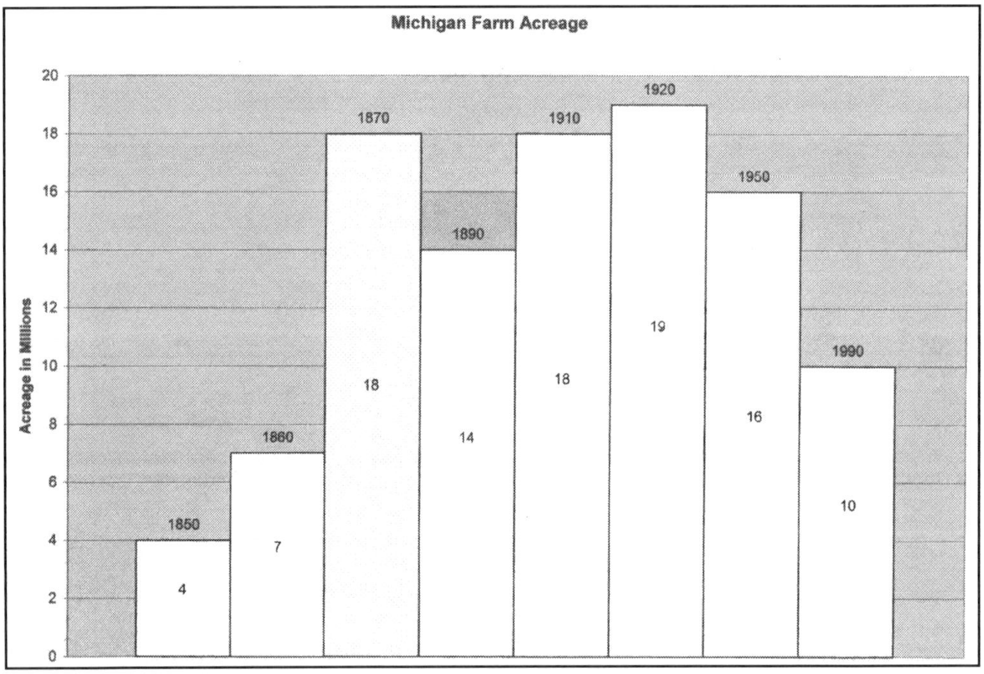

Courtesy John Parlin

The Michigan farm acreage includes both the Upper and Lower Peninsulas.

Not many acres of land in the northern U.P. proved to be suitable for farming in the long term but land on top of Onota Hill was a notable exception. Charles Schaffer planted crops, fruit trees and also raised cattle. In 1910, all the fires for charcoal were extinguished at the Onota kilns accelerating the decline of the community. Farming then became the main source of income for the residents. That is the year that Orvice LaBounty resigned as superintendent of Munising Township schools and began scientific farming on a large scale.

Courtesy Superior View Studios

Old horsepower

Courtesy Superior View Studios

New horsepower

Charles Schaffer eventually owned 77 forty acre plots of land with 325 fruit trees, an 80 acre farm with 60 acres under cultivation, a store and houses which Orvice LaBounty bought. He then added a large wooden barn using some stone from the ruined kilns. This Schaffer venture failed but others took up the challenge. LaBounty sold out to Stacey and Keith about 1918 who had plans to raise cattle on the land. After a short-term of ownership the enterprise was sold to Harry Stoop who then went bankrupt and the property went for tax sale.

Another attempt at agriculture in Onota was Pamona Farms, a fruit farm planned on a grand scale on Onota Hill. After the Alger County Agricultural Society was formed about 1915, George Burtis was hired to manage the farm. George had run a lumber operation in the Tyoga area in the late 1800's with the mill subsequently shipped to Munising. Apple and plum trees, gooseberries, currents, strawberries and raspberries were introduced and potatoes were planted. Because of George Burtis' lack of experience in farming and extravagant business methods, the venture failed. When his lavish home burned in the 1920's, he left before his plantings had a chance to mature. Andy Smith took over for the stockholders who finally lost their investment. J. Longyear bought the property for back taxes but it eventually reverted to the State.

Another farm was developed on the old Burtis property by Bill Wilson who had an extensive truck garden there. Bermuda onions were his specialty. He lived below the Onota Hill in the "Big House" adjacent to the Onota kilns.

Leo Johnson was the next settler in 1932. Before FDR stopped sale of all public land in 1934 he was granted one of the last homesteads and lived there until his death in the 1990's.

Several experienced farmers arrived from other countries and were successful. Julius Gogarn was born along the River Rhine in Germany in 1834. He came to the USA in 1853 and came to Munising in 1872. He had a big farm (365 acres) on East Connors Road, east of Munising. Some of his family settled in Deerton. Another big farm in the Munising area was opened in 1896 by John Robinson, and was later owned by John Lehnen and then Dan Becker.

In 1899 the noted U.P. experimental station near Chatham was established by Michigan State University and to this day has many extension sites in the U.P. The main purpose was to experiment with crops, soils, etc., as to what would do best for farmers in the U.P. An agent came to each small farm frequently during the summer and fall to test cows for TB and give advice on new products and techniques. He also helped with other problems the farmer was having at the time. It was a very valuable way to communicate to every one as no one had a radio or phone.

Also there was a summer camp started in the 1920s that was erected for children from the Upper Peninsula who were considered to be malnourished. It was located just north of Chatham, operated by the school nurse or some agency. It ran for about four weeks in the summer. In 1924 it was named "Camp Shaw" in honor of Robert S. Shaw, a former president of Michigan State University. It continued on into the 1960's. The main experimental buildings of MSU have been moved in recent years to a site about one-half mile south of Chatham and east of M-67.

Sugar Bush Maple Syrup

Maple syrup and sugar were vital staples for Native Americans and early settlers. Charles Schaffer had a sugar bush shack at Deer Lake. It was called "The Antlers" and produced about 200 gallons of syrup annually until the 1930's. With 40:1 ratio, this required 8,000 gallons of raw material. Peter White's sugar bush was managed by Charles Anderson of Marquette. About 3,000 trees yielded 500 gallons of syrup produced annually. Maple syrup shipped all over the world. The Mayflower Hotel in Washington D.C. was one of the more prominent customers.

Nathan Kaufman's lodge on Silver Lake had a huge sugar bush in the area which was tapped for many years. There the trees matured into very large trees and when Penegor Lumber Co. were harvesting this virgin timber about 1942-1943, you could no longer see the metal spikes which were overgrown by at least 8 inches of wood. After the logs got to the saw mill, the spikes were "found" by the saw, creating havoc and ruining many a band-saw's teeth. To get around the problem officials at the mill went to the site and ordered all trees not already cut to be cut off 4 feet above the ground. Any tree already felled had four feet cut off above its base as well. (See Appendix Chapter V, Item 3)

Courtesy Superior View Studios

Young boy helping to collect maple sap in early spring
Forty gallons of sap boiled down to one gallon of maple syrup.

Courtesy Superior View Studios

Top Left: Men and mining car, Calumet and Hecla mine shaft #2, 1912
Top Right: Copper mining Middle right - Keweenaw Amygdaloid copper mine 1900
Bottom: Miners with "sunshine" carbide lamps that burned acetylene produced by dripping water on calcium carbide in a canister below the flame. Mining was cold, dirty and dangerous.

Copper and Iron Mining

To cover the copper and iron mining industries is obviously outside of the scope of this book but this too had a major impact on the timber industry. Copper mining in the Keweenaw began in 1850 and was almost completely shut down shortly after the turn of the 20th century. In the year 1900, it was said that there were more millionaires in Calumet, Michigan, than in New York City. Iron mining in the central U.P. known as the Marquette range, began in approximately 1850 and continues today. These industries used a significant amount of lumber for construction of mining shafts and railways and competed with the timber industry for labor. The mining industry drew heavily on workers from mining countries in Europe, beginning with Cornwall in England, followed by Germans, Scandinavians, and Italians.

Early Settlers and Reasons for Coming

The French were the first Europeans to come to the Lake Superior area as voyageurs, missionaries, and fur traders in the 17th century. They were also our first farmers, among the first lumberjacks and probably our first white fisherman. Most of these people came by way of Canada. Many blended with the Native Americans. Later in time many people arrived by many other routes. The varied ethnic background cultures and nationalities added much diversity to the communities; this impact is still felt today. In many cases there were primary roles played by a particular nationality. Anglo-Saxons were builders; Croatians, Slovenians, and Finns were woodsmen and farmers; the Jewish were traders; Swedes were ironworkers, woodsmen, and farmers. A few black people arrived in the early 1900's with their primary roles as seamstresses, barbers, log jobbers and fishermen. There was even a Chinese laundry in Munsing. In all, 23 different countries representing four continents laid the foundation for the north county as it was in 1885 and still is today.

Reasons for leaving the old country included:

> Most people came seeking greater economic opportunity to provide for their families and greater freedom from oppressive political situations. Finland has a long history of oppression by their eastern foe. After many wars, Finland was ceded to Russia in 1902. Not until December 6, 1917 did Finland once again become an independent nation. During the early years of Alger County, many young men left Finland to avoid conscription in

the Russian army. The saying was "Better an ax in an American lumber camp than a rifle in the Russian army." The Finnish word sisu refers not only to the courage and optimism but to a concept of life that says, "I may not win but I will give my life gladly for what I believe." Not all newcomers were impressed by what they found here and returned to their native land. With the era of socialism in the U.P in the 1930's some were recruited to return to an Utopian society in Europe.

Jo Marsh was born in Virginia in 1866 and came to Everett, Michigan in a covered wagon in 1904. In about 1906 he and his wife Nora Rickman (a Cherokee Native American), their family and his brothers came to Munising. He worked at a saw mill and farmed and then went into commercial fishing and berry picking. He was a state trapper and for some time manager of Munising Tourist Park. Their son, Earl, was well known in boxing circles in the 1930's, both as a fighter and teacher of young men of the area.

Mrs. Matthews was an expert seamstress and sewed for many people plus teaching young girls the craft.

A black barber from Marquette periodically made the rounds to Deerton farms giving the boys the city look for a brief time after which they reluctantly returned to the "pot on the head" method used by busy mothers.

Art Skinner was a popular wood jobber and drayman in the Deerton area in the early 1900's with a large outfit, including many teams of horses plus saddle horses, turkeys, geese, hogs, and a large assortment of equipment. Art built his camp one mile west of Deerton (Whitefish kiln area) and also a large number of houses for his workers. In 1925-1926, he moved his operation to Rock River. There were not many black people in the area but the few added a lot to the community.

A smaller number of immigrants left Europe because of crowded conditions in their homeland. This also happened in Alger County, strange as that may seem. John Mathias arrived in the area from Ontario, Canada in 1883. He didn't like crowds and so he followed a blazed trail along Lake Au Train,

Mud Lake, and then Trout Lake. He made his homestead there. Mathias Township was named after him as he was the first settler. The state wanted to honor him but he was not present for the honor. They tried to locate him with no success because he had quietly moved out of the area in 1887. He said it was getting overcrowded!

Only a few Native Americans made Onota township their home after it was settled. In addition to LaPete and Bayfield there was Longlad, a piecemaker 6 foot 6 inches tall, weighting 200 pounds. He handled the long saw alone cutting cordwood. He shared his traditional medicines which saved lives before "modern" medicine. There was also an Native American midwife at Onota and one or two men who hunted deer for the Tyoga Company.

Chapter VI

RAILROADS AND ROADS OF ALGER COUNTY

by John Parlin and Jim Carter

At the start of the 21st century, roads have become vastly more important than railroads in the U.P.. In the early 1900s, the reverse was true. Railroads were the major mode of transportation for the timber and iron ore industry; and also for passenger service before reliable cars and good roads were developed.

In Theodore J. Karamanski's book, *Deep Woods Frontier: A History of Logging in Northern Michigan*, he pointed out three distinct time periods of the timber industry: 1840 -1900 the white pine era; 1900-1930 hardwood era; and finally the pulpwood and hardwood era which is still ongoing. This is known as the "rubber tire" era where trucking is the primary mode of transport for pulp and hardwood.

The Tyoga Lumber Company (1905-07) logged primarily white pine and hemlock. A dedicated spur from the Duluth, South Shore and Atlantic Railroad (DSS&A) ran from Onota to the Tyoga Lumber Company. It was used primarily to transport timber products to the market as well as personnel and supplies to the company as there were few adequate roads. At that time, logging railroads were not subject to either state or federal regulations so some spurs to logging operations were of questionable quality. This does not seem to have been the case with the Tyoga spur.

As one can imagine, building railroads in the U.P. presented significant challenges because of the unique topography with many streams, swamps, and rock formations not to mention the dreaded black flies and mosquitoes. The hydrology of the U.P. is depicted on the following maps from the Superior Watershed Project. Note that the watershed of the Upper Peninsula runs the expected east and west paralleling the southern shore of Lake Superior. What is striking is that the watershed is less than 20% of the distance between Lake Superior and Lake Michigan between Marquette and the Two Hearted River east of Grand Marais, Michigan.

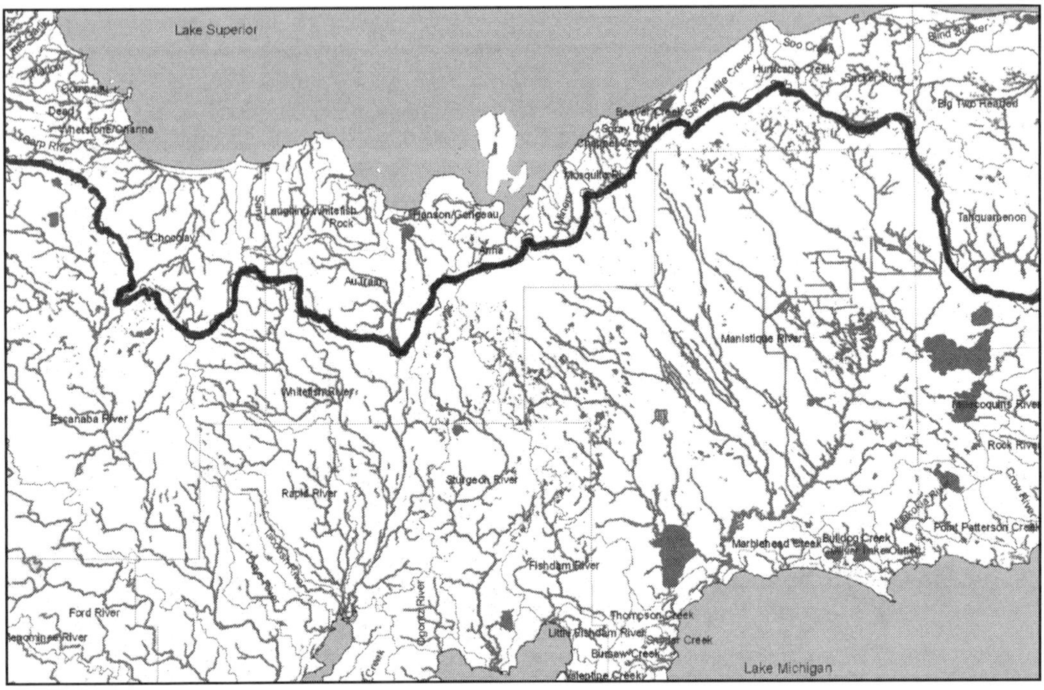

Courtesy Lake Superior Watershed Partnership

Rivers of Upper Peninsula - Note the far north location of watershed between Lake Superior and Lake Michigan. Bold line represents the watershed.

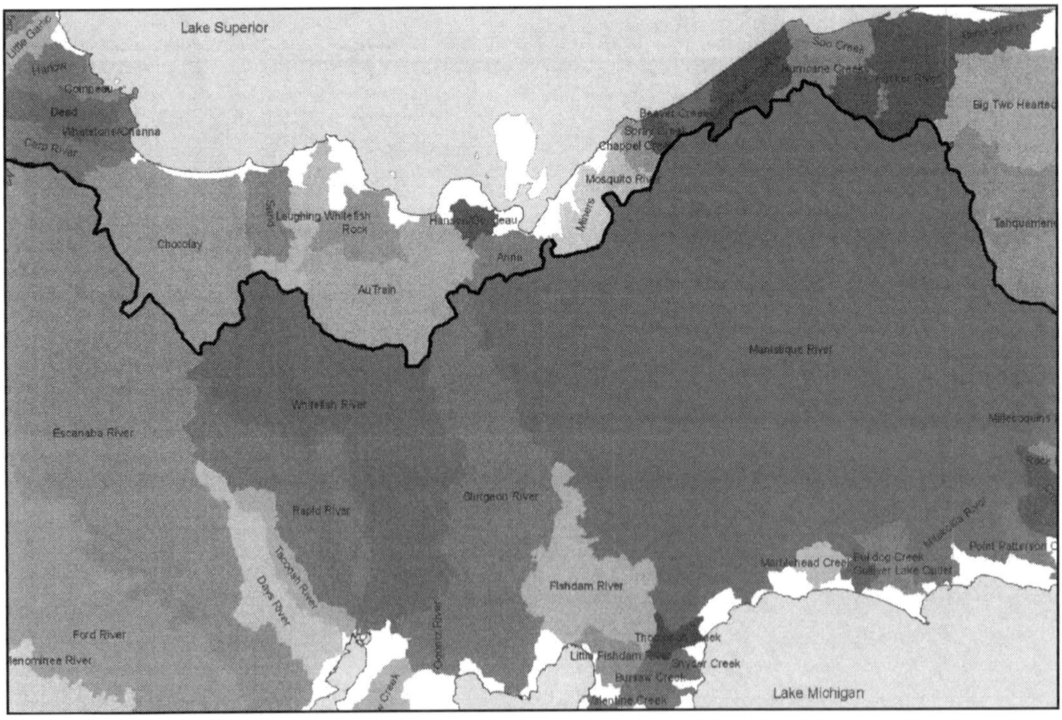

Courtesy Lake Superior Watershed Partnership

Drainage area of rivers of the U. P. Bold line represents the watershed.

As one drives the 23 mile straight Seney Stretch of M-28 between Shingleton and Seney, the numerous streams are all flowing south into the Manistique River and on to Lake Michigan at Manistique. This meant that much of the harvested pine was floated south from areas far north in the Upper Peninsula to mills at the river mouths on Lake Michigan or were gathered on the lake into booms and were towed to mills elsewhere on Lake Michigan or Lake Huron.

By 1860, there was a large network of railroads in the Lower Peninsula and in the northeastern United States but only one line in the U.P. from Marquette southwest to the Marquette range (Ishpeming and Negaunee) and from the Marquette Range to Escanaba. The Iron Mountain Railroad was 14 miles in length between Marquette and Negaunee. The railroad line south from Negaunee to Escanaba was part of the Peninsula Railroad, owned by the Chicago and Northwestern Railway, completed in 1863. This was separate from the Iron Mountain Railroad in both name and ownership.

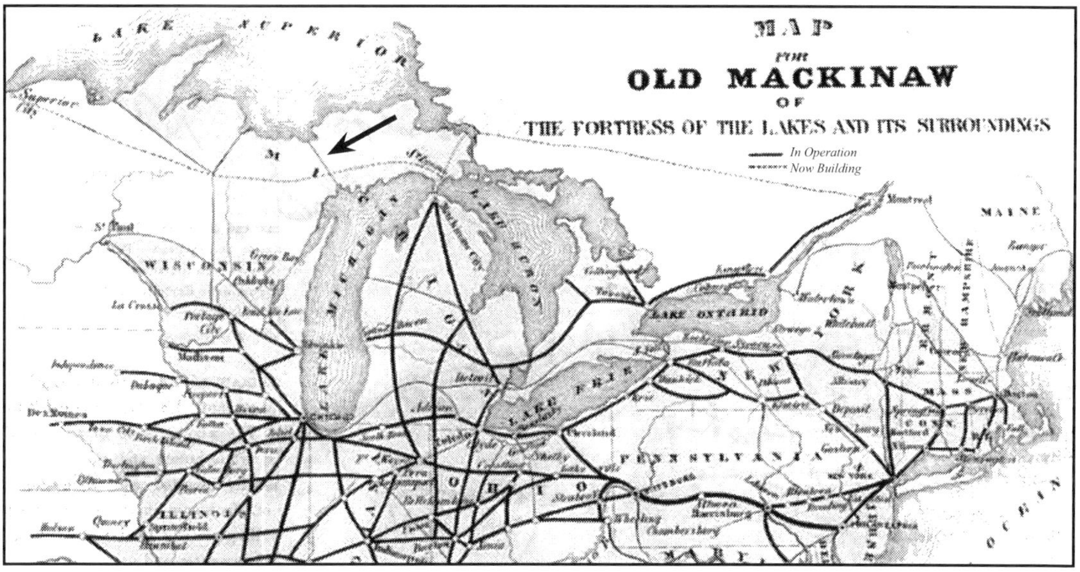

University of Michigan Library

Railroads of Northeastern U.S., 1860 - Railroad line (bold arrow) between Lake Superior and Escanaba should be shown to originate further to the west south of Marquette rather than from the Munising area.

By an act of the U.S. Congress approved on June 3, 1856, certain federal lands were granted to the state of Michigan to aid in the construction of a railroad from Marquette to the Wisconsin state line near the mouth of the Menominee River. On March 18, 1859, the Michigan State Legislature confirmed the grant of land to the Peninsula Railroad Company of Michigan, now the Chicago, and Northwestern Railway Company. (see Appendix Chap. VI, Item 1) Other additional lands were granted to the State of Michigan for the benefit and use of the Chicago and Northwestern Railroad Company after the railroad from Marquette to Menominee was completed and in "running order." These land grants totaled 104,000 acres (162 sections).

Courtesy Michigan Railroad Commission

Railroads of Michigan, 1876
Rail line from Marquette to Escanaba and From Marquette to St. Ignace

Courtesy Superior View Studios

<u>Chief Wawatam</u>, 1911-1984

This was the ferry between Mackinaw City and St. Ignus for passenger, rail, and car. It was a 338 foot coal-fired vessel owned and operated by Mackinac Transportation Company. The steel hull also served as an icebreaker so it could operate year around. Its maximum capacity of 22 fully loaded freight cars was power by three engines rated at 4500 horsepower. The <u>Chief Wawatam</u> was doomed to the scrap yard following the opening of the Mackinaw Bridge in 1957.

Courtesy Superior View Studios

Fleet of boats stuck in ice on Lake Superior near Sault Ste. Marie, April 1905. Shipping on the upper great lakes is now closed between January 15th and March 15th.

By 1880, the Detroit, Mackinac and Marquette Railroad (DM&M) extended east from Marquette to Onota (formerly Glenwood) which is midway between Deerton and Rock River. That same year, transport of trains across the straits of Mackinac was made possible by the ship *Algoma* which was later replaced by the large train ferry, the *Chief Wawatam*. This once proud ship finished out its days as a rusting hulk in the Mackinaw City harbor in the latter part of the 20th century. Commercial transport on the Great Lakes, particularly Lake Superior, has always been problematical. Coast guard icebreakers continue to run between January 15 and March 15 below the soo locks.

In 1886, a merger of several small U.P. railroads created the Duluth, South Shore, and Atlantic (DSS&A), known to many as "darned slow service and abusive". A map of the Deerton/Tyoga area in 1886 shows the railroad between Marquette and Munising. A later map dated 1926 shows the railroad spur north to the settlement of Tyoga. Tyoga is misspelled as Tioga and is erroneously located east of Deerton on the main DSS&A track. In the Appendix, Chapter VI, Item 2, there are a series of maps of railroad expansion and contraction in the U.P. between 1850 and 2005.

Courtesy Superior View Studios

Lake Superior & Ishpeming (LS&I) engine at Birch
Birch was located adjacent to Michigan Highway 550, halfway between Marquette and Big Bay at a junction of Loma Farms Road and the railroad spur.

Although there is no record of the specific railroad equipment used by the Tyoga Lumber Company, Jim Carter, with his long time interest in railroads of the U.P., surmises that in the period 1905-7, Tyoga would be using a steam locomotive similar to the LS&I engine at Birch. It probably would be a lighter "local" work engine not as heavy as those used on the main east-west passenger freight line. There also probably would have been a siding on the main line near the Tyoga spur where cars with lumber from the mill were parked to be picked up for transportation elsewhere or they might have taken lumber directly to Marquette for redistribution. Following a very hard winter in 1906, the company decided to abandon the Tyoga Lumber Mill. The Tyoga station on the main line 1¼ mile east of Deerton and the spur to Tyoga were removed. The overgrown railroad grade can still be followed from Onota to Tyoga.

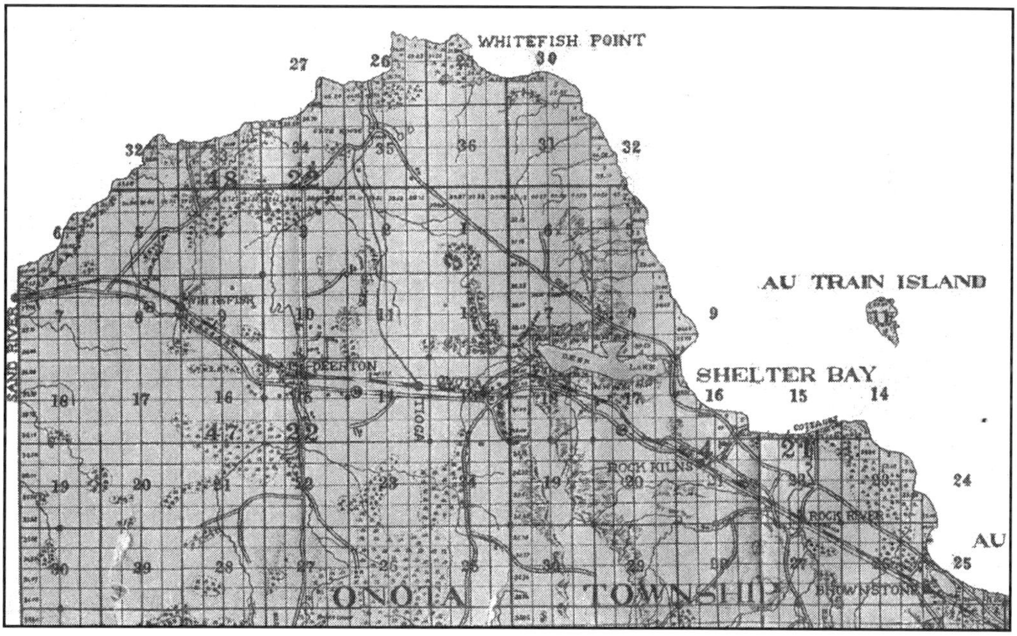

Courtesy Paul Petowsky, Munising, Michigan

The Roads and Railroads of Onota Township – 1926
Tyoga was misspelled and placed between Deerton and Onota on DSS&A line rather than its true location due north in Section 35.

Steam locomotives similar to the one at Birch were used well into the 1920's. The main changes between 1875 and 1920 were an increase in the size and weight of the locomotives and the rolling stock, so that heavier rail had to be laid as the years went by. The mills were often quite busy in the winter when much logging was done by sleigh and iced roads. Large plows mounted on the front of locomotive engines were required so that service was not suspended during the winter months. Otherwise the basic mechanics of steam railroading remained the same but with significant changes and improvements of safety features such as brakes and the comfort and service for the passengers.

DSS&A train with snowplow

The DSS&A and the SOO Line were merged with Wisconsin Central, a division of Canadian Pacific in 1987. By the turn of the 21st century, rails have been removed in many areas of the U.P. with ATV's, snowmobiles and bicycles using the grades for recreation.

Turning to the history of roads in the central U.P., there are currently three roads between Marquette and Munising. To the north is M-28, with construction which began in 1926 and connected Duluth, MN with the Sault Ste. Marie. The second is the gravel Mangum Road between U.S. 41 at Beaver Grove and Rock River ,Chatham Rd HOI. The third road is M-94 which runs easterly from U.S. 41 south of Skandia, to Chatham and on to Munising.

The history of these roads is somewhat confusing as the early road between Sand River and Deer Lake took a more northerly route, crossing the Laughing Whitefish River just north of the Tyoga Lumber Company, the site of the Halfway House. This was the original stagecoach line between Marquette and Onota (Christmas) that carried the mail and an occasional passenger in 1870. This was also called the Old State Road or Route 94, and is now called M-28. The current M-94 extends from M-41 east of K.I. Sawyer Airport, to Chatam and Munising. (A letter from a potential contractor, George Wagner, to Governor J.J. Bagley is included in Appendix Chap. VI, Item 3.)

With stagecoach service available from Marquette to Munising three times a week, a writer for the *Weekly Mining Journal* in 1874 waxed poetic.

> a sleigh ride of forty miles, into and thru a wilderness on a mild December day is not only pleasant but exhilarating. A road built from Chocolay River to Au Train River, permits a steady jog toward Onota. The road, level and smooth, contributes to the comfort and the ever changing scene lures the eye from starboard to larboard – from thin pine plains to dense forest – from ledges of rock to moss valleys fleeced with snow – and from high ridges and abrupt bluffs to the broad stretch of deep rolling sea of the lake and the mad surf of the beach. Nevertheless, as the daylight became more and more shaded, the lead pencil marks on the blazed trees, indicating the distance from Munising, began to be attractive, and were watched with increasing solicitude as they were approached and passed. The lights of the village of Onota [Bay Furnace or Onota before the fire of 1887; current name of Christmas] were not unwelcome to the eyes, nor was the hot supper and warm wood fire ungrateful to the inner and outer man. Rarely is a journey begun with pleasure, continued with delight, and ended with joy. The road had been finished from Harvey to Au Train. The road from Au Train to Munising had been chopped out and put in a passable condition, and will be finished next season to Onota and thence to Munising, thus giving a much needed winter outlet to the people of Schoolcraft County. Messrs. Freeman and Bro. have secured the contract for carrying the mail between Marquette and Onota and Munising thrice each week. A stage line will run daily leaving each end of the route every morning. Stages have been prepared expressly for the route, and passengers will be provided with every possible comfort. The opening of this daily line will be a great accommodation to the traveling public, and everyone may feel assured of safety and dispatch over the route while the line is under the management of these gentlemen.

Further south, an early road between the Chocolay and Au Train Rivers was built in 1874. Early roads were frequently the corduroy style with 8-foot lengths of timber laid parallel to each other and perpendicular to the direction of the roadway. Sand or sawdust were used as fillers. An alternative was using shorter lengths of wood to form a roadway.

The Corduroy Road technique is still used by loggers in swampy areas. – Wikipedia

Courtesy Superior View Studios

A log road was constructed with multiple short lengths of wood.

Between the Tyoga years of 1905 to 1907, there was no good north-south road. Tyoga depended upon the railroad spur. Both of these early techniques of corduroy road and log road building are still employed by logging operations and then left to return to nature after the timber is hauled out. By the turn of the century, some of the roads became designated as improved with the application of oil/aggregate.

Courtesy Superior View Studios

Oil/aggregate roads have almost been completely replaced with asphalt and concrete roadways.

Weather statistics back up the old timers saying that snow loads were significantly greater in the early part of the 20th century and were a significant impediment to public transport and commerce during the winter months. Teams of horses were used to pull large rollers to pack the snow. Mud and flat tires were always a challenge before paved roads.

Courtesy Superior View Studios

Mud and flat tires made travel by early auto a real adventure.

Courtesy Superior View Studios

Team of horses with snow roller

Beginning with the 1600's, Lake Superior became an increasingly important "pathway" for east-west transportation of goods and people. The Great Lakes continue to be the dominant mode of transport for coal, iron ore, and other raw materials. Currently it is more cost effective for coal trains from the Northwest to off-load to 1000-foot transport freighters in Duluth for transport to the Sault Ste. Marie rather than by the longer route around the southern ends of Lake Michigan and Lake Erie. The development of the rail system in the late 1800's and early 1900's made it possible to reach into the interior of the southern shore of Lake Superior to extract its riches of copper, iron ore, and timber for shipment east, west, and south.

In the 20th century, highways have become the dominant mode of transport for goods and passengers because of an aging rail system. The quest to find alternative technologies to eliminate the pollution from coal electric generating power plants has become important as it is estimated that we have a supply of coal adequate to provide electricity for 200 years. Beginning in the 21st century, the reality of limited fossil fuel (oil) has made alternatives of solar, wind and nuclear power more attractive. A "green company" in the U.S. is in the early stages of converting CO_2 emissions into bricks.

The Appendix Chapter VII contains two additional items of interest from the Perron Archives: excursion rates for fishermen and campers of 1905 and the DSS&A annual report of 1907 to the Michigan Railroad Commission.

Chapter VII

CLEVELAND CLIFFS IRON COMPANY

by John Parlin

CCI has been a major economic force in the U.P. since its formation as the Cleveland Iron Mining Company in 1847 by Samuel Mathers. It has had a significant regional impact with the production of iron and wood products. Led by Mathers for 50 years and subsequently by his son, William Mathers, CCI is currently the largest supplier in North America of taconite iron-ore pellets (60% pure) with a 40% share of the market. Jackson Mining Company was the first producer of Iron in the Marquette Range at their site in Negaunee in 1849. It was one of many mines and forges acquired by CCI.

Shipping out of Lake Superior became economically feasible with the opening of the Soo locks in 1855. The design of ships developed to fulfill the function as bulk carriers of lumber and iron ore underwent significant change during the last half of the 19th century.

Courtesy Superior View Studios

Brightie at Bay Point,
was a typical Great Lakes cargo schooner (three masts), one of the early cargo schooners in the mid 19th century which replaced traditional schooners.

Courtesy Superior View Studios

The "Racine", a Port Huron based schooner barge, was wrecked in Chicago 1876.

Courtesy Superior View Studios

"Charles Bradley" (1890-1931)
A hybrid wooden sail/steam cargo ship that burned in 1931 in entry to Keewenaw Waterway

1. The *Brightie* at Bay Point was one of the early cargo schooners in the mid 19th century which replaced traditional schooners.
2. The next generation of a cargo schooner was similar to the *Racine*, towing a barge that was frequently a demasted schooner.
3. The *Charles M. Bradley* (1890-1931) was a hybrid variety combining sail and steam.
4. The *R. J. Hackett* (1869), measuring 211 feet with the capacity of 1200 tons was the first wooden bulk carrier owned by C.C.I. The last wooden bulk carrier was built in 1902.
5. The *Onoka* (1882), was the first iron-hulled freighter which was followed in 1886 with the first steel-hulled freighter, *Spokane*.
6. By 1900, sailing vessels were almost completely replaced by bulk steam-driven freighters such as the *Peter White.* The design of the *Peter White* evolved into the current freighters of 1000 feet like the *Indiana Harbor* which is capable of carrying 79,000 tons with a mid-summer draft of 34 feet.

Courtesy Putin Bay Photos

The "Onoko" (1882-1915)
Was the first iron-hulled bulk freighter, measuring 302 feet and capable of carrying 3,000 gross tons of iron.

Courtesy Library of Congress

"SS Spokane" (1886-1912)
The first steel-hulled bulk freighter on Lake Superior.

Courtesy Superior View Studios

"Peter White" (1937)
A typical freighter of the 20th century.

Courtesy Boatnerd.com

"Indiana Harbor" (1979)
1000 feet long with a beam of 105 feet a draft of 34 feet
and a capacity of 81,000 tons.

By the turn of the century, CCI had subsidiaries in the U.P. and elsewhere that produced coke (replacing charcoal for smelting iron ore at the turn of the century), owned coal mines in Pennsylvania and West Virginia, uranium and shale oil fields, invested in gas and oil drilling, and owned a railroad (the Lake Superior and Ishpeming in 1892). By 1984, they had divested themselves of the Great Lakes shipping business.

In Alger County, CCI had many products based on the timber industry other than lumber, including well known Munising Woodenware, veneer, wood-ethyl alcohol, pulp for paper, cedar and spruce for mining shafts, and hemlock bark for the tannery in Munising. Modern timbering techniques leave little waste.

By 1920, the emphasis in the U.P. was on mines and forest products with CCI remaining the largest landholder in the U.P. By 1930 CCI had amassed 1.5 million acres with the addition of approximately one million acres after 1907. By 1995, CCI had seven iron mines in the U.S., Canada, and Australia. By 2006 they employed 40,000 workers in North America. In 1990, the CCI forest division was sold to the Bishop Estate Trust of Hawaii and managed as Benson Forest Products until its purchase by Shelter Bay Forests in 2000. In 2006, 500,000 acres were sold to Forestland Group. Of that, the Nature Conservancy negotiated a conservation easement on 248,000 acres assuring sustainable forest management by Forestland. In addition, the Nature Conservancy purchased 23,000 acres outright to protect unique areas such as the Two Hearted River between Grand Marais and Whitefish Point NW of Tahquamenon Falls.

By 2000, the closing of mines in the Marquette range was being predicted due to the dwindling supplies of high grade ore. Thanks to a partnership with Kobe Steel of Japan, CCI can use Kobe's patented process to make taconite into nearly pure iron in nugget form which will greatly prolong the life of the mines in the Marquette range. Much of the iron ore ends up in China or India as heavy industry continues to decrease in the U.S.

Although the Tyoga Lumber Company name was lost after its purchase by CCI in 1907, many of the 7,000 acres are currently being logged. A common belief is that CCI had frozen out Tyoga by owning surrounding lands. The CCI annual report for years ending November 30, 1907, gives us a snapshot of the Tyoga Lumber Co. operation between the time of its sale to C.C.I. on March 18, 1907 and its relocation in Munising in the fall of 1907. (Appendix, Chap.VII, Item 1) The report of the CCI Land Department states:

> the remaining bodies of timber are quite closely held and there is very little offering of sizeable compact acreage. There still remains in the market isolated descriptions and some contiguous to our holdings. The prices on these lands have risen from one

hundred to one hundred and twenty-five per cent during the past four years. The purchases for the fiscal year were mainly small parcels with the exception of the Tyoga, Powell, Prickett and Jenson Purchases. The demand for timber lands is keen and there would be very little difficulty in disposing of a very large acreage of our timbered lands at prices in the neighborhood of $25.00 per acre. We are making every effort to acquire small holdings adjacent to our lands and find the owners stiff in their prices and not willing to make concessions. Prices range from $10.00 to $20.00 per acre ($265 to $530 in 2008 dollars. (Onota Township rural wooded land sold for approximately $600 per acre in 2008.) We can conservatively estimate owning 600,000 acres of well timbered forest lands in the U.P.

The report further states that

with the purchase of the Tyoga Lumber Company we acquired a modern single band saw mill. It has up to date machinery with a rated capacity of 40,000 feet per day (pine timber). The mill was only operated seventy days and its machinery was in good condition, and the mill was put up substantially with heavy timbers and good workmanship. We made an effort to sell the mill and might have done so at a sacrifice. It was deemed best, however, to move it to Munising and operate it by utilizing such material as we can conveniently buy and get from our lands. There is a large class of material on all lands that buyers have to scale closely to make sure of coming out even. With our own mill and railroad we are able to save any over-run there might be in the logs, besides utilizing the slabs, etc., as fuel for the Paper Mill, in other words, making use of all the material in the defective log.

The inventory from Tyoga Lumber Company in 1907 included:

80,000 feet of hemlock timber at $9 per foot – $720.00
1200 cedar poles at $1 – $1200
25 thousand feet mixed softwood at $10 – $250,000
One hundred 32 inch mixed lath at $1 – $100
Tyoga store, etc. – Lots of harness, camp tools, and supplies - $30

The receipts for the year end 1907 from Tyoga Lumber Company included

Tyoga store ($37), logs ($10,000), lumber ($1,400), and cedar ($1400). The total receipts for the Land Department were $42,000 with expenses of $23,000, the largest receipts came from the Hotel Williams on Grand Island, the sale of stumpage, and the sale of limestone.

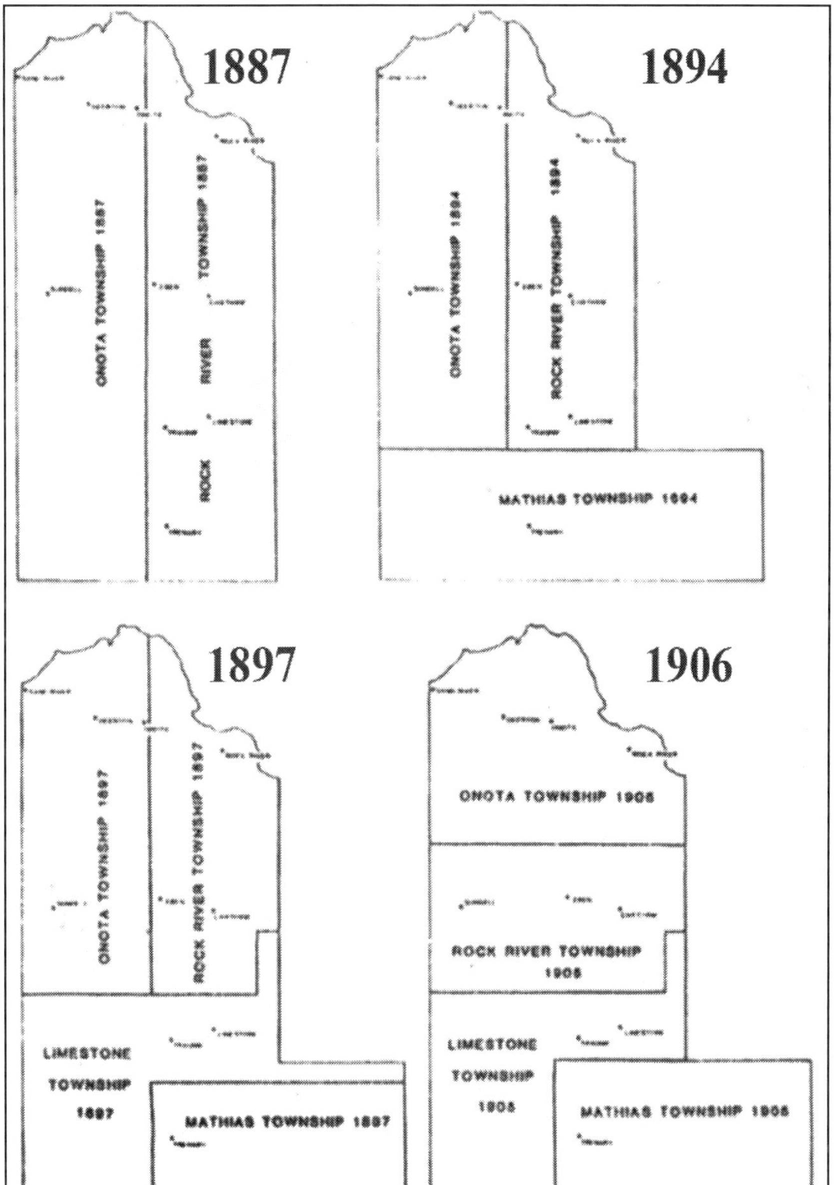

Courtesy Alger County Historical Society

The evolution of townships (1887-1905) with changes in the name, size and shape of Onota, Rock River, Lewiston and Mathias Townships

The 1907 annual report's record of the purchase of Tyoga Lumber Company (entry # 589) earlier in the year indicates that there were 7,141 acres at a price of $9.37 per acre with total purchase price of $140,000. In addition, there was $284 in taxes and legal expenses and $294 was listed as miscellaneous.

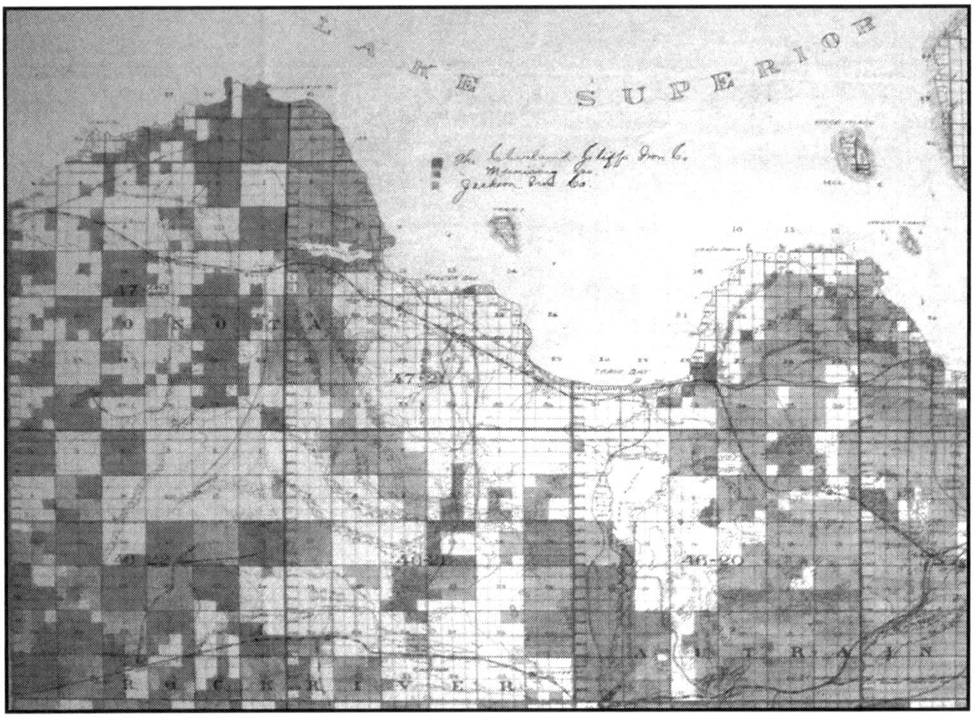

Courtesy Northern Michigan University Archives

Timberland ownership of Alger County pre-1900 embraces CCI (darkest shade), Munising Company (medium shade), and Jackson Iron Company (lightest shade). CCI purchased Munising Company and Grand Island in 1900.

CCI's land holdings in the U.P. did not occur all at one time but in bits and pieces. CCI's holdings at any given date are somewhat difficult to depict as plat maps frequently lag behind transfers of land. To understand local holdings of CCI by township, one needs to follow the evolution of Onota, Rock River, Mathias, and Limestone townships as depicted in the Alger County Historical Society Centennial Edition. By 1905, Onota Township (56,000 acres) and Rock River Township (52,000 acres), had achieved their present configuration. An undated CCI map shows part of Alger and Schoolcraft Counties with the holdings of CCI, Munising Company, and Jackson Iron Company. This map can be dated to about 1900 as Grand Island was purchased by CCI in 1900 (for a grand total of $93,000), the Munising Company in 1900, and the Jackson Iron Company in 1905. On this map, the Jackson Iron Company lands extend south of Deer Lake. CCI is shown as the owner of the region of Tyoga Lumber

Company. As CCI did not purchase Tyoga Lumber Company until 1907, Terry Reynolds, an authority on CCI at Michigan Tech, surmises that CCI would have taken this existing map and colored in the Tyoga purchase of 1907 with the darker shade after that date.

The year-end corporate report of November, 1907, breaks down the CCI holdings in the western, central, and eastern divisions.

	FEE	SURFACE	MINERALS	TIMBER	TOTALS
Western Division	60416.17	14183.53	1431.57		76031.27
Central Division	128360.80	45691.88	1714.97	8396.11	184163.76
Eastern Division	286194.01	640.00			286834.01
	474970.98	60515.41	3146.54	8396.11	547029.04

Courtesy Northern Michigan University Archives

Fee = all rights mineral and timber
Surface = land and timber rights but not mineral rights
Minerals = mineral rights only
Timber = timber cutting rights
all units are in acres

The following table breaks down the CCI acreage in 1906 and 1907 for the townships of Alger County as well as Chocolay Township in Marquette County. Note the 10,000 acre increase in Rock River Township between 1906 and 1907 – which probably reflects the Tyoga purchase.

CCI ACREAGE

TOWNSHIP	1906	1907
AuTrain	14,200	-
Burt	71,000	73,000
Grand Island	5,600	5,600
Limestone	11,800	11,800
Mathias	14,150	14,150
Munising	25,244	25,270
Onota	16,656	13,223
Rock River	5,089	15,459
Chocolay	3,132	7,603

Courtesy Northern Michigan University Archives

The next table shows timber cut in 1906 on C.C.I. lands, which includes holdings in Alger, Marquette, and Baraga Counties.

CCI TIMBERLANDS 1906 *(Wood in thousands of pole cords*)*

TOWNSHIP	ACREAGE	HARDWOOD	% HARDWOOD	SOFTWOOD	CHARCOAL WOOD	TOTAL
Grand Island	16	20	67	10	28	59
Burt Township	236	642	60	162	251	864
Onota, Munising and Marquette County	118	237	65	131	326	695
M & T and Barage County	78	340	65	180	285	805

Courtesy Northern Michigan University Archives

The 1907 CCI report continued:

> The Tyoga mill is equipped with a lath mill, and we shall add a shingle mill and perhaps, several small saws with which to cut box material, crating and other similar stuff. The mill is being located near the west side of the so-called Burtis Property [Munising].

As with any large corporation, executives combine hard work with some play. CCI owned the Grand Island Forest Preserve as well as a fish hatchery and an Alger County farm, nursery, and greenhouse. In 1908, they also introduced a single raccoon. This one then met up with one of the opposite sex and our garbage cans have not been safe since.

The 1907 the CCI annual report continues:

> [On Grand Island] The sharptail grouse are frequently encountered in the vicinity of the old Duffy farm. The native partridge appear to be increasing. The wild turkeys which are kept in an enclosure at Williams Landing, do not increase. The number of eggs have gathered each year and hatched, but have been unfortunate in not being able to raise the chicks. This is also true of pheasants. Our spring season is usually wet, with violent changes in temperature which is very destructive to both birds. The farmers find it almost impossible to raise domestic turkeys... although chickens seem to thrive. The Caribou have caused some apprehension as to their whereabouts. The keepers have been

---

**1 pole cord: 4' x 4' x 8' = 128 cu'*

unable to locate them, although from time to time apparently their tracks are found in different parts of the island. As these animals are great roamers and very alert, it is not strange that they can keep out of sight. Six young caribou were added during the fiscal year. These were put in an enclosure with expectation that they would become tame and could be watched over for a year to two. Two have since died from apparent over-feeding.

Capercailzie (wood grouse – native to Northern Europe with dark plumage and fan-like tail) and Black Game (heathgrouse – also European) have been seen periodically during the summer.

Three wolves made their appearance on the Island at different periods, the first one coming on the West Side near the Lookout and following the road to the North Light near Gull Point. From there it went inland and in less than a mile killed five deer. This wolf was finally driven off after twenty-four hours of continuous hunting. It left the Island at Gull Point and passed in front of some fishermen on the North End of the Island."

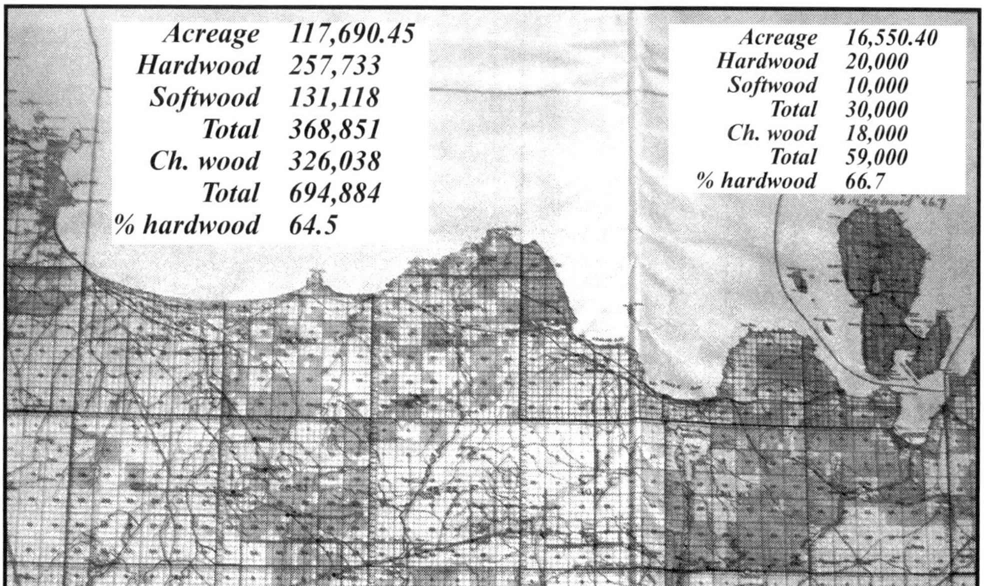

Courtesy Northern Michigan University Archives

CCI holdings, 1912 (darker areas are CCI)

Donations in 1907 included supplies to the poor ($30) and $230 to the highway fund. In the year-end report of November 30, 1908, CCI lists donations to the Munising Fire Department ($25), for the rental of Munising YMCA building ($9), and for the maintenance of the Munising playground ($300).

They had retained the Halfway House on the Laughing Whitefish River to be used as a "club house camp". There were also some residual frame buildings and two log dwellings at Tyoga.

The CCI map from Northern Michigan University archives dated 2/10/1912 shows the acreage and breakdown of categories of wood in cords times 1000. (Fig. 30) With the exception of Grand Island, the red lines that they have used to designate the last three areas are not geographically accurate and only an approximation.

The following two plat maps are of Onota Township. The earlier Alger County plat map is from somewhere between 1910 and 1920. Land holders Le Bounty and Pamona Manor farms are shown south of Onota. Lands south of Deer Lake were subsequently purchased by CCI as shown on a current plat map almost 100 years later. When comparing Onota Township Plat maps of 1920 and 2001, there is relatively little change in CCI land north of M-28.

The Tyoga settlement area of 200 acres was purchased by the state of Michigan for back taxes in 1939 and designated as the Bay DeNoc Forest or Escanaba River State Forest. The Michigan DNR has designated this as the Tyoga Historical Pathway.

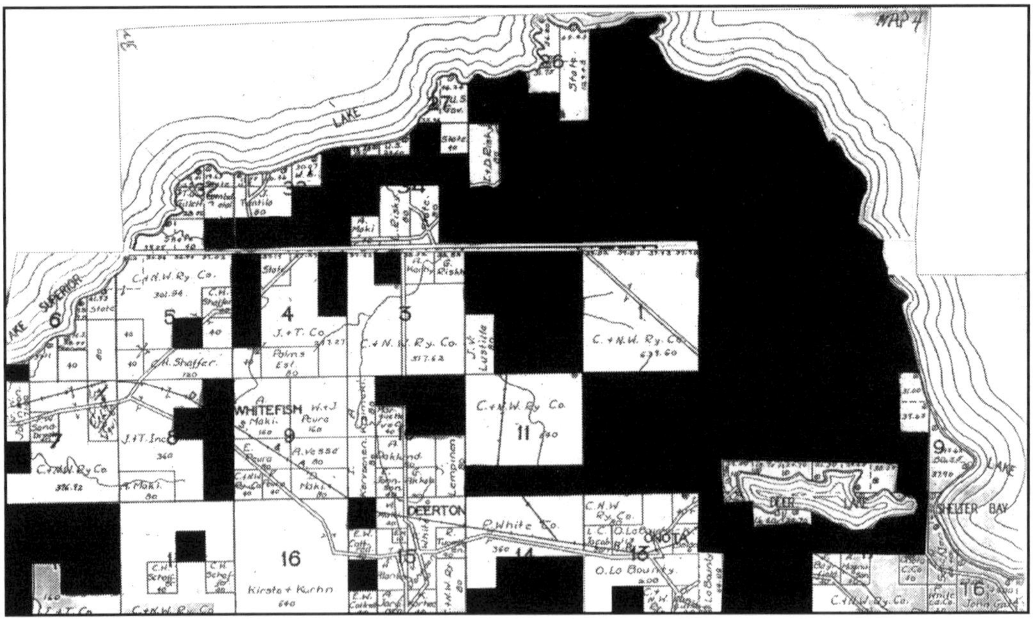

Alger County Plat Map 1920, T.47 – 48N.-R.22W. Darkened areas are CCI

Alger County Plat Book

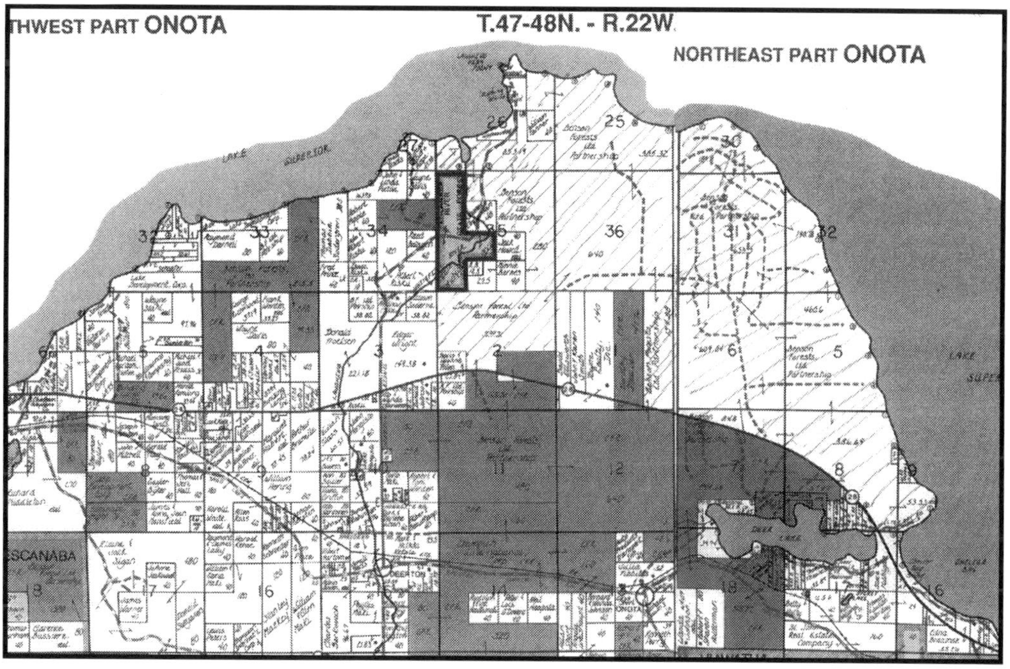

Alger County Plat Book

Alger County Plat Map 2001, T47-48 N and R22 W, the Dark shading is CCI, the light shading is State and Federal lands. Tyoga Historical Pathways perimiter is marked with a bold line.

The Tyoga Lumber Company ceased to exist as a separate entity and was folded into CCI which has steadily increased in size in both the U.S. and worldwide between its founding in 1847 and the present day. For the U.P., particularly the northern part, CCI continues to be the major employer through its iron ore mines. The forest division has been sold with conservation easements to regulate the timber harvest.

Chapter VIII

GEOGRAPHY

by John Parlin

Early Maps of Lake Superior 1400's – 1840

Maps seem to hold a special interest for readers when viewed in sequence as they show the geography of the area during its early exploration. Maps of the "world" as known then date back to Ptolemy in the 6th century B.C. Beginning with the 15th century, early maps began to depict Lake Superior as the best prospect for finding a route between the Atlantic and the Pacific, the ultimate goal for early explorers of North America. An early unauthenticated Chinese world map of 1418 shows the North American continent with a St. Lawrence outlet and a continental divide.

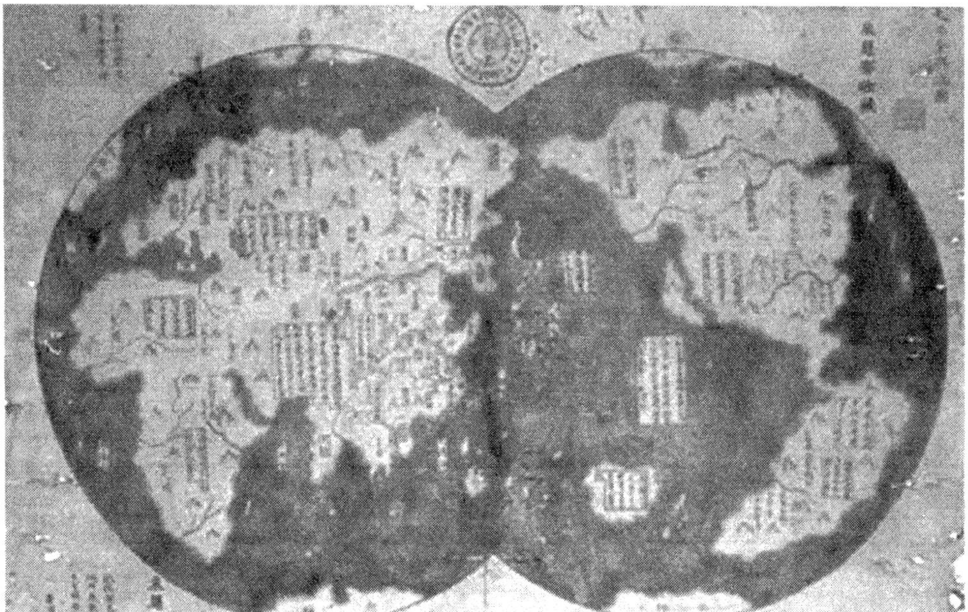

Courtesy BBC news

A Chinese map of the world, 1418, which is regarded by some as bogus. The American Continent is to the right.

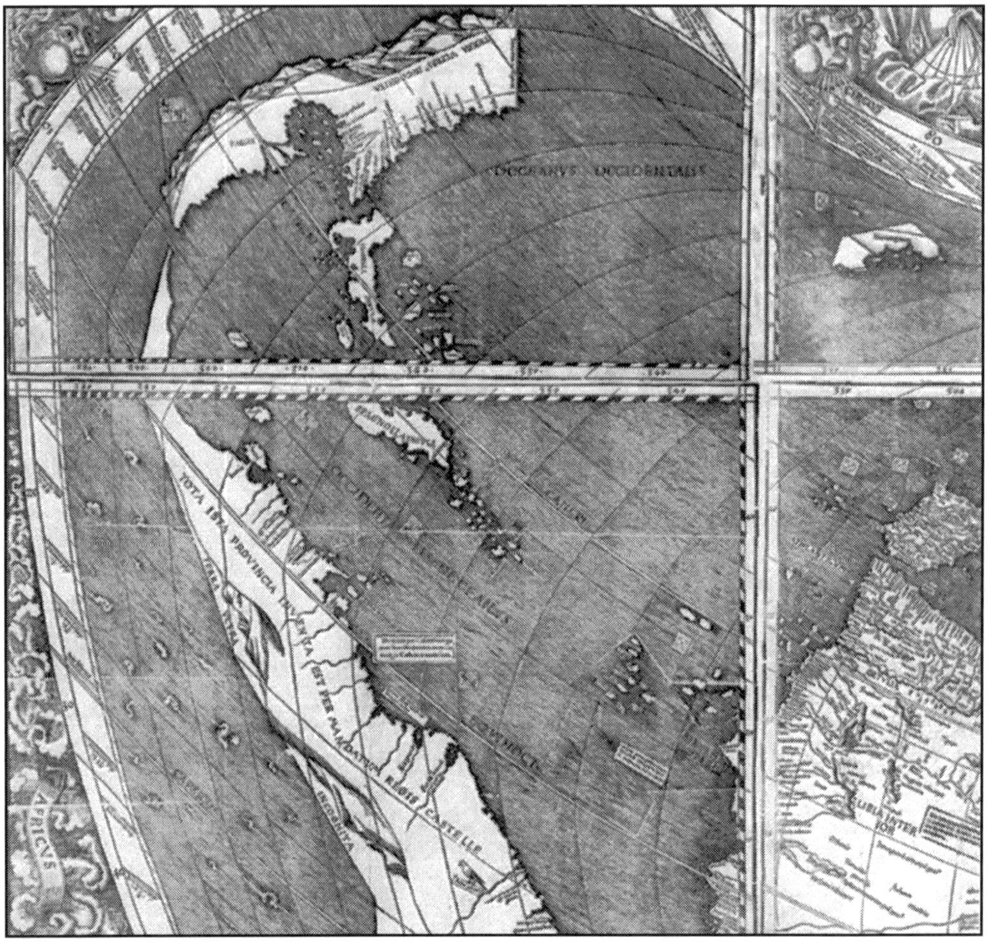

U.S. Library of Congress

Pictured above is the Martin Waldseemueller World Map – 1507. The western third of North America is shown as mountains labeled as "Terra Ultra Incognito" (translation a territory beyond our knowing).

The Martin Waldseemuller "flat map" of the world is the only surviving copy of the original. The original of which there were 1,000 copies. was based on the 1507 Voyage of Amerigo Vespucci and is the first map to depict North America as being separate from the European-African continent. Note that this includes the area east of the Alleghenies with the Terra Incognita to the west. This does not include the northern portion of United States or lower Canada. Compare this to a current "flat map" of the world.

U.S. Library of Congress

A portion of the Matteo Ricci World Map, 1602
The St. Lawrence and Great Lakes are drawn as a transverse waterway in the upper right. An Italian born Jesuit Priest drew this map during his mission years in China.

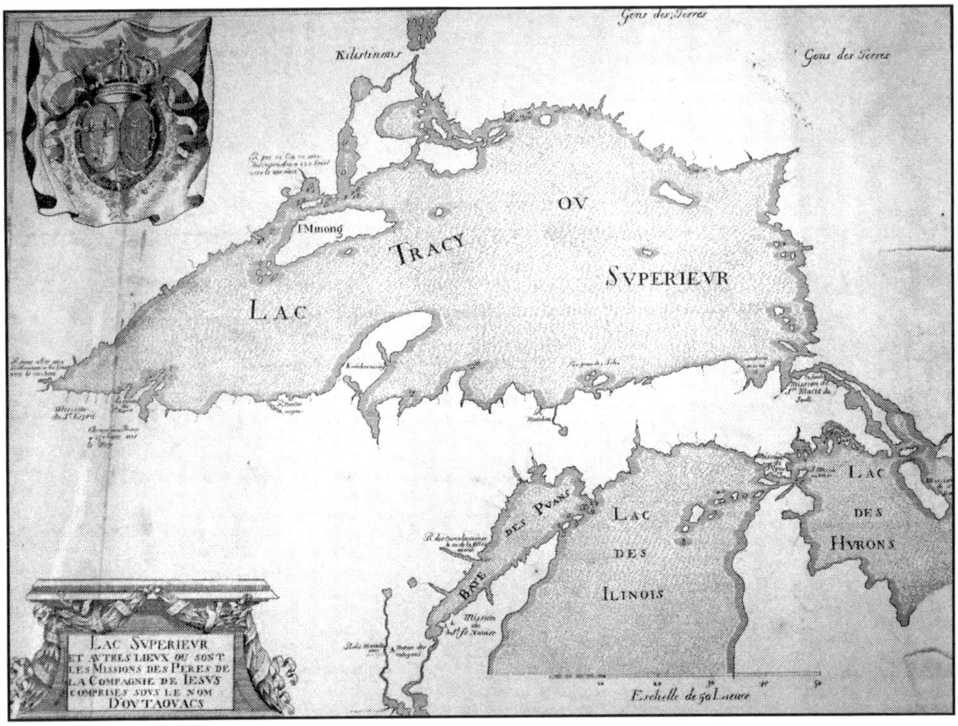

Courtesy Peter White Library, Marquette

This Jesuit map of Lake Superior, ca. 1620, is remarkably accurate when compared to modern maps.

Early Jesuit explorers developed a series of maps of the region between 1507 and 1740. Other maps of the area were from geological expeditions beginning in the 1800's. On the Jesuit maps from 1620, River Mataban (AuTrain) is the end of a portage between the waters of Lake Michigan and Lake Superior. The *Carte de la Nouvelle France* by Champlain in 1632 includes the eastern 80% of Grand Lac. Sanson in 1656 showed the Americas ending at the mid point of the southern shore of Lac Superieur and excluded the northwest one-half of the United States. Green Bay is labeled as Lac de Puans. The first maps to include all of Lake Superior in some detail were by the Jesuit Allouez in 1671 and Father Jacques Marquette in 1673. On these early maps, that of Allouez is the first to draw the Laughing Whitefish River although the label is unreadable. Lake Superior is sometimes referred to as Lac Tracy. Tracy was the Viceroy of New France in the 1600's. It was also called Lac de Conde after a French prince who lived from 1588 to 1647.

---

*Puans is French for "stinking persons"

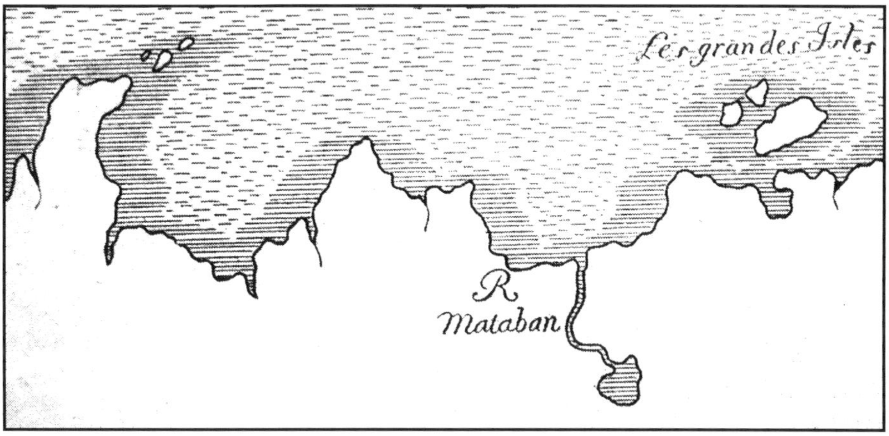

Courtesy Peter White Library, Marquette

A portion of the Jesuit map of Lake Superior dated ca. 1620
Autrain Lake and Autrain River (labeled here as R Mataban)*
The full map covers the southern shore of Lake Superior between the Keewenaw Peninsula and Grand Island.

Courtesy Wisconsin Historical Society

Samuel de Champlain map of New France, 1632
Lake Superior (Grand Lac) in lower left corner

---

*Mataban is Arabic for "Clay Pot" or French for "barrel shaped jar"

Courtesy Pennsylvania Regional Maps

French Cartographer Nicholas Sanson map of New France, 1656

The map of New France of 1791 summarizes the maps of the eastern half of the United States and the south-east quarter of Canada. Very little has changed on maps of today.

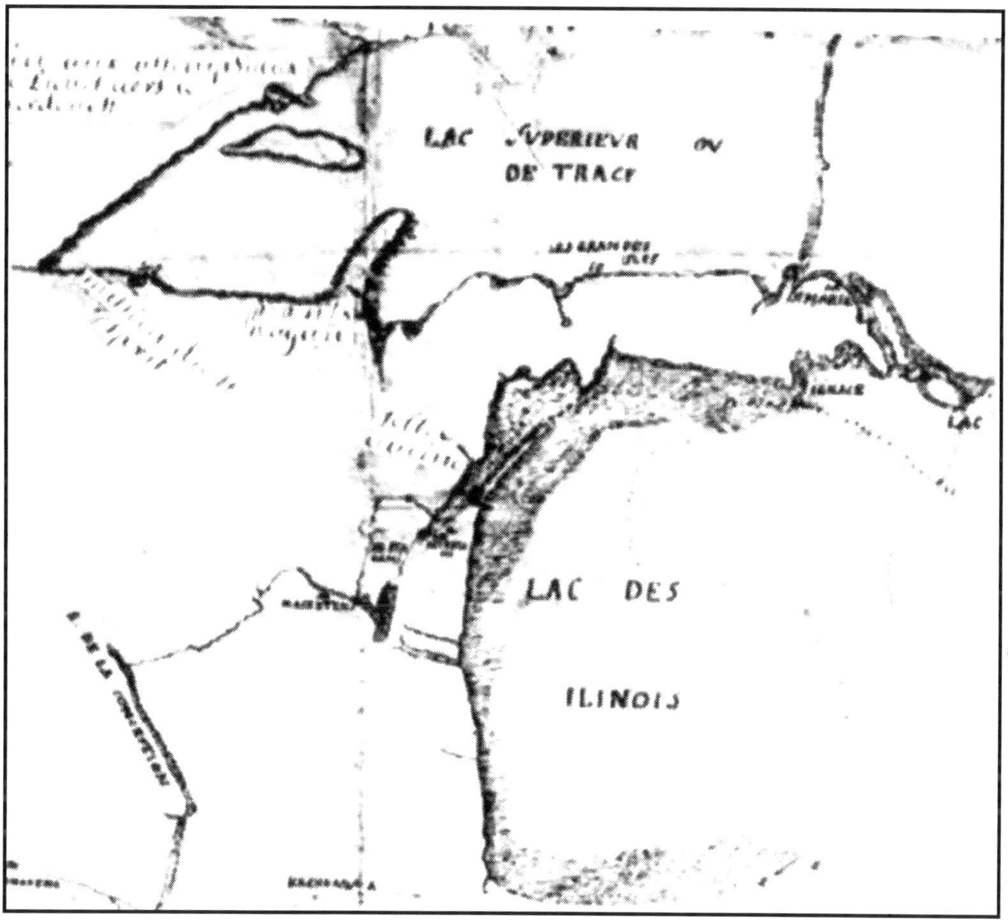

Courtesy Wisconsin Historical Society

The Father Jacques Marquette map of 1673 is rather primitive but does show Grand Island and Au Train River and Lake.

Prior to the surveying of the Upper Peninsula in the 1840's by William Burt, the Bradford map of 1835 named some of the rivers of the southern shore of Lake Superior, including the Chocolay. The Chocolate (Chocolay) River and Au Train River (River Mantaban) are included on maps from this time forward as these were frequent stopping points for travelers along the southern shore of Lake Superior. The Sand, Laughing Fish, and Rock Rivers were added to maps following the Burt surveys of the 1840's. *Lake Superior Place Names* by Bernard Peters contains a thorough explanation of the names of the rivers of the south shore of Lake Superior between the Sault Ste. Marie and Duluth.

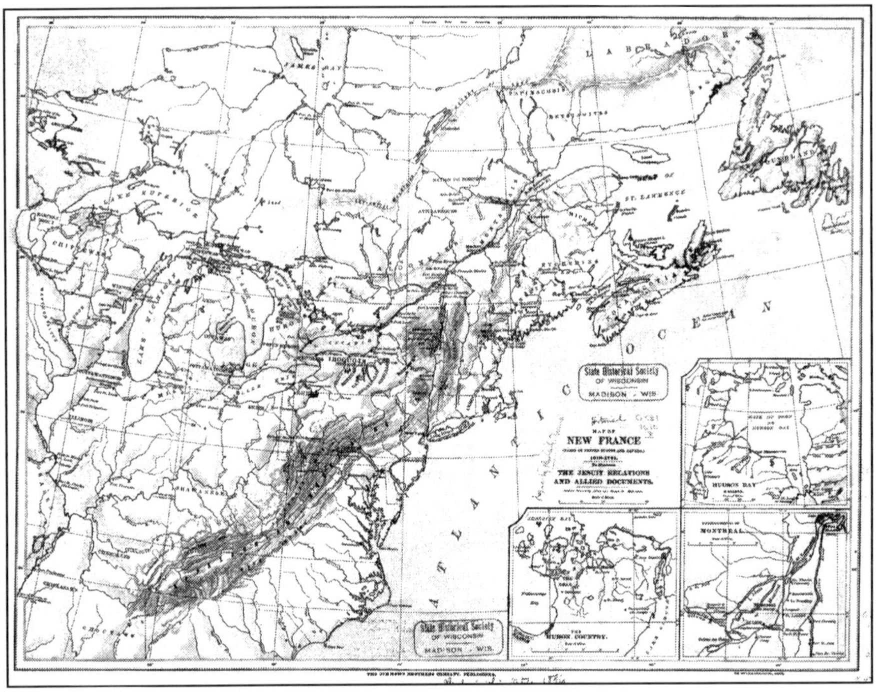

Courtesy Wisconsin Historical Society

Map of New France, 1791

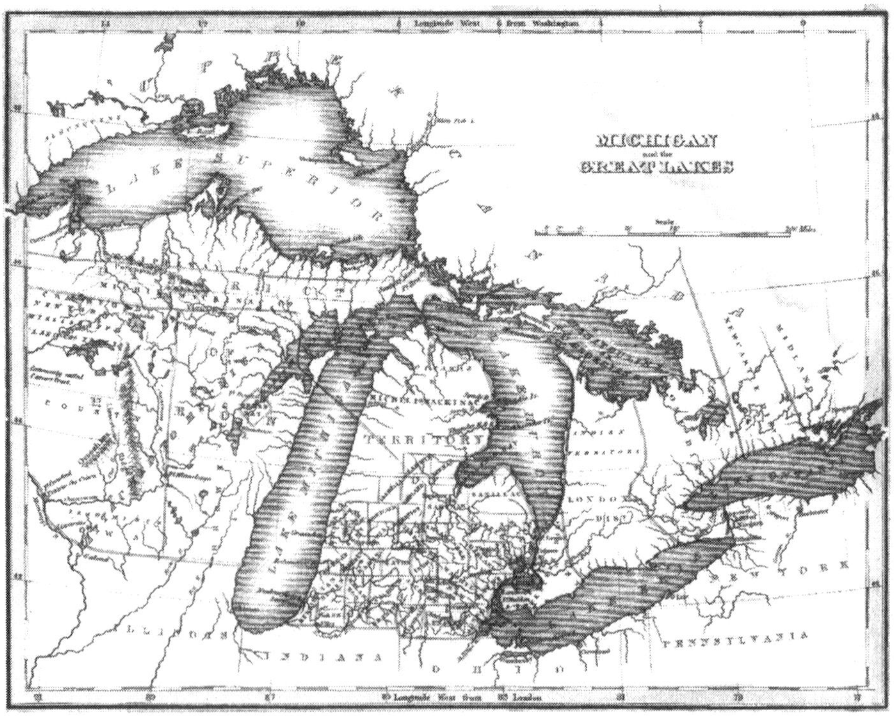

Courtesy Michigan Digital Map Library

Boston Cartographer Thomas Bradford Map of Michigan and the Great Lakes, 1835

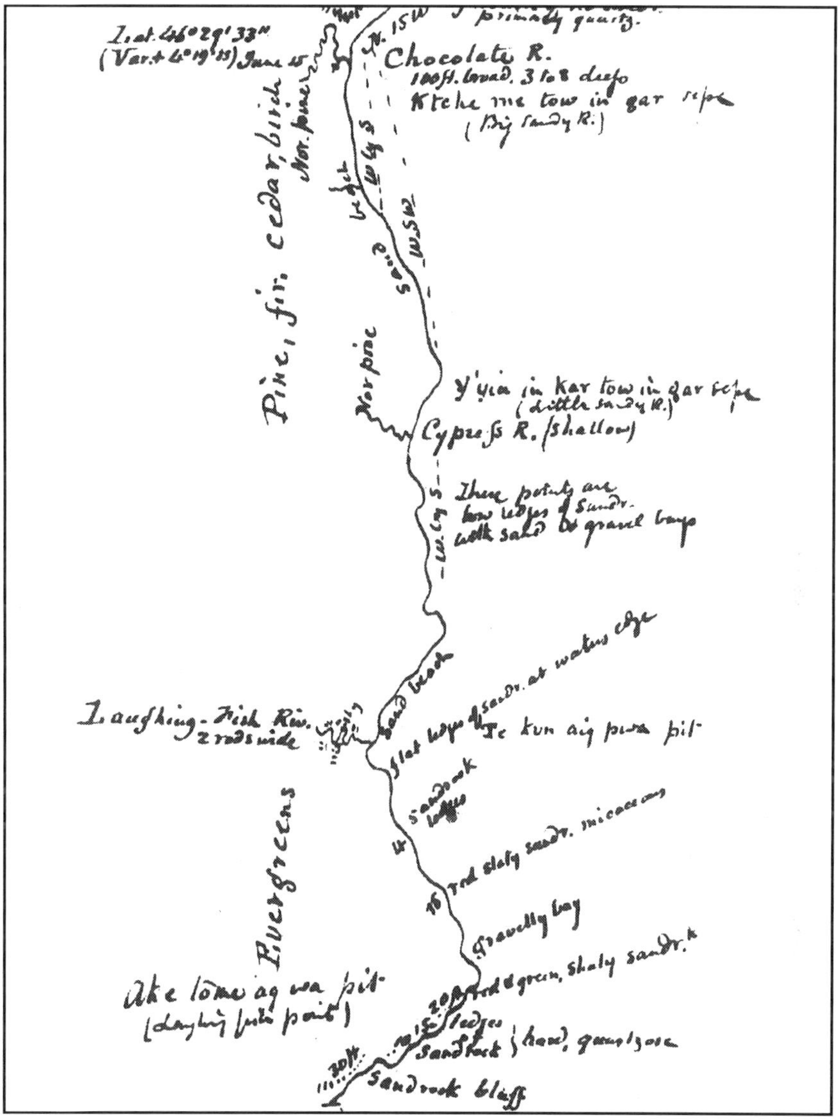

Bela Hubbard – Lake Superior Journal

Bela Hubbard map of southern shore of Lake Superior between Marquette and Shelter Bay – 1841. The map includes Chocolate River (Chocolay), Cypress River (Sand River), and Laughing Fish River (Laughing Whitefish River).

The last map prior to Burt survey maps is that of explorer and geologist Bela Hubbard in his *Lake Superior Journal*, who was traveling with Douglas Houghton and William Penney in 1840. Bela Hubbard's drawing in 1841 is the first detailed drawing of the Lake Superior shoreline between the Chocolay River and Grand Island. On it he described the Chocolay River as 100 feet broad and 3 to 8 feet deep, the Sand River as Cypress River (shallow), the Laughing Whitefish River as Laughing Fish River (2 rods or 32 feet wide) and the Au Train River as the Dog-Train River.

Honorable William Burt, 1792 - 1858

SURVEYING OF THE UPPER PENINSULA 1841 – 1848

Maps of the Upper Peninsula prior to 1841 include only the terminal end of streams flowing into Lake Superior with little attention to the relatively unexplored interior. The traffic on Lake Superior in the 17th and 18th centuries was a combination of Jesuits establishing missions and the Voyageurs plying their fur trade. The coming of logging operations in the southern 80% of the Upper Peninsula after 1800 and the northern 20% after 1840, followed by iron and copper mining beginning in the 1840's, mandated the need for accurate maps.

Thomas Jefferson had spearheaded the Land Ordinance Act of 1775 which established the legal basis for public land surveys:

1. Land must be purchased from Native Americans before it could be surveyed.

2. Survey must precede sale to sellers by auction.

3. Lines measured by chains with trees blazed. Lines were to be described on a flat map to include post at each section corner as described by the size of each tree species.

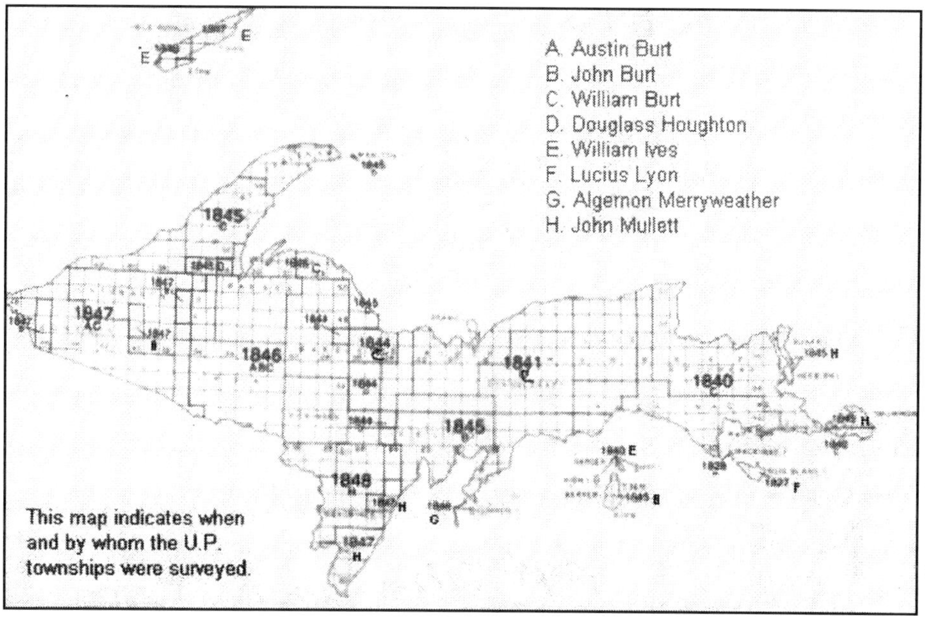

Courtesy State of Michigan magazine – 1982

Upper Peninsula Township survey 1841-48
It was laid out along township and range lines.

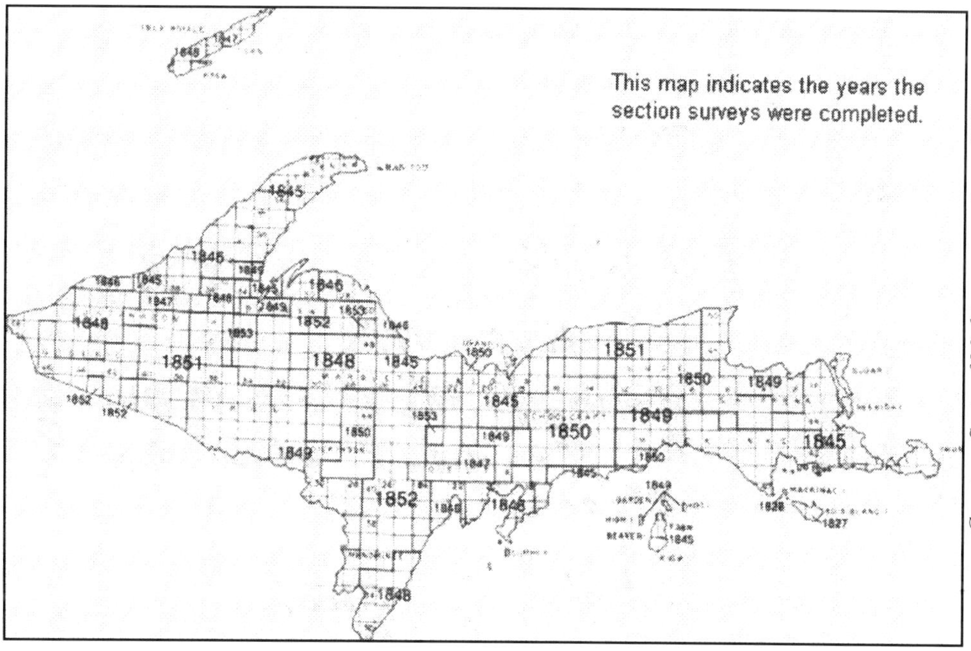

Courtesy State of Michigan magazine – 1982

The Upper Peninsula section line survey of 1845-55, was a survey of 17,000 square miles. It was completed at a cost of 2 cents per acre.

The process was detailed by the Continental Congress in 1784 (Appendix Chapter VIII, Item 1).

Surveying the lower peninsula of Michigan had been in progress since 1815, creeping northward at 10 miles per year, until the Straits of Mackinaw were reached in 1840 with the meridian of 84 degrees 22 minutes extended across the straits to Sault Ste. Marie. Between 1841 and 1848, William Austin Burt and associates laid out the areas depicted on the township/range and section map of the Upper Peninsula. William had four sons one of whom, John Burt, settled in Marquette and played a large role in the development of that area. Burt laid out the township and range lines of Alger County in 1841. A township was defined as a 36 square mile unit of land divided into 36 one square mile sections and a range was defined as a north-south column of townships, east or west of a reference longitudinal Meridian. The section lines were subsequently laid out by William Ives in Alger county. A section and its divisions are shown in the land description diagram.

Surveyors relied on the use of chain, axe, and compass. In 1835, William Burt made a great contribution to surveying with the development of the solar compass which functions similarly to the sextant used in navigation. This instrument required skies without clouds, sometimes a luxury in the Upper Peninsula. Even in the GPS era, surveyors continue to use chain and compass because of the superior accuracy.

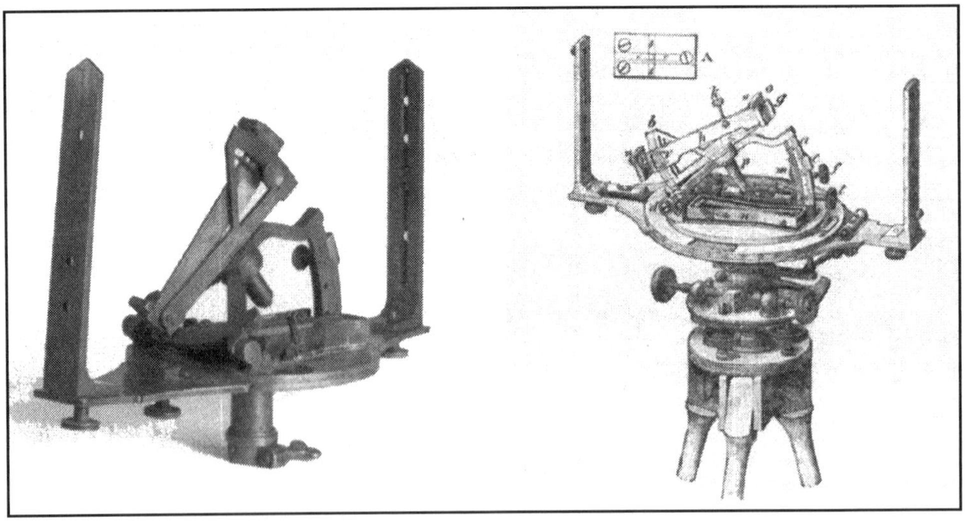

Courtesy Wisconsin Historical Society

William Burt's solar compass of 1855 used the sun instead of a magnetic needle to determine true north. The sextant used for celestial navigation had been invented in 1757.

LAND DESCRIPTION

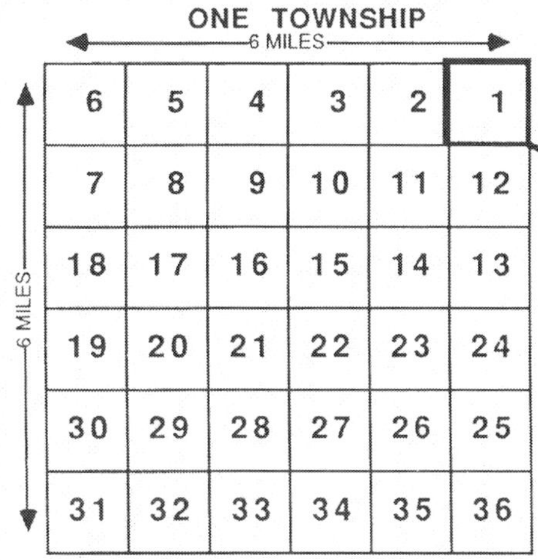

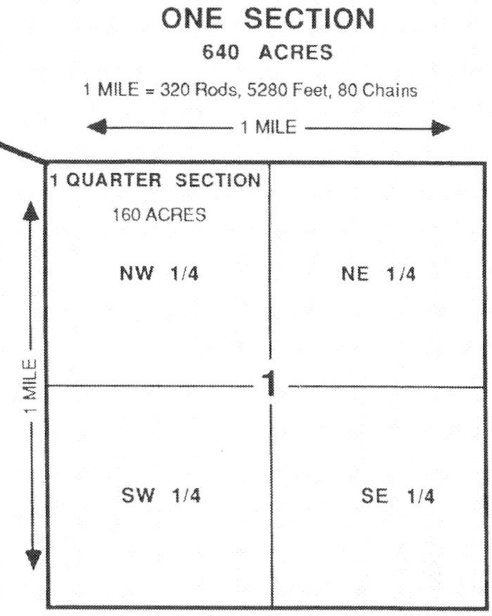

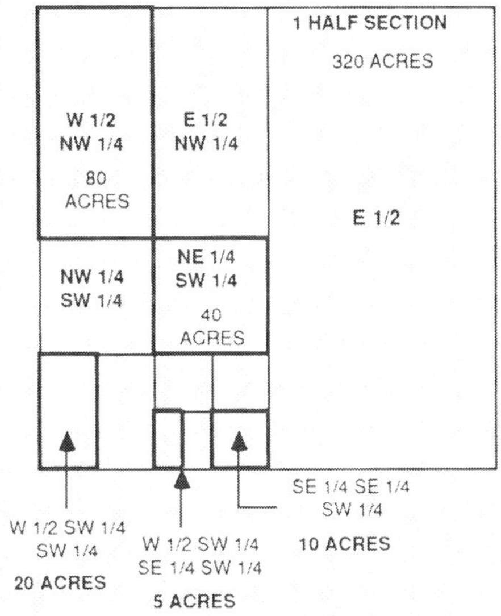

METES & BOUNDS is a method of describing land which identifies a parcel of land by its shapes and boundaries. Permanent markers placed in one corner of a parcel are used as starting points. The land is described in terms of distance and direction from this point. Distances are measured in feet, direction is shown in degrees, minutes & seconds. Each circle is 360 degrees, 60 minutes (') in each degree and 60 seconds (") in each minute.

Another method of land description is the **RECTANGULAR SURVEY SYSTEM.** This method does not use physical markers. Instead it is based on imaginary lines --**LATITUDE LINES** are the east-west lines and **LONGITUTE LINES** are the north-south lines. Certain longitute lines are known as **PRINCIPAL MERIDIANS**. For each of these there is an intercepting latitude line known as a **BASELINE**. In the U.S. Survey System, there are 36 principal meridians with their intersecting base lines. **RANGE LINES** are longitude lines lying every 6 miles east and west of each principal meridian and are consecutively numbered east and west of each principal meridian and are consecutively numbered east or west of the principal meridian. **TOWNSHIP LINES** intersect with the range lines and run every six miles north or south of a baseline. Thus each 6 by 6-mile imaginary square is called a township. Each 36 square mile township is divided into 36 one-square mile units called **SECTIONS**. Each section contains 640 acres and each acre is 43,560 square feet. 1/4 sections contain 160 acres and dividing that quarter section into another quarter results in four 40-acre parcels. i.e. the NE 1/4 of the NE 1/4 contains 40 acres. Secitons are numbered 1 through 36 starting in the upper right hand corner of a township.

MEANDERED LAKE OR STREAM - Meandered water is generally known as public water, meaning the land owner pays the taxes on the land only and not the land under the water. The land is divided into divisions known as government lots which is the legal description for that piece of land. If there are government lots adjoining the body of water it is meandered, otherwise it is privately owned land and the owner is paying taxes on the land under the water as well as adjoining the water. The landowner also owns fishing, hunting, etc., rights on that piece of land within the regulations of the federal and state governments except when the water is navigable.

SQUARE MEASURE

144 sq. in	1 sq. ft.	160 sq. rds.	1 acre
9 sq. ft.	1 sq. yd	4,840 sq. yds.	1 acre
30 1/2 sq. yds	1 sq. rd.	43,560 sq. ft.	1 acre
16 sq. rds.	1 sq. chain	640 acres	1 sq. mile
1 sq. rd.	272 1/4 sq. ft.	1 sq. mile	1 section
1 sq. chain	4,356 sq. ft.	36 sq. miles	1 township
10 sq. chains	1 acre	6 miles sq.	1 township

LINEAR MEASURE

7.92 in.	1 link	66 ft.	1 chain
3 ft.	1 yd.	80 chains	1 mile
25 links	16 1/2 ft	320 rds.	1 mile
16 1/2 ft.	1 rd	8,000 links	1 mile
100 links	1 chain	5,280 ft.	1 mile
5 1/2 yds.	1 rd.	1,760 yds.	1 mile
4 rds.	100 links		

Alger County plat book

Land description of range/township and divisions of sections

In letters to his wife, Burt vividly describes the rigors of the life of a surveyor: "The heavy brush took its toll on clothing" Burt wrote to his wife as he neared the Superior shore. "Coat and pantaloons are most gone," he requested that she make him a new outfit from the "strongest kind of bed-ticking." Years later, when he published a manual for surveyors, Burt recalled his own Upper Peninsula experiences and suggested that work in the field called for stout boots, heavy trousers and "flannel for underclothes…for all seasons and all kinds of weather." He added that a large silk handkerchief "to tie over the ears and neck, is a good protection from flies and mosquitoes."

The distances between towns in the U.P. forced surveyors to hire as many men to pack in supplies and equipment as to clear brush and sight town lines. Operating costs soared. Since the deputy surveyor paid all help from his own pocket, his chances of making much on a given contract were greatly reduced. Burt and his sons were willing to take on the work, often accepting contracts that others refused. Their expertise with the solar compass and their long experience in rough country were severely tested, but each contract was accurately completed and a small profit credited to the accounts.

In September of 1844, Burt and his survey team made the momentous discovery of iron ore in Negaunee which started an era of iron ore mining of the Marquette range that continues today. Alan Brown's article in *Michigan History* magazine describes the event:

> As though weather and terrain were not enough to contend with, Burt's men were almost out of supplies and subsisting on three porcupines they had been able to take the previous day. But Burt was thorough. His lines and field notes must be accurate; there would be no shoddy work done under his name. He and his crew would not go back to base camp for supplies until the township lines were properly run.
>
> Bad weather, difficult terrain, short rations – surely enough for the party to cope with. But there was more. The compass needle was fluctuating continually! An obvious solution was to use the highly accurate solar compass that Burt had perfected, which the United States government urged it surveyors to employ. But clouds threatened to obscure the sun.
>
> 'On the morning of the 19th of September, 1844, we started to run the line south between ranges 26 and 27. As soon as we reached the hill to the south of the lake [Teal Lake in Negaunee] the compassman began to notice the fluctuation in the variation of the magnetic needle. We were of course using the solar

> compass, of which Mr. Burt was the inventor, and I shall never forget the excitement of the old gentleman [age 52] when viewing the changes of the variation – the needle not actually traversing alike in any two places. He kept changing his position to take observations, all the time saying, 'How would they survey this country without my compass? What could be done here without my compass?' It was the full and complete realization of what he had foreseen when struggling through the first stages of his invention. At length the compassman called for us to come and see a variation which will beat them all. As we looked at the instrument, to our astonishment the north end of the needle was traversing a few degrees to the south of west. Mr. Burt called out, 'Boys, look around and see what you can find.' We all left the line, some going to the east, some going to the west, and all of us returning with specimens of iron ore mostly gathered from outcrops. This was along the first mile from Teal Lake. We carried out all the specimens we could.'

The accuracy of the Burt survey was confirmed by a registered survey completed by Al Pierce of Ishpeming near Sand River in 1990. One hundred and fifty years later, Burt's original landmarks of 1841, as recorded in his small leather-bound field notebooks, were used to identify the corners with descriptions such as "clump of five maples". By 1847 the township lines had been completed by John Burt. In the 17 years beginning with 1841, 5700 miles of township boundaries and 31,000 miles of section lines were drawn at a cost of $33,000 (at 58 cents per township mile) and $145,000 (at 21 cents per section mile) respectively. 17,000 square miles of U.P. were surveyed at a cost of 2 cents per acre. By 1887, it was estimated that 800,000 square miles of land had been surveyed using the solar compass. John Burt, great, great, great grandson of Wm A. Burt and author of *They Left Their Mark*, estimates the solar compass saved the U.S. government $15,000,000 between 1849 and 1878. The solar compass continued to be used by the Bureau of Land Management into the late 20th century. William Burt's patent on this instrument expired and he received only $80 total for the invention. Earlier, Burt had invented the first American typewriter ("typographer") but sold his patent for a pittance after three months.

Neither Burt nor a single member of his survey crew ever attempted to capitalize on their discovery. Perhaps that is not surprising. The geologist, Douglas Houghton, had warned copper speculators in the Keweenaw Peninsula that instant wealth was unlikely, and that profit from iron ore would be even more difficult to obtain. Unlike copper, which exists in a pure state, the heavy iron had to be extracted from the rock. It is costly and time consuming to market heavy iron ore. In fact, it was 1862 before the Jackson Iron Mine, established on the site of Burt's discovery, returned a dividend to its stockholders.

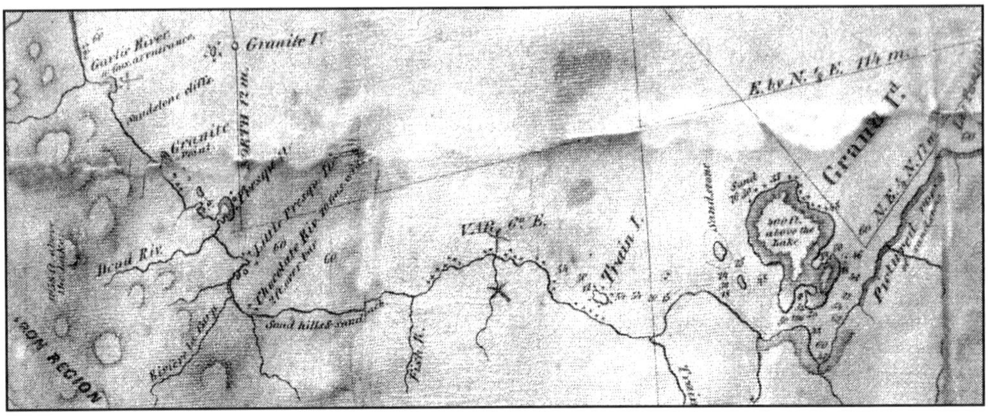

Courtesy Peter White Library, Marquette, MI

The 1847 map of English Cartographer T. W. Bristol includes Fish River (Sand River), the unnamed Laughing Whitefish River, and the Train River (Au Train).

The chart of Lake Superior by T. W. Bristol in 1847 does not reflect Burt's surveys just prior to this time. As one compares these maps of the Laughing Whitefish River area, the variation in the streams included as well as the relative size of the streams become apparent. On the Bristol map, the Chocolay is drawn as being quite small with the Fish River (Sand River) being larger than the Chocolay. Rock River is not drawn and Au Train Lake is omitted.

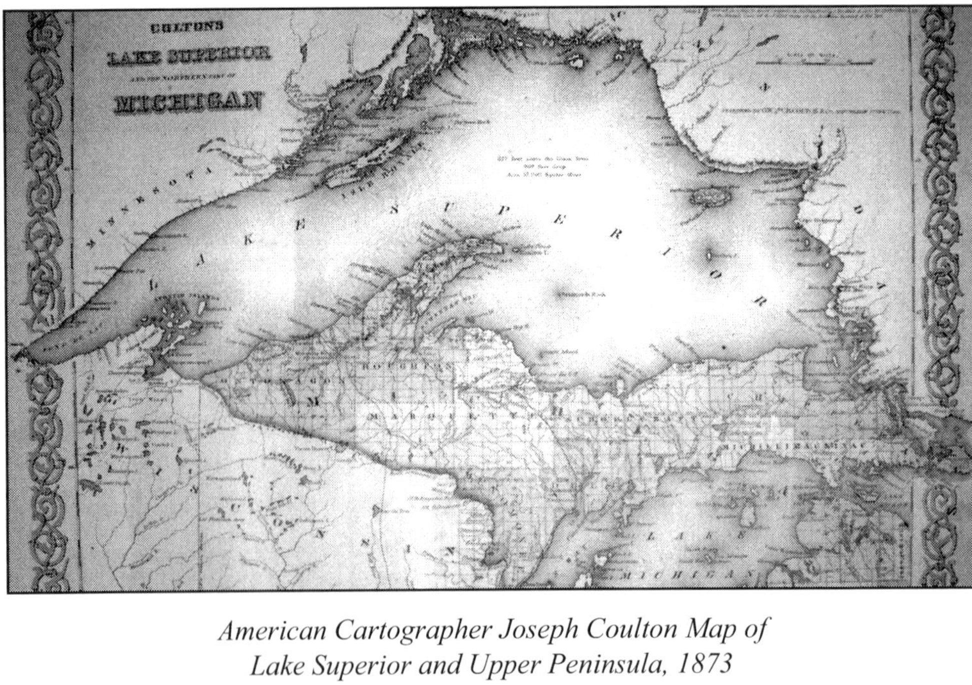

Courtesy Historical Maps of Michigan

American Cartographer Joseph Coulton Map of Lake Superior and Upper Peninsula, 1873

The images of the William Burt 1841 survey maps are included in the Chapter IX appendix along with those from the Marquette law firm of A. H. Matthews between 1880 and 1903. The plat maps of Alger County by C. A. Baldwin are ca. 1880.

THE MODERN (POST BURT'S SURVEY) ERA 1872 – 2008

The excellent map of Lake Superior in the U.P. by Coulton in 1872 breaks the Upper Peninsula into nine counties. The divisions evolve over the years to the current configuration of fifteen counties in 2008 (Alger was separated from Schoolcraft in 1885). At that time Lake Superior was believed to be 627 feet above ocean level, 900 feet deep, and 32,000 square miles. These figures are slightly different in a current profile of the Great Lakes. Current statistics are: 350 x 160 miles, 3,000 miles of shoreline, 600 feet above sea level and 1330 feet deep. Without the weight of glaciers, the north shore of Lake Superior is rising at the rate of 1 cm. per year.

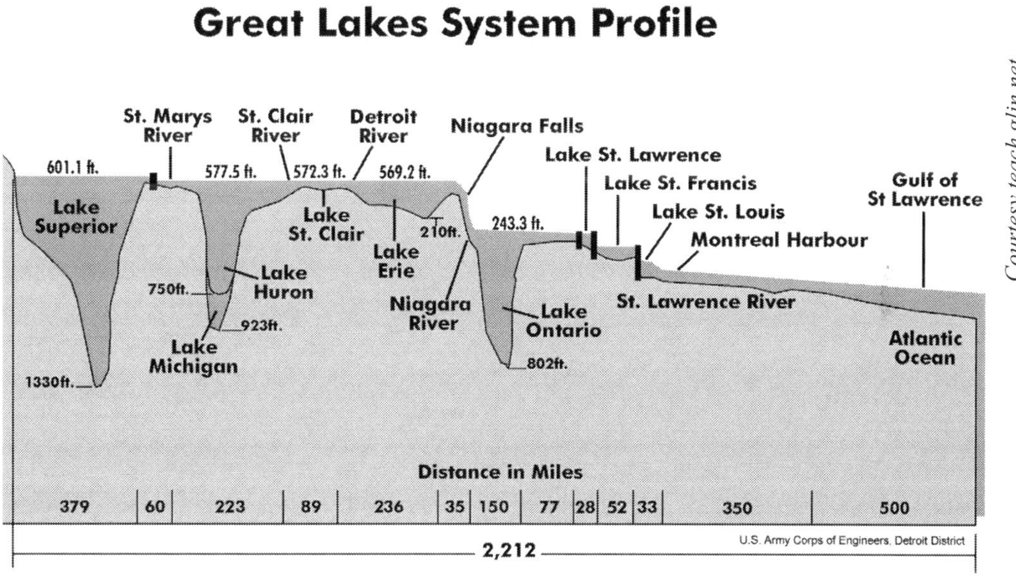

Lake Superior measures 350x160 miles with 10% of the world's fresh water. Lake Superior has 3,000 miles of shoreline, and is 600 feet above sea level and 1,330 feet deep.

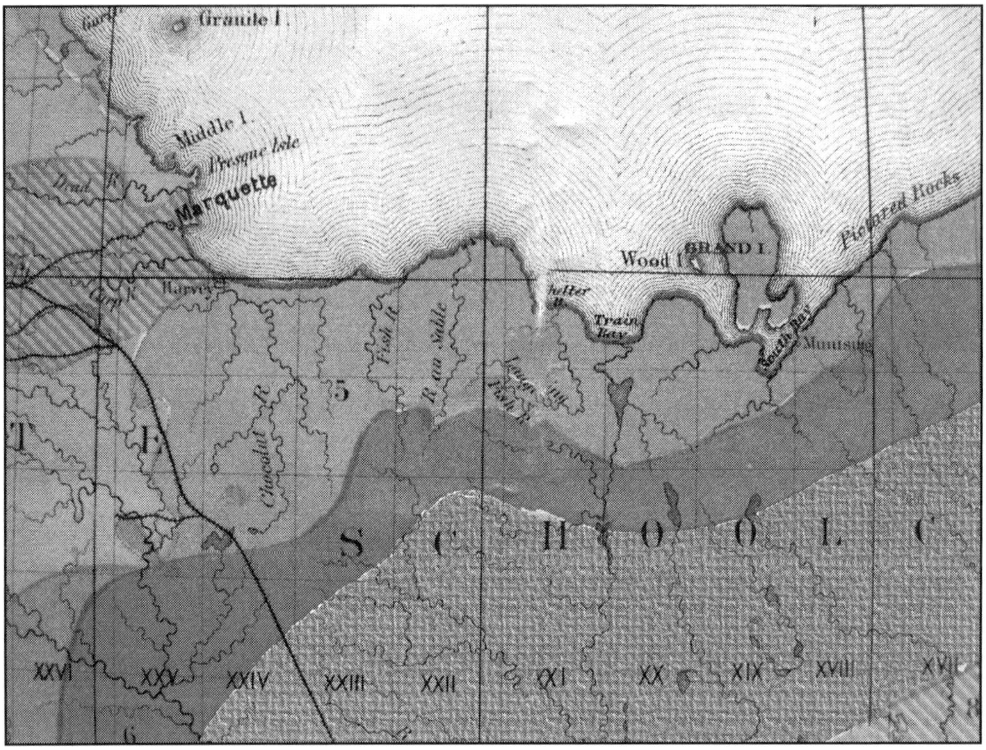

Peter White Library, Marquette

T. B. Brooks Map of Southern Shore of Lake Superior, 1873
Note: Laughing Whitefish River is referred to as River Au Sable. The Laughing Fish River is shown flowing into Shelter Bay.
Light: St. Mary's (Cenozoic era 5-25 million years ago)
Darker: Calciferrous (Paleozoic era 410-440 million years ago)
Crosshatch: Trenton (Paleozoic era 410-440 million years ago)

In 1873, T. B. Brooks, a number of geological survey of Michigan, published a map of the central portion of the Upper Peninsula. The purpose of the map with colors was an early effort to define the geologic era of the predominant rocks. His second map of 1873 reflected the surveying by Burt and his colleagues. Note the names given to the rivers with the Laughing Whitefish River labeled as Ausable and Rock River labeled as Laughing Fish River. These are labeled correctly on the second 1873 map. The map showed the stagecoach route for mail going from Sand River northeast to the Halfway House located in Section 35, then took a hard right and went to Shelter Bay. A proposed line of the Mackinaw Railroad had been drawn. The size and distribution of the Sand, Laughing Fish, and Rock Rivers are a reflection of their growing importance as the primary route for the moving of cut logs to Lake Superior.

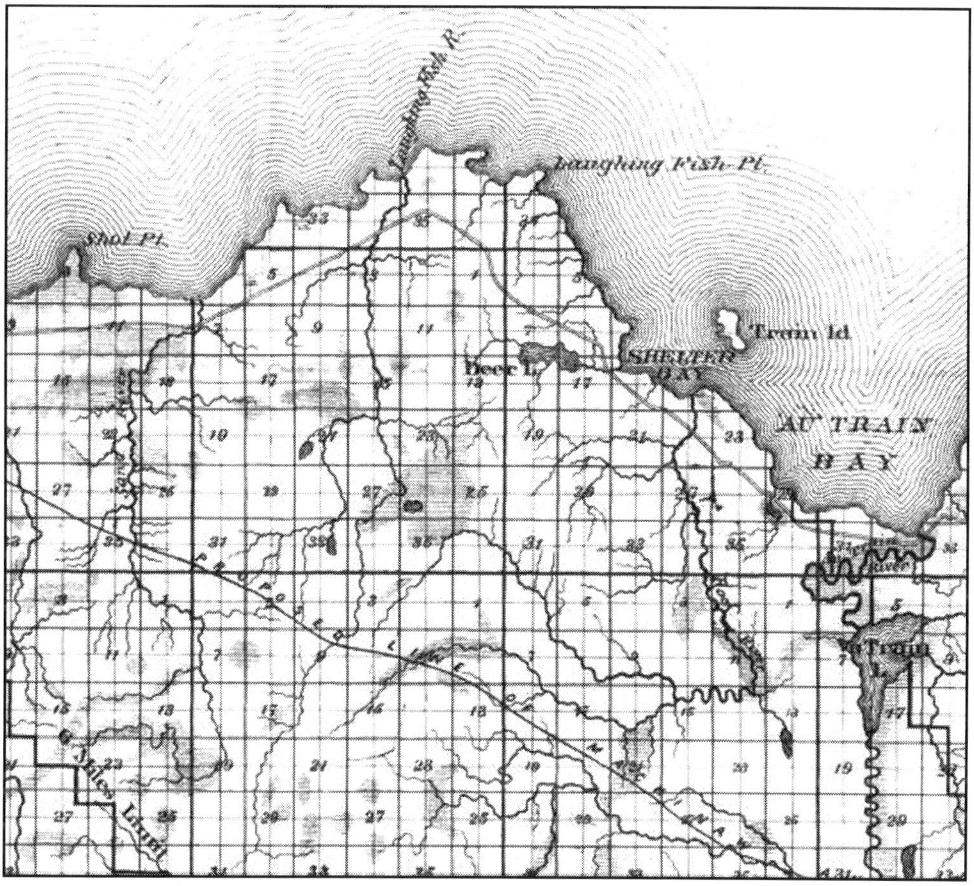

Courtesy Peter White Library, Marquette, MI

Second Brooks Map of Southern Shore of Lake Superior – 1873
The Laughing Fish River is now correctly labeled.

The 1896 map by Cadarette shows the road and railroad coming from Marquette, through Sand River, Whitefish (labeled Kilns), Deerton, Onota, and on to Rock River and Au Train. The stream that extended from the Whitefish Kilns (on border between Section 8 and Section 9) and which flowed north into Lake Superior midway between the Sand River (Section 6) and the mouth of the Laughing Whitefish River (Section 35) is locally known as the Whitefish Creek. The Halfway House is labeled as Club House.

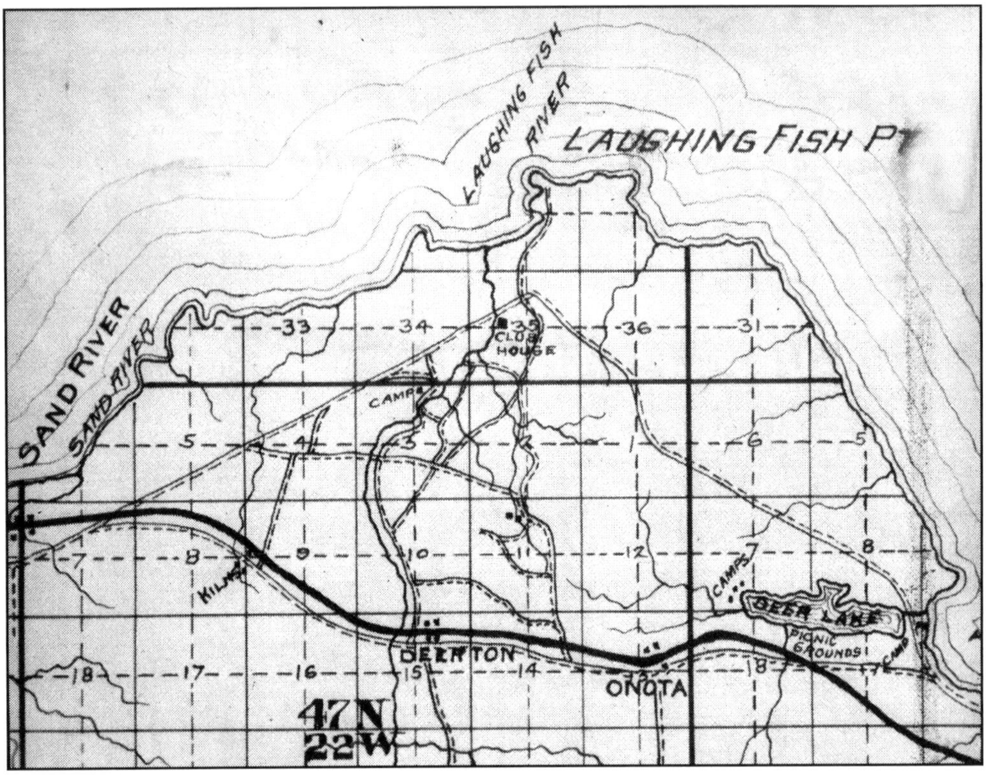

Courtesy J. M. Longyear Library

Cartographer Cadarette Map, 1896. The Club house (Halfway House) sat at the junction of the Laughing Whitefish River and the old road between Sand River and Deer Lake.

The 1928 Department of Conservation Land Economic Survey map shows some of the camps of the area. Note also the railroad spur from the DSS&A labeled Tyoga originates in Section 14, courses northward and ends in the Tyoga lumber camp in Section 35. On the Hinsdale map of 1931, Peter White Lake is incorrectly labeled as being a lake west of the Laughing Whitefish River. In reality, the "Whitefish Lake" just south of the Peter White camp is currently known as Laughing Whitefish Lake. Peter White Lake on the 1931 map is now known as Howe's Lake.

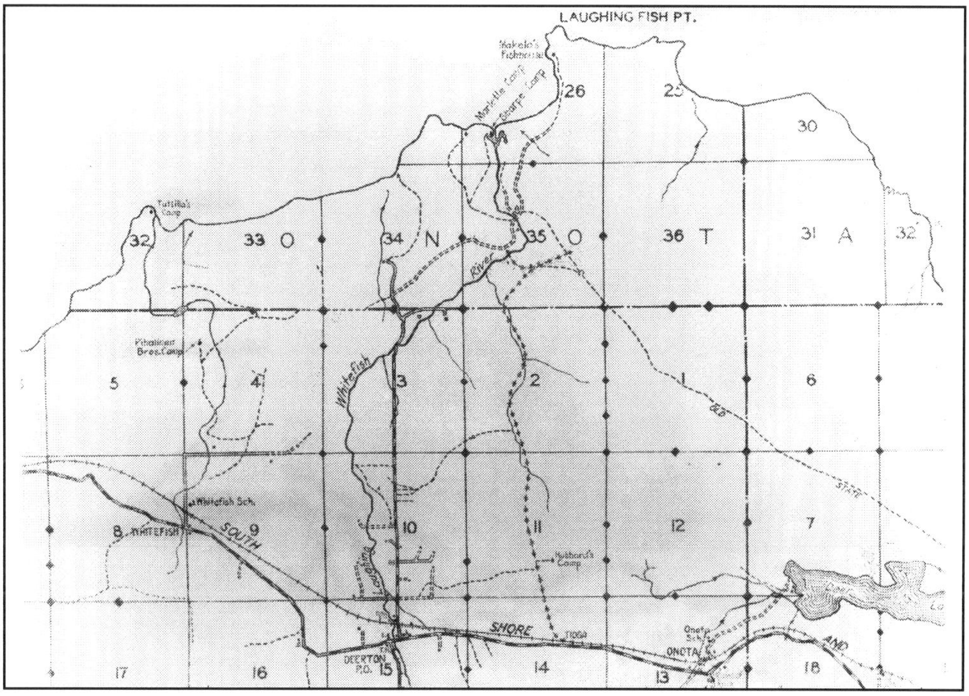

Courtesy J. M. Longyear Library

Department of Conservation Land Economic Survey, 1928 Map includes several camps, some of which are still in existence. The Railroad spur for DSS&A is incorrectly labeled as originating at Tioga.

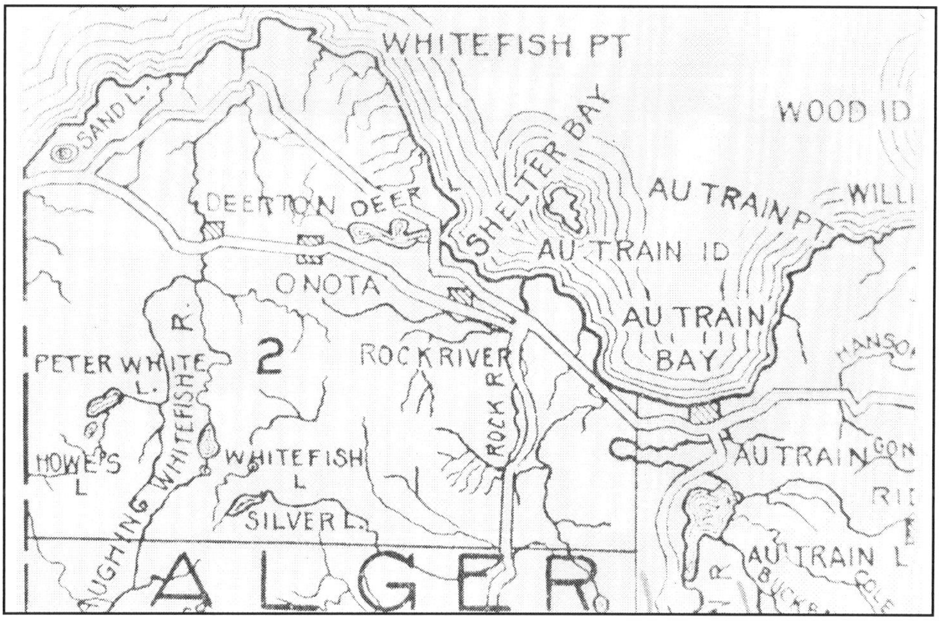

Courtesy Peter White Library, Marquette

Wilbert Hinsdale Map, 1931,
Wilbert was dean of Homeopathic College at the University of Michigan with a life long interest in archeology.

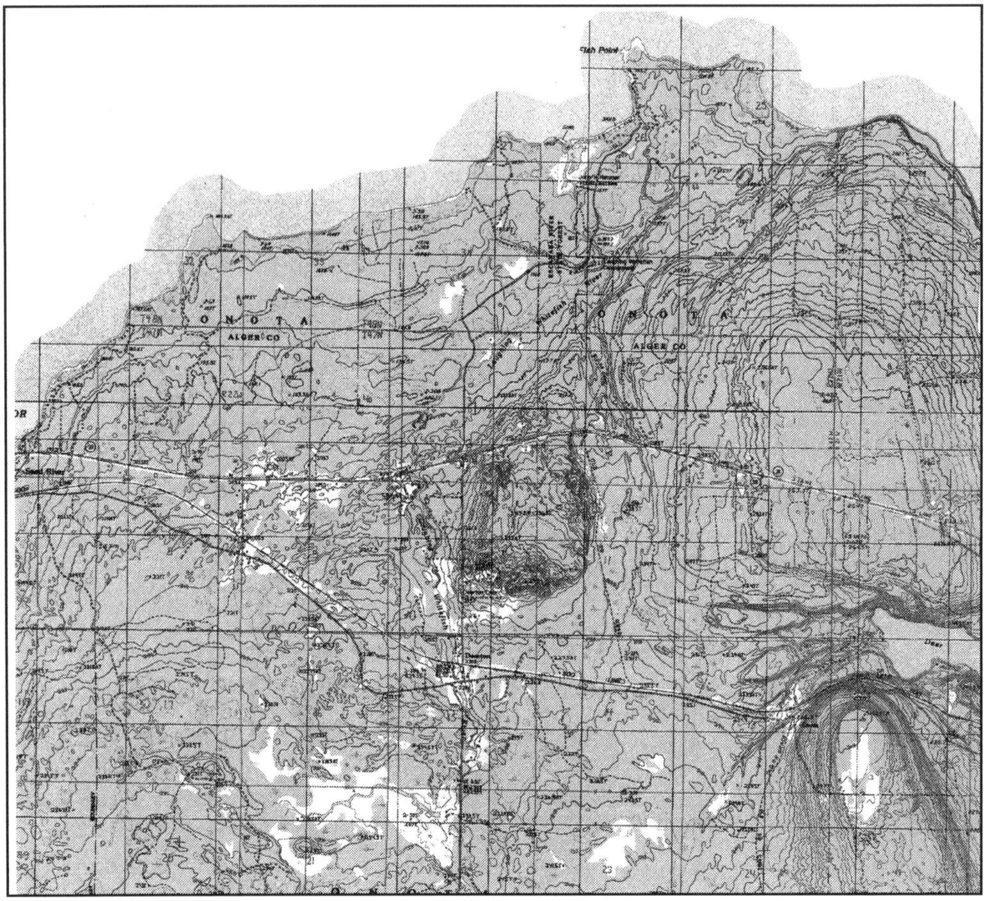

Courtesy Peter White Library, Marquette

U.S. 1:24,000 topographic map, 1985

The 1985 1:24,000 topographic map shows the current appearance of the area. The streams of the central U.P. between Lake Superior and Lake Michigan are well delineated on a map from the Lake Superior Watershed Project. (Chap. VII, pg 112) The watershed between Lakes Superior and Michigan in the upper 20% of the central U.P. is well depicted on this map showing the drainage area of the rivers. The watershed of the Laughing Whitefish River is just south of M-94, west of Chatham.

There is a set of maps by the U.S. Army Corps of Engineers of 1900, 1943, 1944, and 1983. In the 1983 version (NOAA) the Laughing Whitefish River is labeled as the Sable River flowing north of Deerton, taking a hard right and emptying into Deer Lake. You can forgive the Federal government for these inaccuracies as these charts are for navigational purposes and not land travel.

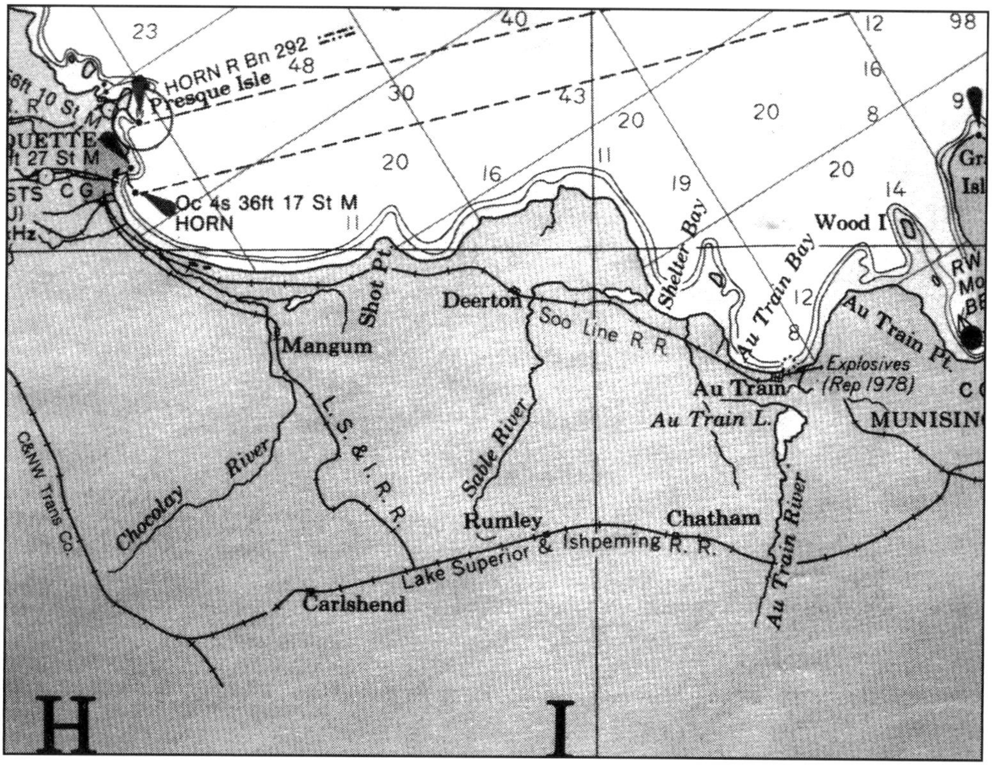

Courtesy NOAA

NOAA Map, 1983 This map was intended for nautical use. It shows The Sable (old term for Laughing Whitefish River) emptying into Shelter Bay via Deer Lake.*

For those of us who love the UP, the most accurate map is that by Eugene Sinervo, a long time employee of the *Mining Journal*. This map served as the place setting in restaurants throughout the area for a number of years. Note that the center of the United States, if not the universe, is the Sand River. Adjacent to the mouth of the Sand River was *Mean Gene's Lakeside Inn*, a well-known watering hole for many years that was torn down in 2007.

*Sable is the French word for "sand" .

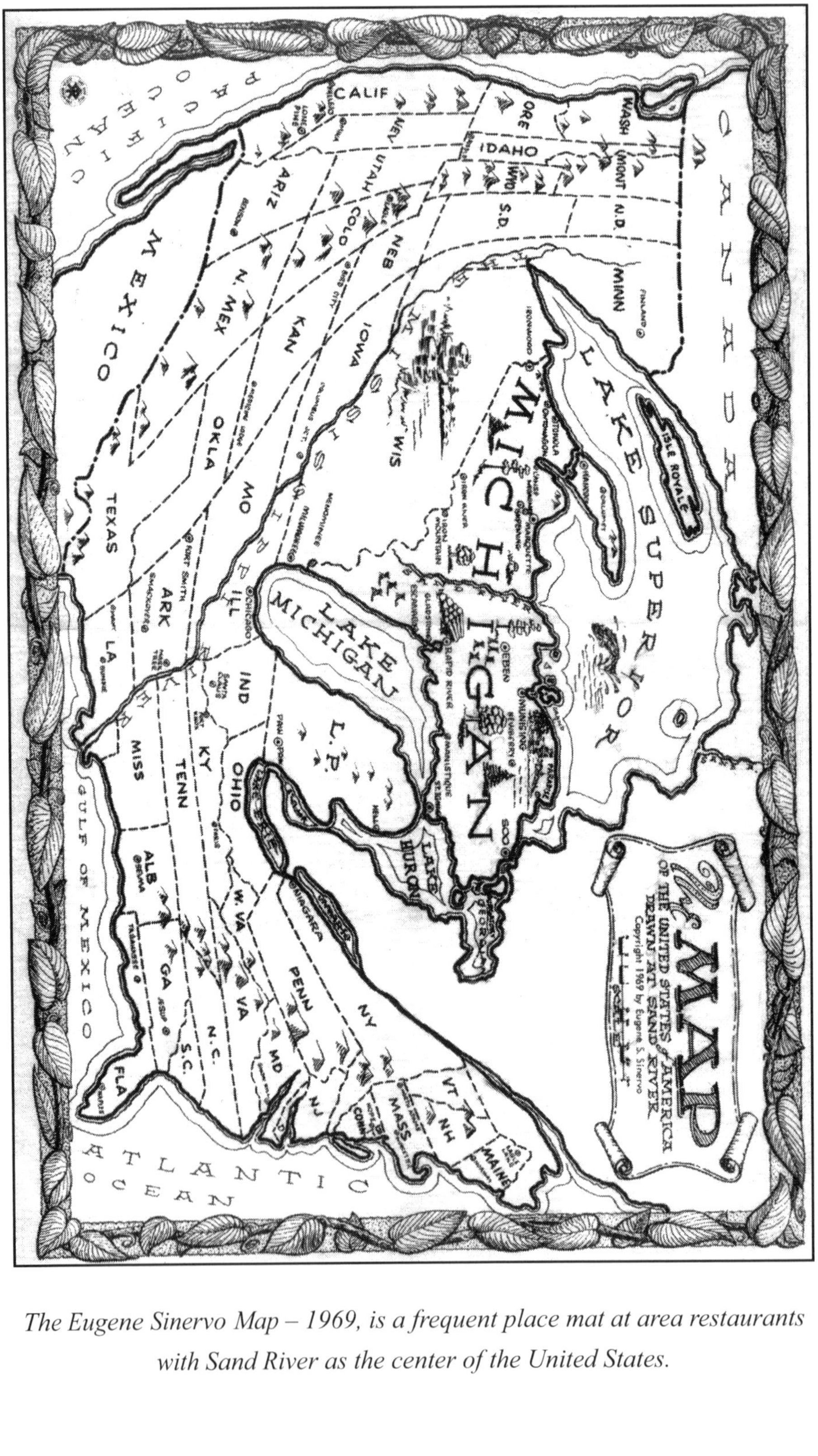

Courtesy Vincent Sinervo

The Eugene Sinervo Map – 1969, is a frequent place mat at area restaurants with Sand River as the center of the United States.

Chapter IX

DEMOGRAPHICS AND STATISTICS

1800 - 2000

An article in the *Daily Mining Journal* of 1965 paints an alluring portrait of Alger County for both newcomers and life-long residents:

> "Without a doubt, Alger County in the Upper Peninsula of Michigan, contains more attractions per square mile than any other section of the state. The world famous Pictured Rocks of sandstone...Miner's Castle, from the top of which Father Marquette preached to the Native American in their birch bark canoes below...caves of the chiefs where the Native Americans placed their prisoners of war, with such thing as escape impossible...the gigantic caves on Grand Island...the remarkable grandeur of the shifting Grand Sable Dunes of Grand Marais...remains of ancient iron furnaces near Munising and the ancient charcoal kilns near Rock River...the enormous meteor at Star Siding on M-28, just a few miles from Shingleton...the 200-foot glacial pothole on Perch Lake Trail out of Munising... Cox's Trout Pond at Wetmore – the only inland body of water in Upper Michigan that flows both into Lake Superior and Lake Michigan...ancient Native American burial grounds on Sand Point...home of George Shiras, famous wildlife photographer, with almost all of his night photography of wild life...
>
> The ancient fur trading posts of the American Fur Trading Co., still standing, with fur derrick intact...Au Train, the resting point for the dog teams that carried the mail on one of the first mail routes to be established in the Upper Peninsula. Peter White was the mail man...the beautiful Adam's trail near

Grand Marais, one of the finest drives in Upper Michigan… the Hiawatha National Forest, one of the government's most ambitious efforts in Michigan; it has considerable wilderness that is untouched by either fire or the ax of the lumberman… the vast pine plantings in this forest which are a study in themselves."

Courtesy Linda Sturvist

This photograph of a doe and fawn was taken by George Shiras III about the yaer of 1892. Shiras was the son-in-law of Peter White and was first to take outdoor night time photographs. He began his career in the area noe known as the Peter White Camp on the Laughing Whitefish Lake, south of Deerton

What this does not mention are the people, the area's most appealing feature – bright, energetic, and, yes, delightfully eccentric. A wide range of employment opportunities guarantees both ethnic and professional diversity.

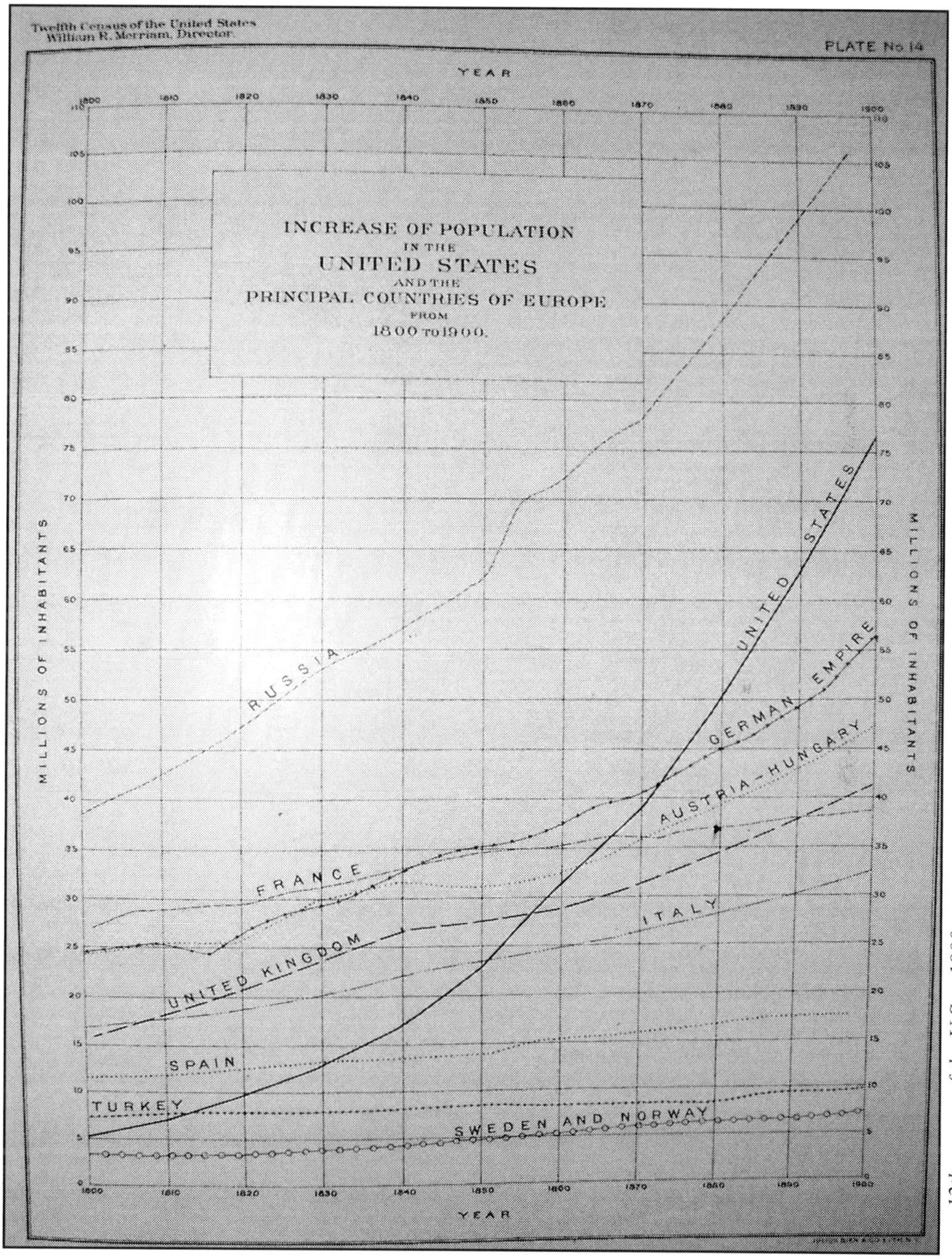

12th census of the U.S. – 1900

Increase in population of United States and European countries, 1800-1900

A recent trend has been the arrival of "snowbirds" coming to the area for retirement, most of whom come only for the warmer months but many hardy souls stay year-around. Although this chapter deals primarily with census statistics of 1900 and 1910 bracketing the short lifespan of Tyoga Lumber Company from 1905 to 1907, it will also look at trends of area, nation, and world between 1800 and 2000.

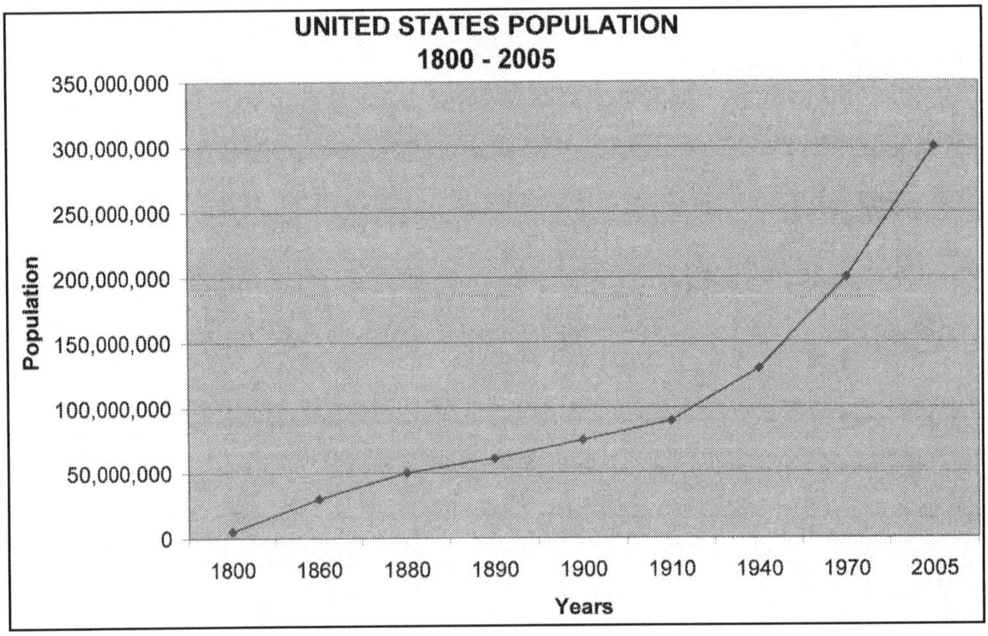

Courtesy John Parlin

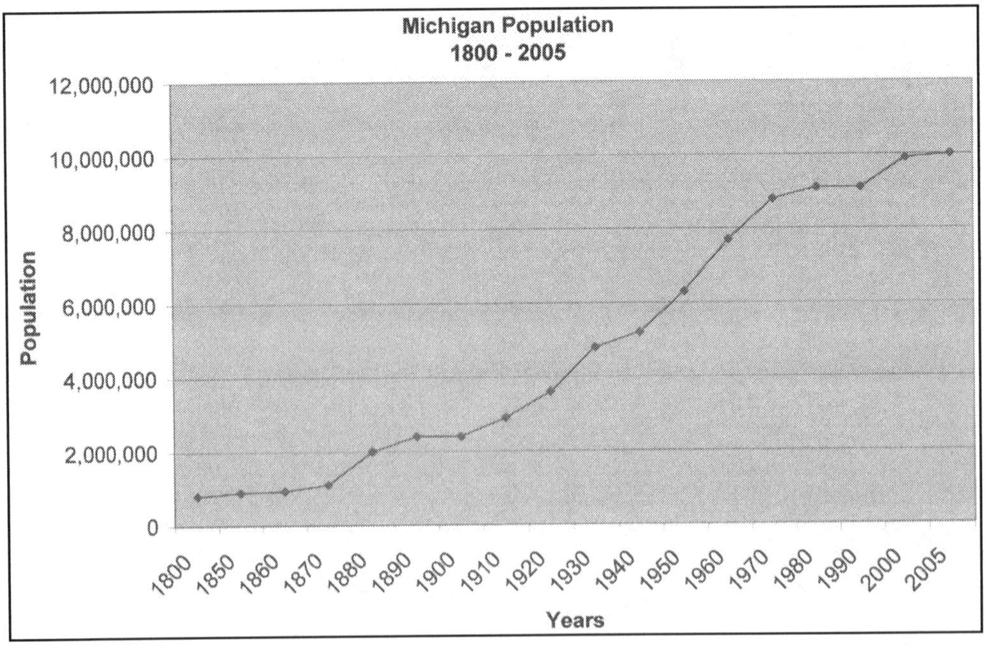

Courtesy John Parlin

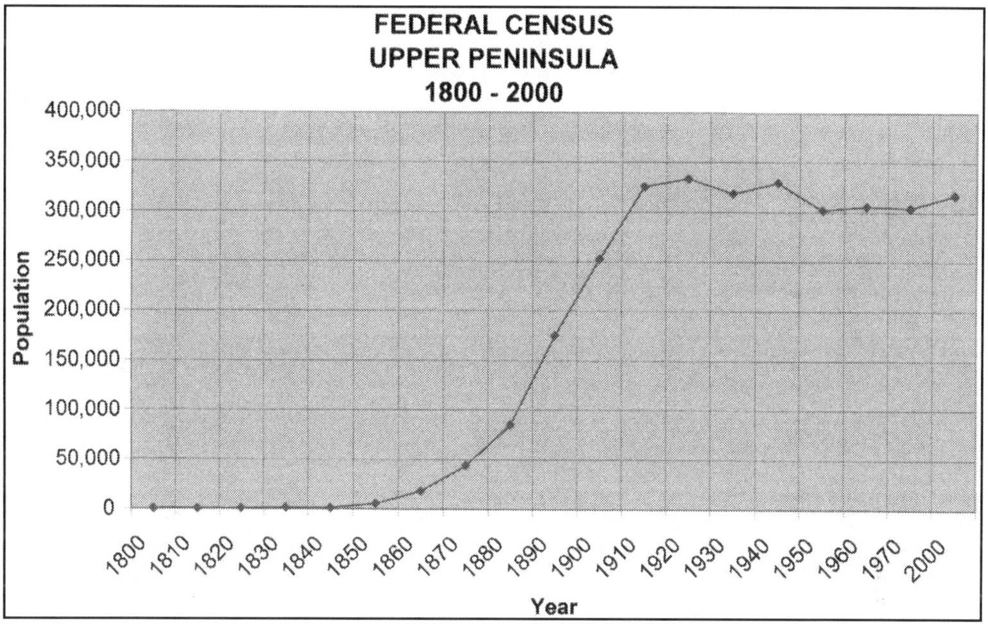

Courtesy John Parlin

FEDERAL CENSUS UPPER PENINSULA
1800 – 2000

Year	Population	Year	Population
1800	601	1900	252,628
1810	615	1910	325,628
1820	819	1920	333,556
1830	1,503	1930	318,616
1840	1,359	1940	329,544
1850	5,745	1950	302,258
1860	18,085	1960	305,952
1870	43,700	1970	304,347
1880	85.085	2000	317,000
1890	175,226		

While the U.S. and Michigan population has increased 300% between 1900 and 2000, the population of the U.P. is virtually unchanged at 317,000. The U.P. experienced its largest population growth between 1855 and 1915 to satisfy the need for labor in the timber, copper and iron industries.

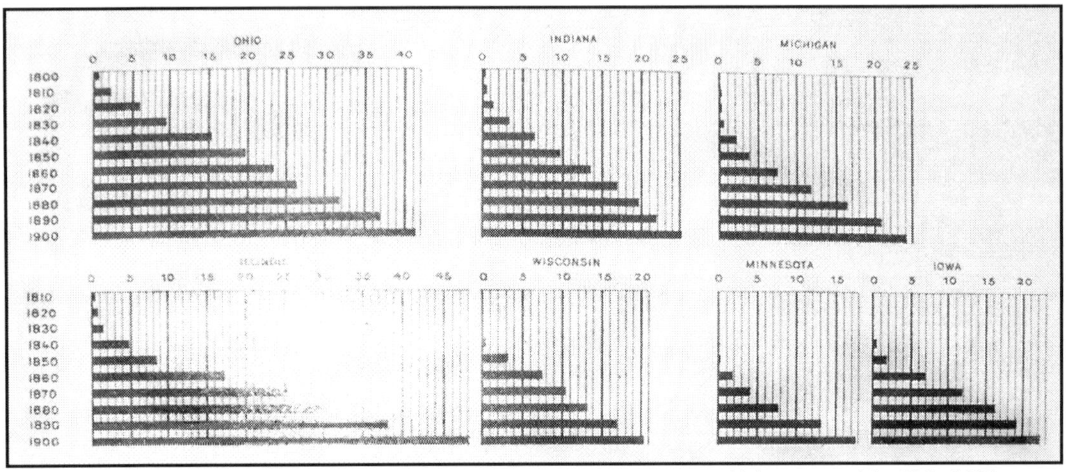

12th census of the U.S., 1900

Total population of each state at each census – 1800-1900

The above bar graph shows that the greater population growth in Michigan between 1800 and 1900 was somewhat different from other midwest states in that the growth began slowly in the early 1800's and accelerated in the late 1800's. A breakdown of state populations by race (on the following pages) shows Michigan to have a relatively smaller component of native White born of native parents and a relatively larger component of native White of foreign parents and foreign White. The leading nationalities in each state were of German and Canadian ancestry Michigan when you include both the Lower and Upper Peninsula. Also shown is the U.S. census at ten year intervals between 1850 and 1900 with the proportions of origin of foreign born. When comparing 1850 to 1900, there was a significant decrease in Irish-born citizens, minimal change in German or British, with most significant increases in Italians and Canadians and in particular of Scandinavians and Slavs. Ethnic groups of the U.P. from 1870 to 1920 are listed in the table on page 173.

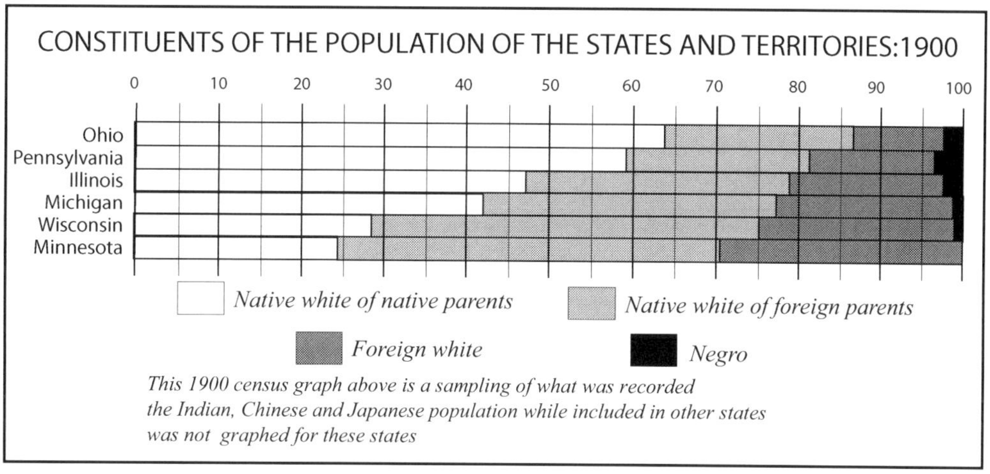

12th census of the U.S., 1900

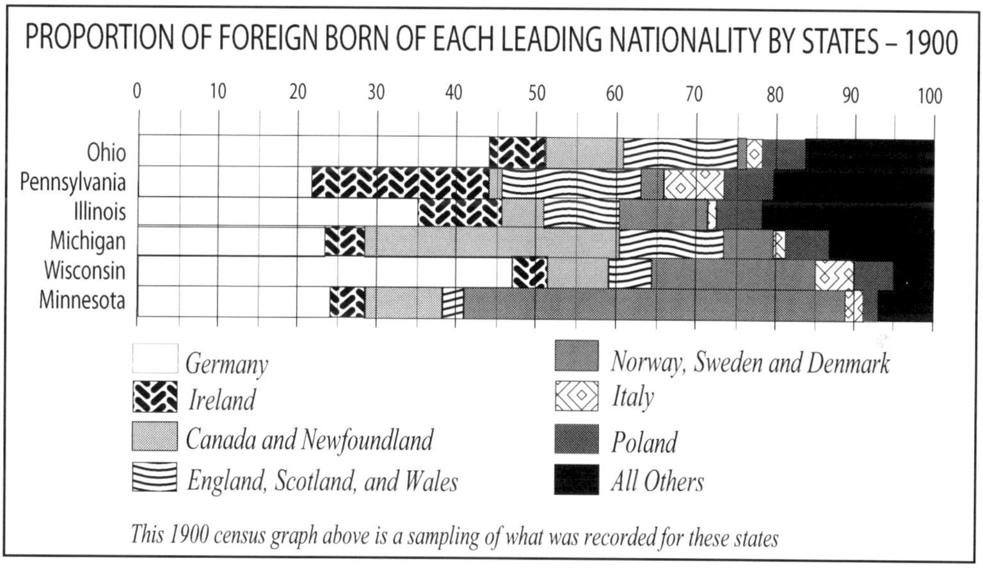

Proportion of foreign born of each leading nationality by states, 1900

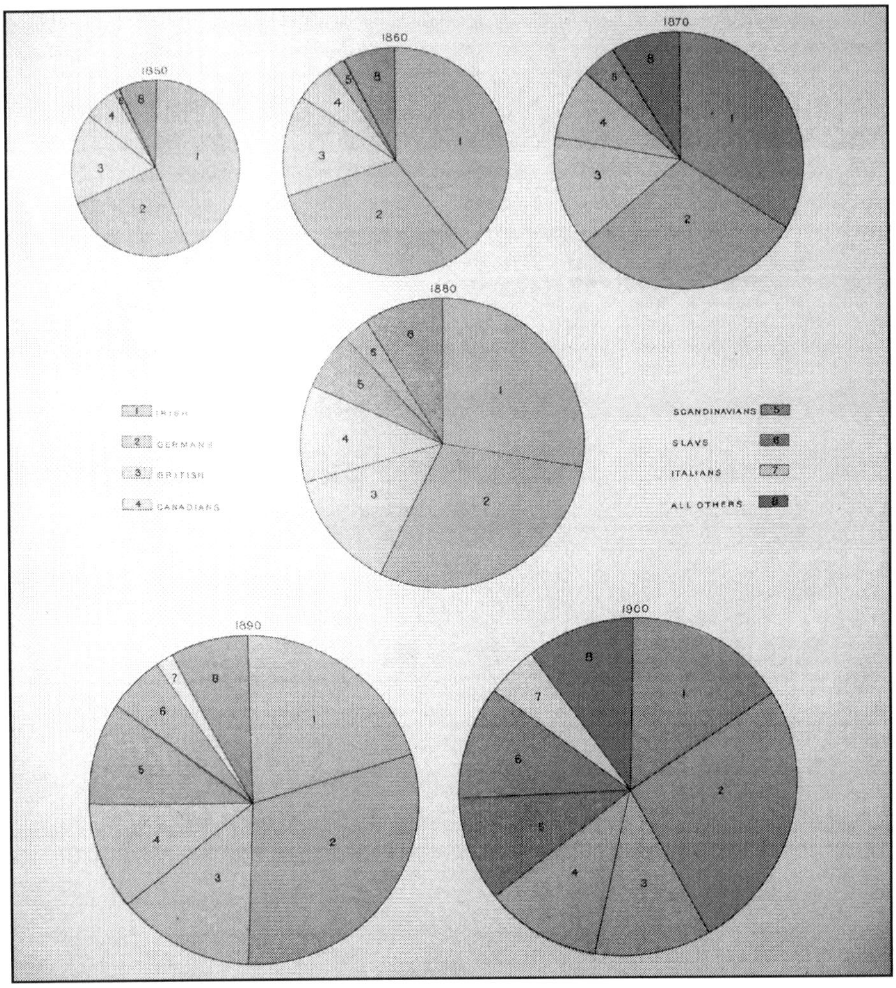

12th census of the U.S., 1900

A proportion of foreign born of each leading nationality – 1850-1900

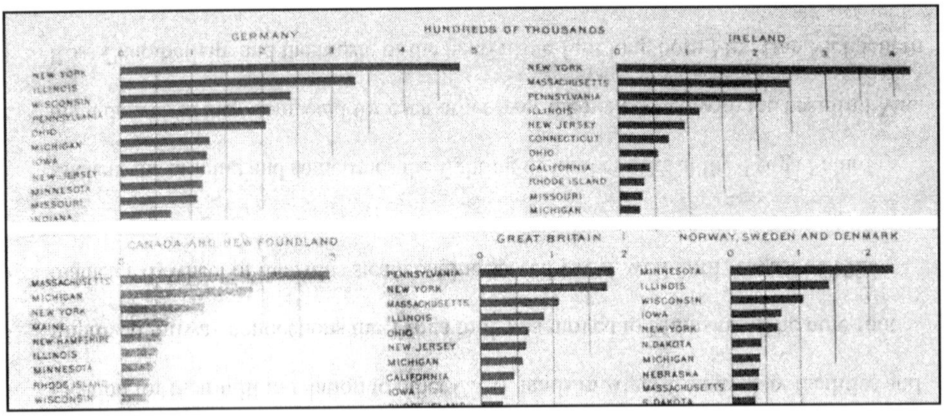

12th census of the U.S., 1900

This illustration shows the distribution of foreign born and their countries of origin in 1900.

ETHNIC GROUPS Upper Peninsula 1870 - 1920

Nationalities	1870	1880	1890	1900	1910	1920
Afro-Americans	147	173	417	272	289	285
Armenians						100
Asians (except China)						
Australians			48	23		
Austrians			2,074	4,540	9,759	2,493
Belgians			785	945	1,401	1,124
Bohemians	62		705	511		
British Americans	6,766	15,466				
Bulgarians						124
Canadians, English				18,986	13,012	9,233
Canadians, French				13,310	11,610	7,481
Canadians & Newfoundlanders			27,003			
Chinese			54	122	?	61
Cubans & West Indians			5	52		
Czechoslovakians						1,299
Danes			1,001	1,187	1,054	827
English			11,396	10,699	10,705	7,176
English & Welsh	4,468	7,291				
Finns				18,103	29,619	26,542
Germans	3,292	3,638	7,626	7,302	6,884	3,782
Great Britain (not specific)		9				
Greeks				31	169	289
Hollanders	53	94	140	150	230	230
Hungarians			162	385	1,806	660
Irish	5,775	5,808	5,484	4,099	2,681	1,442
Indians, Chinese, Japanese, etc.					2,874	
Italians			2,386	4,589	8,278	7,874
Lithuanians						306
Mexicans			11			
Norwegians			3,415	3,839	3,730	2,680
Norwegians and Swedes	1,564	5,728				
Poles (Austrian)				358		
Poles (German)				1,072		
Poles (Russian)				801		
Poles (Unknown)				332		
Poles		239	1,333			
Rumanians						117
Russians			8,857	849	4,157	1,674
Scots	470	672	1,216	1,196	914	572
South Americans			7			
Swedes			14,289	15,121	14,191	11,011
Swiss	231		265	195	237	184
Syrians						212
Turks (Ottoman Empire subjects)					294	
Welsh			168	145	140	80
Yugoslavs						3,398
Miscellaneous			116	262		499
Total (Foreign born)	22,828	39,118	88,963	109,562	124,034	95,683

Courtesy John Parlin

The following tables and graph are population figures for the U.S., the State of Michigan and for Alger, Marquette, and Schoolcraft counties, as well as Onota Township.

COUNTY POPULATION 1850 – 2000

YEAR	Alger County	Marquette County	Schoolcraft County
1850		136	16
1860		2,821	78
1870		15,033	
1880		25,394	1,575
1890	1,238	39,521	5,815
1900	5,868	41,239	7,889
1910	7,700	46,739	8.681
1920	10,000	45,786	9,977
2000	9,862	64,000	8,903

Schoolcraft County split from Chippewa County in 1871.
Alger County split from Schoolcraft County in 1875.

Courtesy John Parlin

POPULATION

YEAR	UNITED STATES	MICHIGAN	ALGER COUNTY	MARQUETTE COUNTY	ONOTA TOWNSHIP
1900	76 million	2.2 million	5,686	41,239	82
1910	9.2 million	2.9 million	7,675	46,729	150
1920	10.6 million	3.5 million	9,983	45,786	
2005	296 million	10 million	9,662	64,000	310

Courtesy John Parlin

POPULATION % Increase

PERIOD	U.S.	Michigan	Alger Co.	Marquette Co.	Onota Twp.
1800 - 1850	360	8,000			
1850 – 1900	230	700			
1900 – 1910					82
1900 – 2000	290	300	78	56	300
1920 – 2000				42	

Courtesy John Parlin

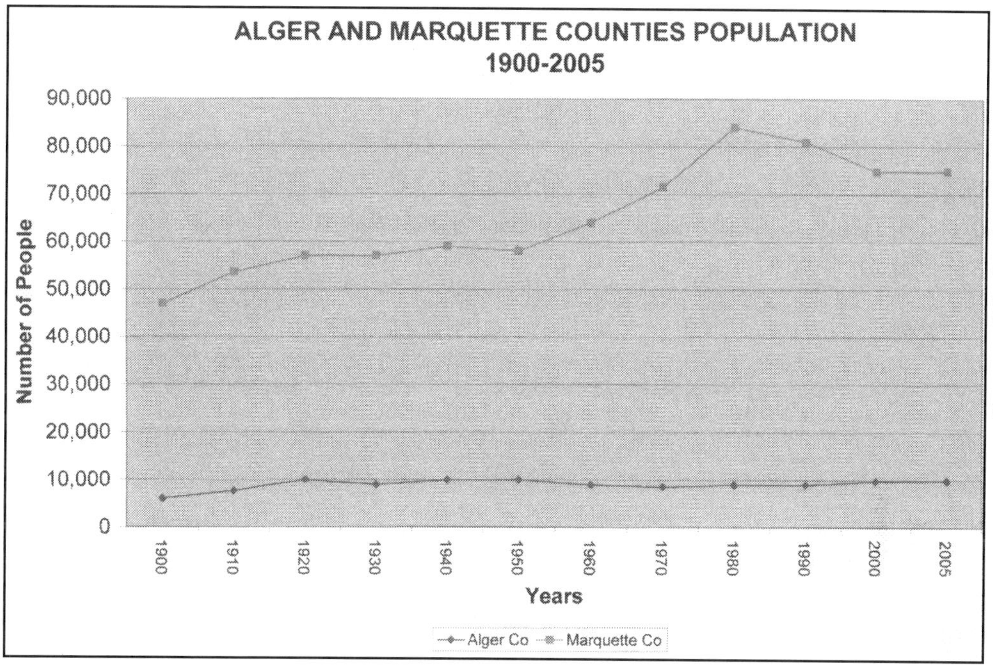

Courtesy John Parlin

Life expectancy, population over 75 in 1900, and death rate in four areas of the U.P. show interesting trends. The worldwide average life span increased from 18 to 66 (267%) over the past 5000 years and doubled between the early 20th century and 2000. It would appear that Michigan was far behind the curve in 1900 with only 15 individuals over age 75 (15 out of 2.2 million), reflective of a young working class. The lower rates for death from infectious disease tend to be in the western half of the U.P. and probably reflect an overweighting of Native American population in the eastern half of the U.P. at that time. Cancer is reported at the tip of the Keweenaw but not elsewhere. This may well be due to a combination of exposure to the pollutants in the copper mining industry and the effects of smoking.

LIFE EXPECTANCY	**(Average Life Span)**
Bronze Age (3000 BC)	18
Pre-Columbus North America	25-30
Early 20th Century	30-40
World in 2000	66
U.S. in 2000	77
Highest (Andorra)	84
Lowest (Mozambique)	42

Over Age 75 in 1900

RACE	Michigan	United States
White	8	82,741
Indian	7	229
Foreign-born	0	1,846
Negro	0	4,954

Courtesy John Parlin

1900 DEATH RATE

DISEASE	Tip of Keweenaw	Western U.P.	Central U.P.	Eastern U.P.
Influenza	<15	30-44	20-29	10-19
Diphtheria & Croup	15-24	5-14	15-24	>35
Cancer & Tumors	>90	-	-	-
Typhoid Fever	<15	20-24	25-34	20-24
Tuberculosis	<15	95-114	115-144	115-144

Courtesy John Parlin

Alger County population by townships between 1800 and 2000 shows the largest increases for Au Train and Onota Townships and the striking decrease for Burt Township which had been the first area to undergo intensive logging beginning in the 1870's.

POPULATION OF ALGER COUNTY TOWNSHIPS

TOWNSHIP	1890	1900	1910	2000
Au Train	248		400	1172
Burt	177		1,400	480
Grand Island			111	45
Limestone			500	407
Mathias			407	571
Munising *Village* *Munising Township*	288		3,000 550	3,125
Onota	161	82	150	310
Rock River	328		1205	1213
TOTALS	1,238	5,868	7,700	9,862
Population Per Square Mile	0.2	1.2	1.5	11

Courtesy John Parlin

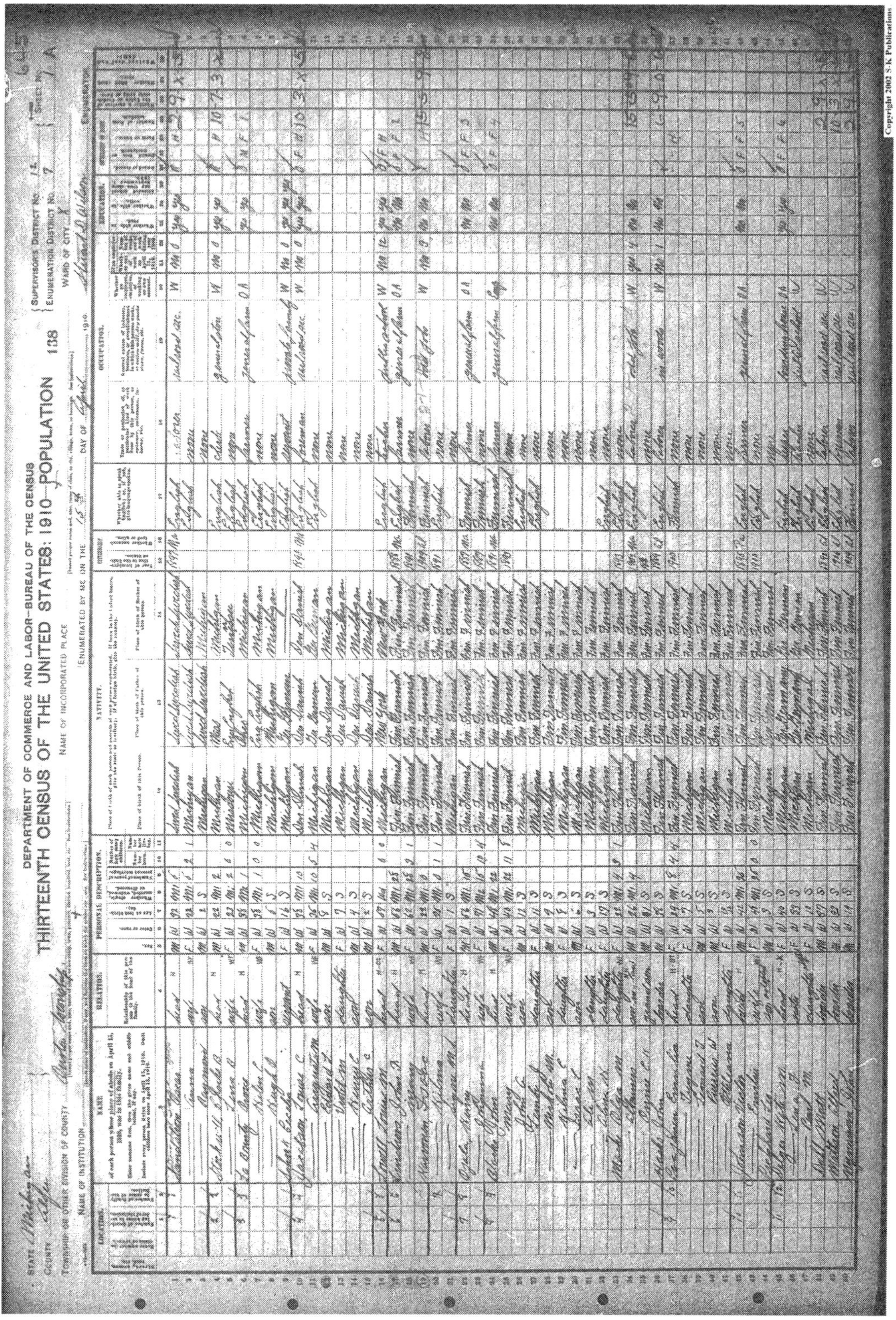
DEPARTMENT OF COMMERCE AND LABOR—BUREAU OF THE CENSUS

THIRTEENTH CENSUS OF THE UNITED STATES: 1910—POPULATION

STATE Michigan

COUNTY Alger

TOWNSHIP OR OTHER DIVISION OF COUNTY Onota Township

SUPERVISOR'S DISTRICT NO. 12

ENUMERATION DISTRICT NO. 7

SHEET NO. 1 A

13th census of the United States, 1910

Onota Township Census

Many of their descendants are familiar names in the area Onota Township area in 2010

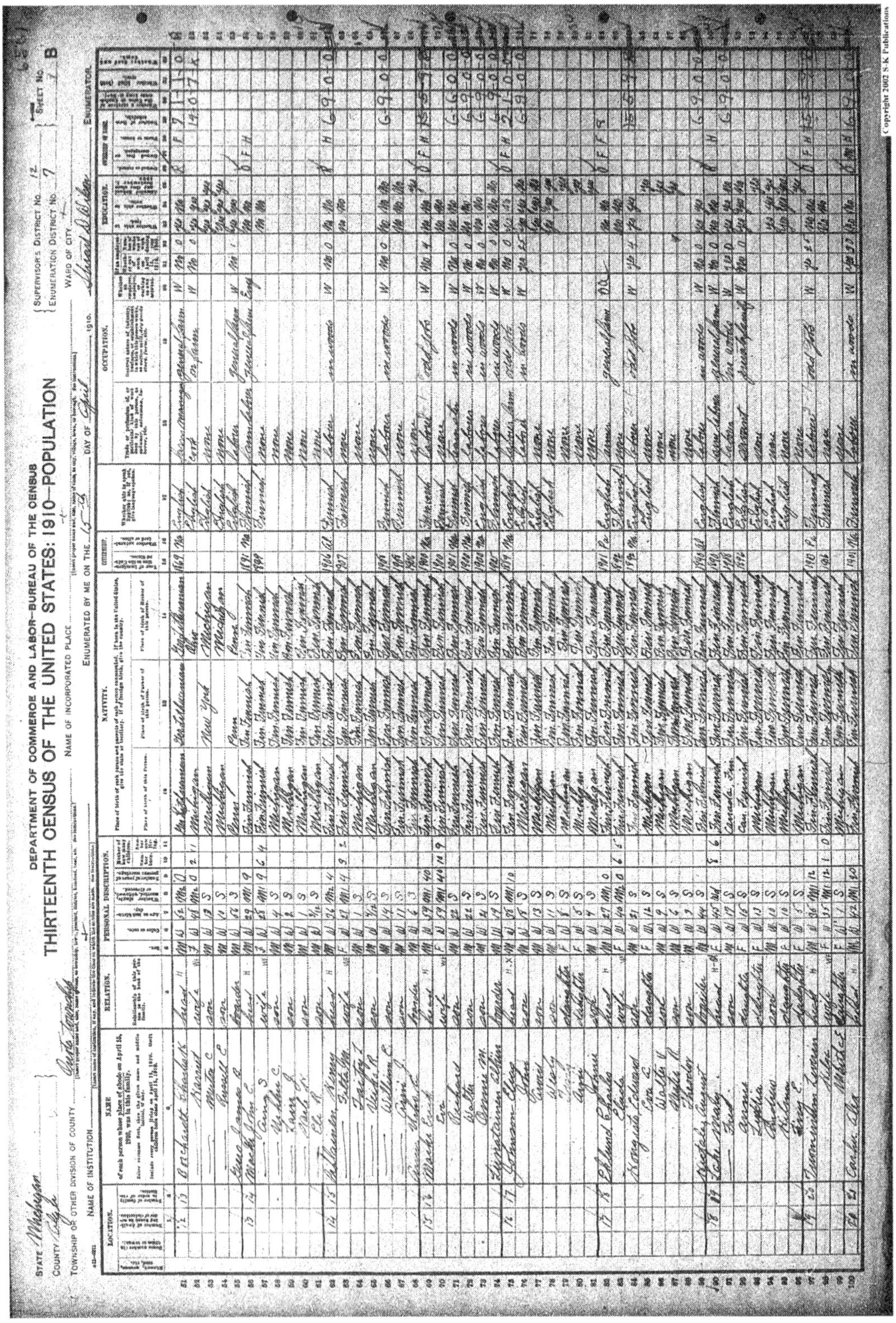

13th census of the United States, 1910

Onota Township Census 1910

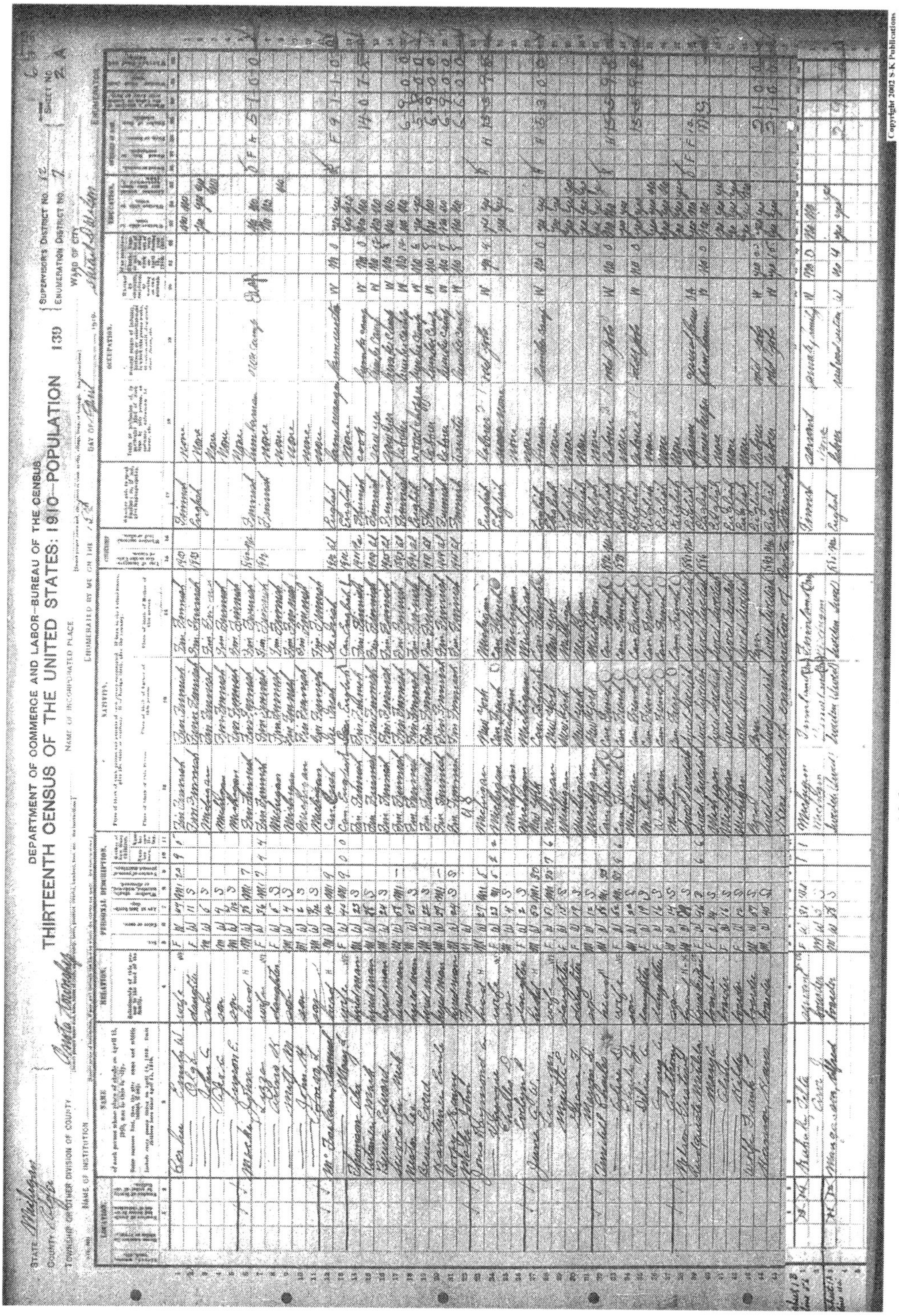

DEPARTMENT OF COMMERCE AND LABOR—BUREAU OF THE CENSUS

THIRTEENTH CENSUS OF THE UNITED STATES: 1910—POPULATION

13th census of the United States, 1910

Onota Township Census

The Population between 1910 and 2010 has increased from 154 to 310.

The 1910 U.S. census for Onota township tells us a great deal about the makeup of the township for three years after the closure of the Tyoga Lumber Company. Two-thirds were male and all were white. Native Americans may not have been in the area at the time of the survey and therefore not included. Fifty percent were born in Michigan, of that number 80% were of Finnish heritage with the remainder primarily French-Canadian, U.S., and Scandinavian. By age distribution, 75% were less than age 40 and two over the age of 60. Their occupations reflect an economy based on the timber industry with a lesser number in farming.

Onota Township Occupations, 1910 Census

Teacher	2
Railroad	7
Clerk	1
Farmer	12
Lumber Camp	11
Servant	2
Odd Jobs	9
"In Woods"	4
Boarders	20

Attendance at the meeting of the Onota Township Board meeting in 1910 lists the leadership of the community, 50% of which were of Finnish heritage. Note the change between the 1905 and 1906 Onota Township Board meeting in 1910.

Onota Township Board Meetings at Schaffer's
Onota Charcoal Manufacturing Company 1905 and 1906

4/3/05 Onota Township Meeting
at Onota Charcoal Manufacturing Company
Those in attendance were:

Frank Bayfield
W. J. Ellison
Louis Jacobson
John LaRue
Arthur Marlow
Edward Penny
C. Schaeffer
Frank Seymour
Charles Trudell
Alfred Wilcher

4/06 Onota Township Meeting
Those in attendance were:

John W. Bailey
John Blake
W. J. Ellison
Louis Jacobson
Elias Johnson
Heming Kellin
Arthur Marlow
Charles Roberts
John Swanson

Total Tax $700
Schools $1,800
Highway tax ½% of assessed value

1907 Township Tax	$1,000
Schools	$1,800
Highway Tax	1%

$230 to Tyoga
Kate Delger and Henry Maki added to the Board

1908 "Highway workers almost finished"
Also firefighters

ONOTA TOWNSHIP BOARD MEETING, 1910

MEMBER	NATIONALITY
Charles Schaeffer	English
Magnasum	Swedish
John Mackie	Finnish
Albert Makie	Finnish
Sandstrom	Swedish
Marleau	French Canadian
Mat Makie	Finnish
Karku	Finnish
Richard Makie	Finnish
Nelson	Swedish or Finnish
Hansen	Danish
Trudell	French Canadian
Hongisto	Finnish
Blake	English
Erick Makie	Finnish

DEMOGRAPHICS 1900-2005

	ONOTA TOWNSHIP			ALGER COUNTY				MARQUETTE COUNTY			
	1900*	1910	2000	1900	1910	1920	2000	1900	1910	1920	2000
Total Population	82	150	310	5,900	8,000	10,000	9,862	41,000	48,000	46.000	64,000
Population/Sq. Mi.	0.9	1.7	3.5	6.4			10.7		2.2		36
Male/Female	77/5	93/57	152/158								
Age							% <24 28				34
% <40	75	75	40				% 25-45 29				27
% 40-60	24	24	38				% 46-65 25				24
% >60	1	1	22				% >65 17				3
Race											
White			98				88				95
Black			0				6				1
Native Indian			2				3.3				1.5
Asian/Hispanic			0				3				3
Total Native Born				3,600					23,000		
Total Foreign Born				2,300					18,000		
Native White/Native Parents				1,200					2,700		
Native White/Foreign Parents				1,300					19,600		
Foreign White				2,200							
**Total Colored				70					148		
Negro				7					64		
White Ancestry											
English	4		6				6				11
Finnish	10	66									
French	32						10				9
French Canadian		4	6				6				
German	1		20				10				12
Irish			12				6				
Italian											8
Native Amer. Indian											
Polish			9				8				
Swedish/Norwegian	10	4									
?	24	69									
African American											

Courtesy John Parlin

*Did not include children

**Persons of Negro, Chinese, Japanese, or Indian descent

DEMOGRAPHICS 1900-2005 (continued)

	STATE OF MICHIGAN					UNITED STATES					
	1850	1900	1910	1920	2000	1800	1850	1900	1910	1920	2000
Total Population	397,000	2.42 mil	2.9 mil	3.5 mil	10 mil	5 mil	23 mil	76 mil	92 mil	106 mil	281 mil
Population/Sq. Mi.	7	41			179						80
Male/Female											
Age					35						
% <40					28						% <18 25
% 40-60					22						% 20-64 60
% >60					15						% >65 12
Race											
White					81%						75%
Black					15%						21%
Native Indian					1.30%						1%
Asian/Hispanic					2%						16%
Total Native Born											
Total Foreign Born											
Native White/Native Parents	>395,000										
Native White/Foreign Parents											
Foreign White											
**Total Colored	2,500	22,500									
Negro	0	15,800									
White Ancestry											
English		15			10						8%
Finnish											
French		10%									
French Canadian											
German		25%			20%						20%
Irish		8%			11%						11%
Italian											6%
Native Amer. Indian		6,350									
Polish					9%						3%
Swedish/Norwegian		10%									4%
?											
African American					14%						12.3%

Courtesy John Parlin

*Did not include children

**Persons of Negro, Chinese, Japanese, or Indian descent

2005 CENSUS DEMOGRAPHICS

	Onota Township	Alger County	Marquette County	Michigan	United States
Total Acres	96	5,049	3,425	98,000	
Land	88	918	1,821	58,000	
Water	8	4,131	1,604	40,000	
Population	310	9,662	64,000	10 mil	296 mil
Population/Sq Mile(land acres)		11	35	175	79
% chg 2000-05		2		1.8	5.3
% chg 1990-2000		9.9		6.9	13.1
Language other than English		5.2%	5%	8.4%	17.9%
Native American	2%	3.7%	1.6%	1.3%	1%
High School Grads	90%	81%	88%	83%	80%
2 yr College	5%				8%
4 yr College	14%				15%
Grad Degree	17%				7%
Median Household Income	49,000	36,000	37,000	46,000	46,000
Median Age	51	41	38	40	36
Cost of Living	19% < US				
Median Home Cost	122,000				217,000
Homes Owned	37%				64%
Housing Vacant	61%				15%

U.S. Census 2005

This table lists demographic data for Onota Township, Alger and Marquette Counties, State of Michigan, and the United States.

As can be seen from the statistics and demographics collected here, Alger County has a population whose origins can be traced back to those hardy souls who came to this wild land to work hard and create a better life. They worked in the mines, in the timber industry, as commercial fishermen and in the ships that moved so much of the ore and timber across the Great Lakes to the cities and industrial plants hungry for the raw materials. Many of those who came were from northern countries such as Finland, Germany, Canada and England where the weather could be as harsh and cold as in the U.P. and the work was familiar. But what the census figures also show is that people from many other countries came and settled and brought with them their customs and traditions adding to the richness and variety of a population that from a distance might look more homogeneous than it is. The U.P. is a beautiful place rich with wild beauty, natural resources and people who love where they live and who accept the hard work necessary to make a living in what many consider to be God's country.

APPENDIX TIMELINE

Formation of the Great Lakes following the retreat of the last glacier in 10000B.C.
Leverett and Taylor, Geology Library of MSU

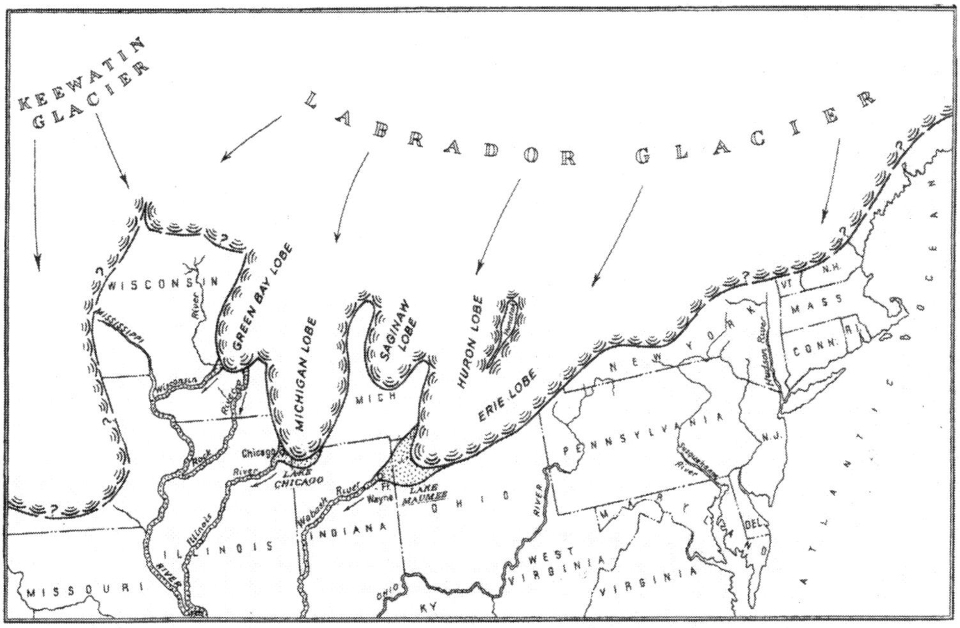

Courtesy Leverett and Taylor

FIRST OF THE GLACIAL LAKES—MAUMEE AND CHICAGO

EARLY LAKE MAUMEE, WITH OUTLET TO THE WABASH RIVER; INITIAL STAGE OF LAKE CHICAGO, WITH OUTLET TO THE DESPLAINES AND ILLINOIS RIVERS.

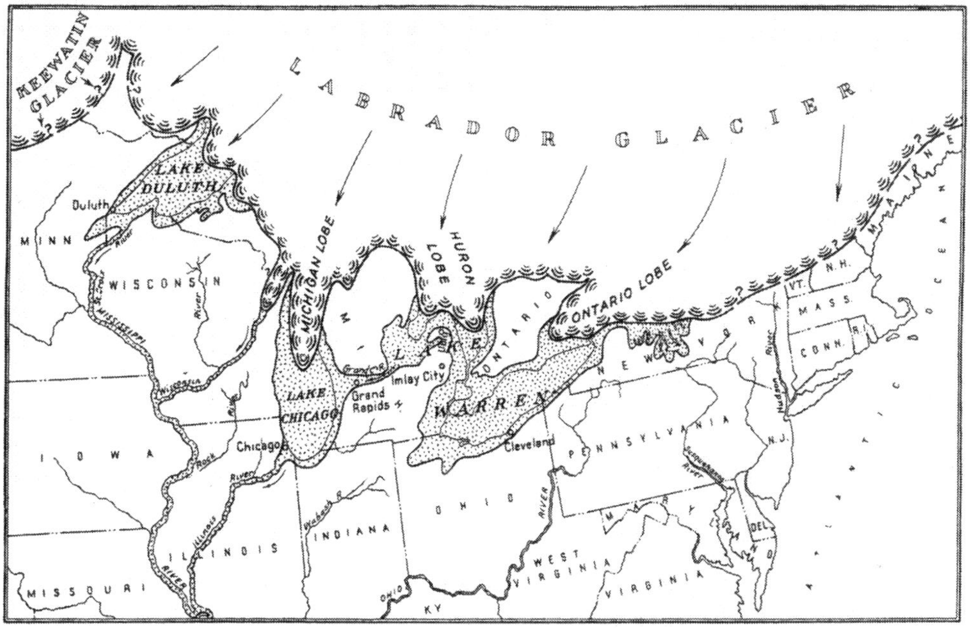

Courtesy Leverett and Taylor

GLACIAL LAKES WARREN, CHICAGO AND DULUTH

GRAND RIVER OULET OF WARREN; ILLINOIS RIVER OUTLET OF CHICAGO; ST. CROIX RIVER OUTLET OF DULUTH; WISCONSIN RIVER DRAINAGE OF GREEN BAY LOBE.

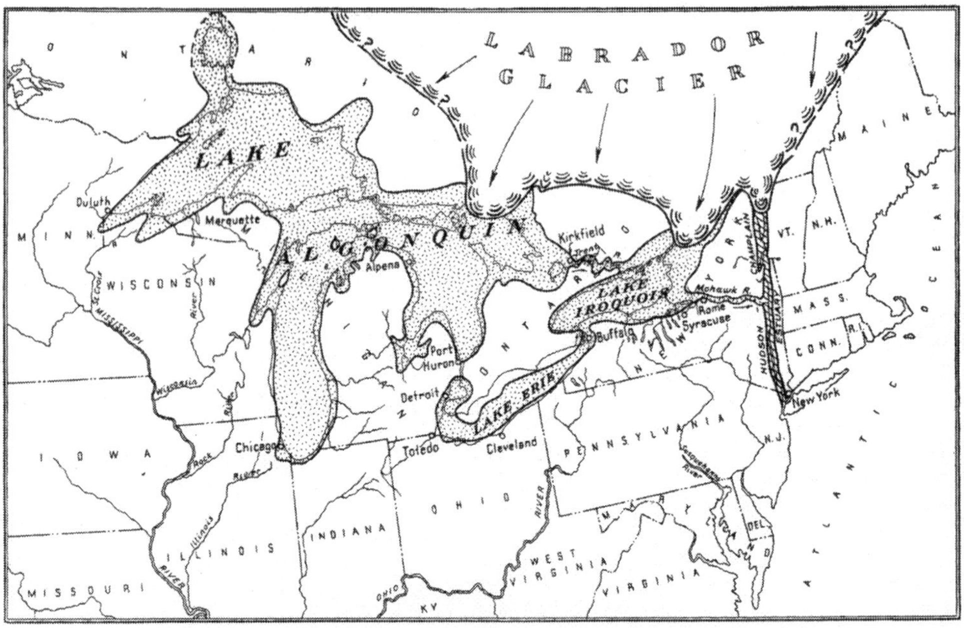

Courtesy Leverett and Taylor

GLACIAL LAKES ALGONQUIN AND IROQUOIS; LAKE ERIE

TRENT RIVER OUTLET OF ALGONQUIN; MOHAWK RIVER OUTLET OF IROQUOIS-ERIE TO THE HUDSON-CHAMPLAIN ESTUARY OF THE ATLANTIC OCEAN.

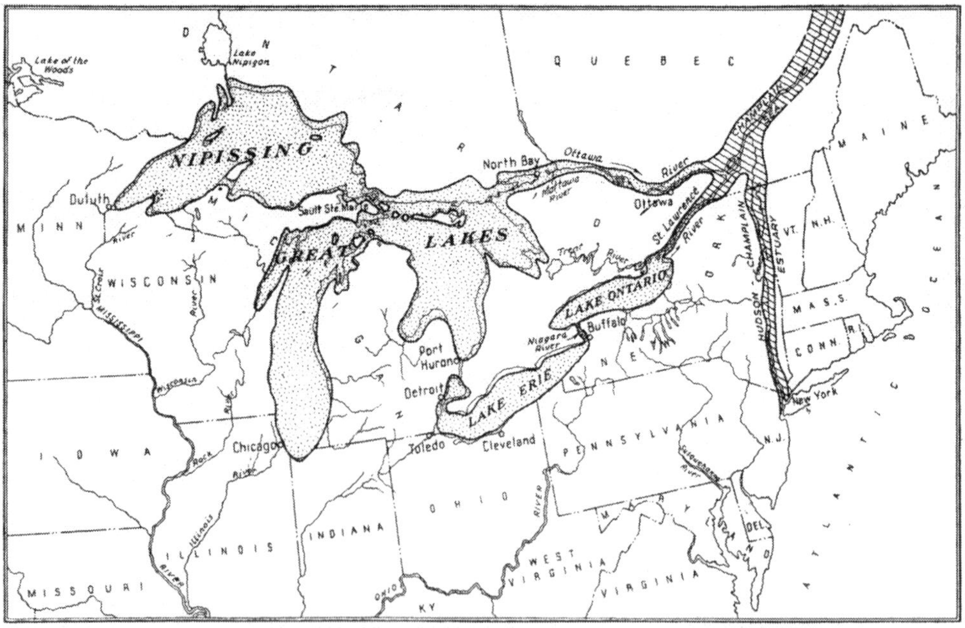

Courtesy Leverett and Taylor

NIPISSING GREAT LAKES, LAKE ERIE, AND LAKE ONTARIO

THE NORTH BAY OR OTTAWA RIVER OUTLET OF THE NIPISSING LAKES; THE ST. LAWRENCE RIVER OUTLET OF ONTARIO-ERIE TO THE CHAMPLAIN SEA.

APPENDIX
Chapter I
Item 1

Treaties of 1826, 1836, 1842, 1847 and 1855

1. Treaty with the Chippewa 1826 – Lewis Cass at the "Fond du Lac of Lake Superior" (Duluth area) –

Whereas a Treaty was concluded at Prairie du Chien in August last, by which the war, which has been so long carried on, to their mutual distress, between the Chippewas and Sioux. The Chippewas give their formal assent thereto, that the peace which has been concluded may be rendered permanent, therefore. . .

Article IV. It being deemed important that the half-breeds, scattered through this extensive country, should be stimulated to exertion and improvement by the possession of permanent consideration of the affection they bear to these persons, and of the interest which they feel in their welfare, grant to each of the persons described in the schedule hereunto annexed, being half-breeds and Chippewas by descent, six hundred and forty acres of land, to be located, under the direction of the President of the United States, upon the islands and shore of the St. Mary's river, wherever good land enough for this purpose can be found; and as soon as such locations are made, the jurisdiction and soil thereof are hereby ceded. It is the intention of the parties, that, where circumstances will permit, the grants be surveyed in the ancient French manner.

Article V. In consideration of the poverty of the Chippewas, and of the sterile nature of the country they inhabit, unfit for cultivation, and almost destitute of game, and as a proof of regard on the part of the United States, it is agreed that an annuity of two thousand dollars, in money or goods, as the President may direct, shall be paid to the tribe, at the Sault St. Marie. *But this annuity shall continue only during the pleasure of the Congress of the United States.*

Article VI. With a view to the improvement of the Indian youths, it is also agreed, that an annual sum of one thousand dollars shall be appropriated to the support of an establishment for their education, to be located upon some part of the St. Mary's river, and the money to be expended under the direction of the President; and for the accommodation of such school, a section of land is hereby granted. But the payment of the one thousand dollars stipulated for in this article, is subject to the same limitation described in the preceding article.

Article VII. The necessity for the stipulations in the fourth, fifth and sixth articles of this treaty could be fully apparent, only from personal observation of the condition, prospects, and wishes of the Chippewas, and the Commissioners were therefore not specifically instructed upon the subjects therein referred to; but seeing the extreme poverty of these wretched people, finding them almost naked and starving, and ascertaining that many perished during the last winter, from hunger and cold, they were induced to insert these articles. *But it is expressly understood and agreed, that the fourth, fifth and sixth articles, or either of them may be rejected by the President and the Senate, without affecting the validity of the other articles of the treaty.*

Article VIII. The Chippewa tribe of Indians fully acknowledge the authority and jurisdiction of the United States, and disclaim all connection with any foreign power, solemnly promising to reject any messages, speeches, or councils, incompatible with the interest of the United States, and to communicate information thereof to the proper agent, should any such be delivered or sent to them.

2. Treaty of 1836 – Washington

During the period from 1819 to 1855, this treaty was perhaps the most detrimental to the Chippewa Native Americans of Grand Island. With the 1836 treaty, approximately 150,000 acres of land in both lower Michigan as far south as Grand Haven and in the Upper Peninsula east of the Chocolay River were ceded to the Federal government. Included in this were 640 acres on the southern end of Grand Island and 2,000 acres on the mainland south of Grand Island. The Grand Island Native American land was purchased by the government over the next five years at 25 cents per acre and sold to the white settlers.. During the 1830's, Native Americans had been used as labor in the logging industry. With this treaty, the Native Americans moved first to Bay Furnace and subsequently to Sand point, north of Munising on the east side of the bay. Native American Town east of Sand Point was created in 1883 with those descendents of the Grand Island band who were living there at that time.

The 1836 treaty stated that the land ceded in the U.P. would be east of a north/south line extending between the mouth of the Chocolay River which empties into Lake Superior east of Marquette and the mouth of the Escanaba River which empties into Lake Michigan. At the end of the treaty, Schoolcraft adds: "**GENERAL NOTE.** – After the selection by Mr. Schoolcraft of the 20,000 and 70,000 acre reserves under this treaty, he was advised that the U.S. might conclude to allow the Native Americans to remain on the other reserves after the expiration of the five years. He was therefore instructed, Nov. 5th, 1840, that the boundaries of all the reserves under this treaty ought to be marked. Aug. 23,

1844, the Native American Office advised the General Land Office that these reserves ought not to be surveyed as public lands, the Native Americans having been tacitly allowed to remain thereon since the treaty.

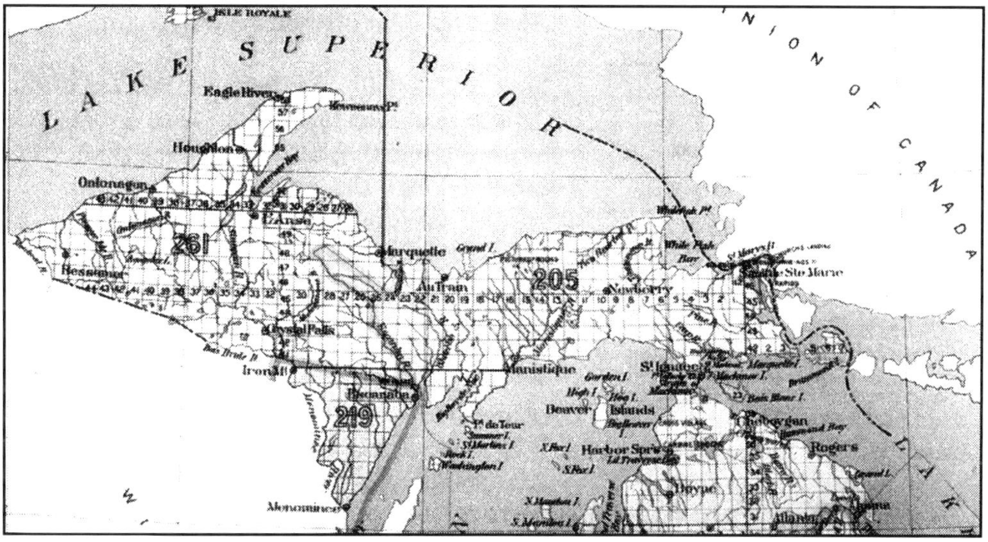

Courtesy Fred Smoot Digital Archives Map Project

In 1836. Chippewa (Ojibwe) ceded land east of the Chocolay River and northwest one-third of Lower Peninsula to Federal government.

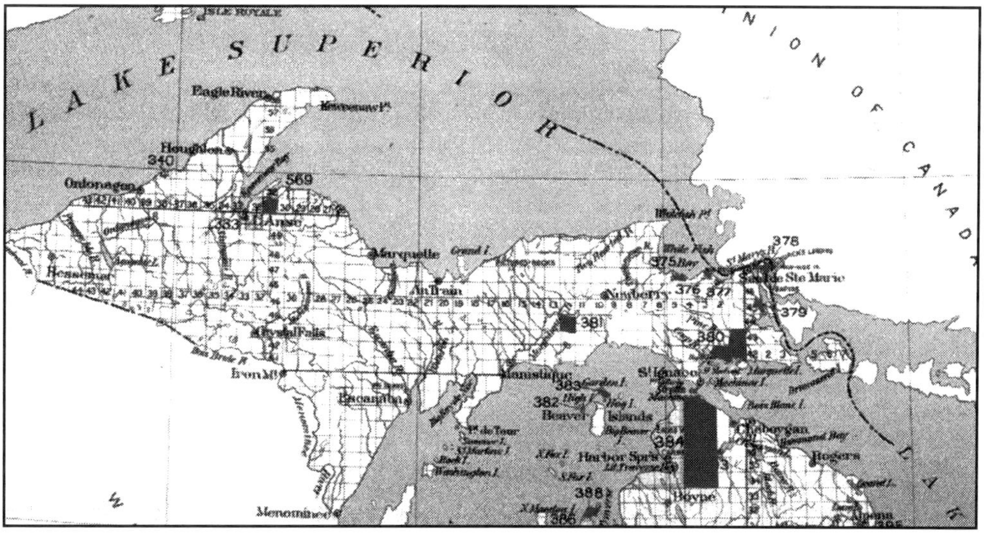

Courtesy Fred Smoot Digital Archives Map Project

Chippewa land ceded to Federal Government in 1842 Reservations are shown in black. The lands ceded in the Treaties of 1836 and 1842 included the entire U.P. and lands in the Lower Peninsula with the exception of the reservation lands shown on the map of Indian Reservations, 1996.

3. Treaty with the Chippewa 1842 – at La Pointe of Lake Superior

(one of the Apostle Islands of Northern Wisconsin).

Article 1. The Chippewa Indians of the Mississippi and Lake Superior, cede to the United States all the country within the following boundaries; viz: beginning at the mouth of the Chocolate river of Lake Superior;

Article 2. The Indians stipulate for the right of hunting on the ceded territory, with the other usual privileges of occupancy, until required to remove by the President of the United States, and that the laws of the United States shall be continued in force, in respect to their trade and intercourse with the whites, until otherwise ordered by Congress.

Article 5. Whereas the whole country between Lake Superior and the Mississippi, has always been understood as belonging in the common to the Chippewas

Article 6. The Indians residing on the Mineral district, shall be subject to removal there from at the pleasure of the President of the United States.

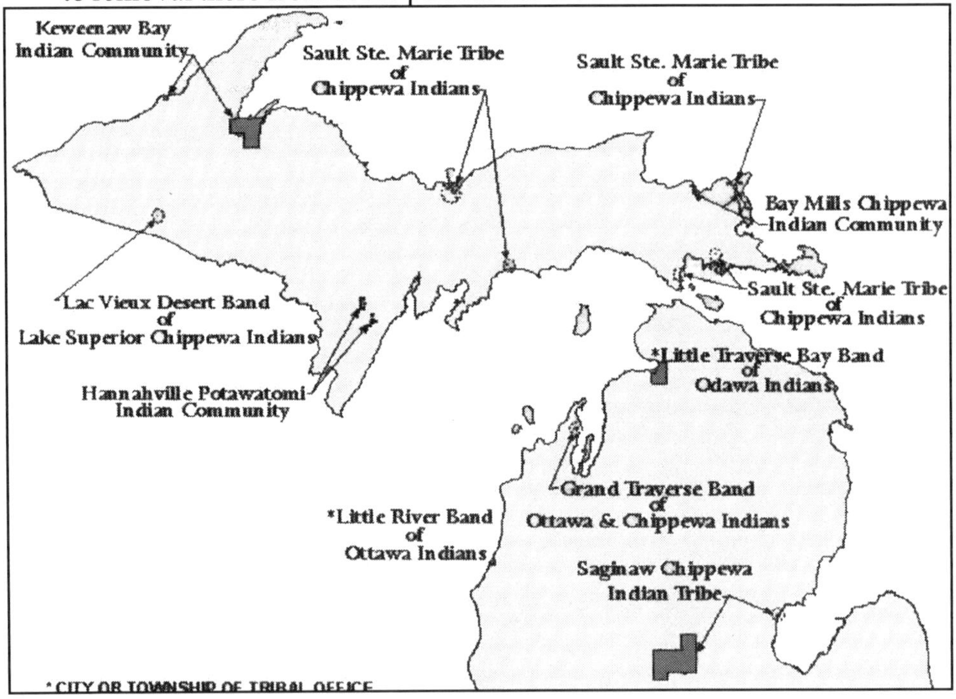

Map of Michigan Indian reservations, 1996
The Chippewa ceded all lands in Michigan to the Federal government retaining only small areas for reservation (shown here in black).Native American reservations have changed only slightly since the map following the treaty at LaPointe in 1842. Since that time, new reservations have been added in Munising, Gogebic, Southern Delta, and Luce Counties. (Appendix Chap. I, Item 8) There has also been a change in the size of the Native American land south of Mackinaw City.

4. Treaty with Chippewa of the Mississippi and Lake Superior 1847 – at Fond du Lac of Lake Superior. This cedes land east of the Mississippi to the Federal government.

It is agreed that the peace and friendship which exists between the people of the United States and the Chippewa Indians shall be perpetual.

The Chippewa Indians of the Mississippi and Lake Superior cede and sell to the United States all the land east of the Mississippi.

In consideration of the foregoing cession, the United States agree to pay to the Chippewas of Lake Superior seventeen thousand dollars in specie, and to the Chippewas of the Mississippi seventeen thousand dollars in specie; the above sums to be paid at such place or places , and in such manner, as the President shall direct, and to be paid within six months after this treaty shall be ratified by the President and Senate of the United States; and the United States further agree to pay to the Mississippi Indians the sum of one thousand dollars annually for forty-six years; but it is agreed that whenever the Chippewas of the Mississippi shall agree as to the schools to be established, and the places at which they shall be located, the number of the blacksmiths and laborers to be employed for them, and shall request the United States to expend, from year to year, the annual payments remaining unpaid, in the support of schools, blacksmiths, and laborers, the same shall be expended by the United States for such purposes; and that Chippewas of full or mixed blood shall be employed as teachers, blacksmiths, and laborers, when such persons can be employed who are competent to perform the duties required of them under this and all former treaties.

5. Treaty with the Chippeway of Sault Ste. Marie 1855 – Chippewa Native American lands east of the Mississippi ceded to federal government in treaty of 1847. Treaty of 1855 addresses specific issues of the St. Mary's River. (Fig on page 23)

Article 1. The said Chippewa Indians surrender to the United States the right of fishing at the falls of St. Mary's and of encampment, convenient to the fishing-ground, secured to them by the treaty of June 16, 1820. (in perpetuity)

Article 2. The United States will appoint a commissioner who shall, within six months after the ratification of this treaty, personally visit and examine the said fishery and place of encampment, and determine the value of the interest of the Indians therein as the same originally existed. His award shall be reported to the President, and shall be final and conclusive, and the amount awarded shall be paid to said Indians, as annuities are paid, and shall be received by them in full satisfaction for the right hereby surrendered: Provided, That one-third of said award shall, if the Indians desire it, be paid to such of their half-breed relations as they may indicate.

Article 3. The United States also give to the chief, O-shaw-waw-no, for his own use, in fee-simple, a small island in the river St. Mary's, adjacent to the camping-ground hereby surrendered, being the same island on which he is now encamped, and said to contain less than half an acre; Provided, That the same has not been heretofore otherwise appropriated or disposed of; and in such case, this grant is to be void, and no compensation is to be claimed by said chief or any of the Indians, parties hereto, in lieu thereof.

APPENDIX
Chapter I
Item 2

Johan Kohl, a German ethnologist, spent the summer of 1855 with the Ojibway in Northern Wisconsin and described the resignation expressed by a tribal elder at yet another treaty. "But for some years they have been asking land from us. For ten years they have asked from us nothing but land, and ever more land. We give unwillingly the land in which the graves of our fathers rest. But for all that we have given land in our generosity. We knew not that we were giving so much for so little. We did not know that such great treasures of copper were hidden in our land."

"Father, the point is this. In our former treaties – yes, in all former treaties – it was settled that a certain sum should be deducted from our tribute for blacksmiths' shops, schools, and other establishments among us. We have *heard* of those moneys. But we have *seen* nothing of these works. They have not come to us. We know not where those moneys are gone, or where they went off in smoke. We beg thee, examine into this closely. This we beg thee much. I could say much more on this point. But I will now sit down. For I am not accustomed to wear these new European trousers which have been given me, or to stand long in them. They annoy me. Hence I will cease to speak and seat myself."

APPENDIX
Chapter I
Item 3

DRAWINGS FROM HENRY R. SCHOOLCRAFT'S - HISTORY OF THE INDIAN TRIBES OF THE U.S., 1857

Henry R. Schoolcraft

Ancient copper mining on Lake Superior, using stone hammer and iron rod Was this miner Native American or European? The Bonnze era of 3000 B.C. was followed by the Iron age in 1200 B.C. in Europe.

Henry R. Schoolcraft

A seer attempting to destroy an Indian girl with a ray of sunlight.

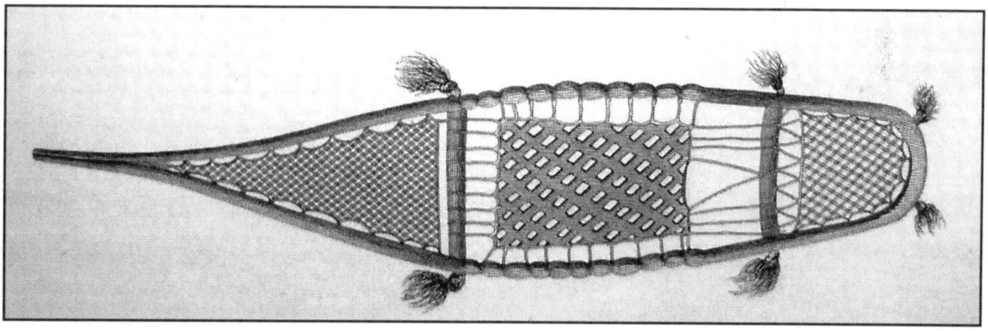

Henry R. Schoolcraft

Chippewa female snowshoe
The design and materials have seen little change in 150 years. The long tail helps the wearer track in a straight line.

APPENDIX
Chapter I
Item 4

CENSUS OF THE INDIAN TRIBES OF UNITED STATES – 1857

This census of the Indian tribes of the United States published in 1857 is included in a five volume publication by Roger Schoolcraft mandated several years earlier by the Congress of the United States. This extensive study of Indians throughout the United States is titled History of Indian Tribes of the U.S. The survey includes 174 questions and answers that provided information that included the usual demographics as well as the Native American life style. Their crops, their livestock, and their hunting were details important to this census.

CENSUS OF THE INDIAN TRIBES OF UNITED STATES – 1857

(Alphabetically arranged,) who have ceded Territory since the Establishment of the present Federal Government, showing the amount paid, &c.

Names of Tribes.	Date of treaty.	No. of acres of land ceded.	Aggregate number of acres ceded.	No. of acres given in exch'ge or reserved.	Aggregate of acres given in exchange, or reserved.	Value of land given in exchange or reserved, at $1 25 per acre.	Aggregate value of land given in exchange, or reserved.	Amount of consideration in money, goods, &c.	Aggregate of consideration in money, goods, &c.	Aggregate of consideration of each treaty.	Aggregate of consideration.
Appalachicolas	1833	5,120	5,120	None				$13,000 00	$13,000 00	$13,000 00	$13,000 00
Caddoes	1836	1,000,000	1,000,000	5,440	5,440	$6,800 00	$6,800 00	80,000 00	80,000 00	86,800 00	86,800 00
Cherokees	1806	1,209,600		None				44,000 00		44,000 00	
Cherokees	1816	1,395,200		"				65,000 00		65,000 00	
Cherokees	1819	566,400		a							
Cherokees	1836	7,882,240	11,053,440	800,000	800,000	e 500,000 00	e 500,000 00	6,224,279 00	6,333,279 00	6,724,279 00	6,833,279 00
Chickasaws	1805	345,600		None				22,000 00		22,000 00	
Chickasaws	1833	6,422,400	6,768,000	"	None	f		3,046,000 00	3,068,000 00	3,046,000 00	3,068,000 00
Chippewas	1807	7,862,400		"				100,400 00		100,400 00	
Chippewas	1819	4,321,280		10,240		12,800 00		41,200 00		54,000 00	
Chippewas	1820	10,240		None				g			
Chippewas	1836	8,320		"				h			
Chippewas	1838	7,000,000	19,202,240	"	10,240		12,800 00	870,000 00	1,011,600 00	870,000 00	1,024,400 00
Chippewas, Ottawas, and Pottawatomies (united nation)	1816	1,418,880		None				12,000 00		12,000 00	
Chippewas, Ottawas, and Pottawatomies (united nation)	1830	4,160,000		16,640		20,800 00		369,801 00		390,601 00	
Chippewas, Ottawas, and Pottawatomies (united nation)	1835	5,104,960	10,683,840	5,000,000	5,016,640	6,250,000 00	6,270,800 00	1,374,289 00	1,756,090 00	7,624,289 00	8,026,890 00
Choctaws	1801	2,641,920		None }				2,201 00		2,201 00	
Choctaws	1802	853,760		" }							
Choctaws	1805	4,142,720		"				108,000 00		108,000 00	
Choctaws	1820	4,500,000		b5,030,912				i			
Choctaws	1831	7,796,000	19,934,400	c15,000,000	d20,030,912	20,750,000 00	20,750,000 00	2,178,529 00	2,288,730 00	22,928,529 00	23,038,730 00
Creeks	1814	14,284,800		None				120,000 00		120,000 00	
Creeks	1826	4,132,480		4,140,000		5,175,000 00		763,000 00		5,938,000 00	
Creeks	1828	1,221,120		None				42,491 00		42,491 00	
Creeks	1832	5,128,000	24,766,400	11,158,400	15,298,400	13,948,000 00	18,123,000 00	1,861,080 00	2,786,571 00	15,809,080 00	21,909,571 00
Delawares	1830	5,760	5,760	None				3,000 00	3,000 00	3,000 00	3,000 00
Delawares, Shawnees, &c.	1803	2,038,400	2,038,400	"				4,000 00	4,000 00	4,000 00	4,000 00
Delawares and Piankeshaws	1804	2,038,400	2,038,400	"				4,000 00	4,000 00	4,000 00	4,000 00

a Acre for acre, and other benefits that cannot now be computed. *b* Lands given in exchange afterward ceded, therefore value not computed. *c* And reservations not known.
d And reservations. *e* Valuation made by the treaty. *f* $3,000,000, estimated nett proceeds of sales of land ceded, to be paid to them.
g Merchandise; amount not specified in the treaty. *h* Nett proceeds of sales of the lands ceded. *i* Amount of benefits cannot now be computed.

STATISTICAL TABLES OF THE INDIAN POPULATION

Comprised within the boundaries of the Consolidated Agency of Sault Ste Marie and Michilimackinac in the year 1832, *together with the number of Trading Posts established under the Act of Congress of May* 26, 1824, *and other facts illustrating the condition and operations of the Fur Trade. Prepared under instructions of the War Department of the* 3d *May*, 1832, *for visiting the sources of the Mississippi.*

NATURE AND GEOGRAPHICAL DISTRIBUTION OF THE POPULATION.

GEOGRAPHICAL DISTRICT.	Name of the Village or Periodical Encampment.	No. of men.	No. of women.	No. of children.	No. of persons of the mixed blood, all ages and sexes.	Population of villages, &c.	Population of geographical districts.
	Michilinackinac (see note *A.*)						
	St. Martin's Islands	18	29	51	140	238	
N. W. coast of Lake Huron	Chenos						302
	Drummond Island	16	18	23	7	64	
	Mineeshco river (*B.*)						
	Muscoda Sagi (*C.*)						
St. Mary's river, American side	Sugar Island	58	73	144	161	436	436
	Little Rapids Kinibitunoong (*D.*)						
	Sault Ste Marie (*E.*)						
	Misconabies Creek						
	Tacquimenon						
	Heart's Blood Lake						
	Manistic river	42	46	98	...	186	
	White Fish Point (*F.*)						
	Shelldrake River (Onzig)						
	Two-hearted River						
	Grand Marais						
	Miner's River and Pictured Rocks						
Southern shores of L. Superior	Grand Island	7	6	23	14	50	
	Presque Isle and Granite Point	4	4	12	...	20	
	Huron Bay	4	6	4	...	14	
	Keweena Bay (*G.*)	31	38	43	28	140	
	Ontonagon (*H.*)	29	32	76	15	152	
	Mouth of Montreal river						
	Mauvais River	32	40	113	28	213	
	Lapointe or Chegoimegon (*I.*)						
	Fond du Lac (*K.*)	44	46	103	38	231	1006

The arrow indicates the Grand Island Tribe

CENSUS RETURNS OF THE INDIAN TRIBES OF THE UNITED STATES.

II. ALGONQUIN GROUP.

Number of Tribes or Geographical Sub-Groups.	Main Tribes, or Geographical Sub-Groups.	Number of Bands, or recognised Divisions.	Names of the Divisions of the Group, as known to the Laws, or recognised by Treaties or Usage.	Geographical Position.	1 Number of Families in the Tribe or Band.	2 Whole Number of Souls of all Ages and both Sexes.	3 Number of Males under the Age of 18.	4 Number of Females under the Age of 16.	5 Number of Males between the Ages of 18 and 60.	6 Number of Females between the Ages of 16 and 60.	7 Number of both Sexes between the Ages of 60 and 100.	8 Number of European or white Heads of Families, male.	9 Number of European or white Female Heads of Families.	10 Number of Male Children of Half or Mixed Blood.	11 Number of Female Children of Half or Mixed Blood.	12 Number of Marriages in the Family during the Year.	13 Number of Births, Male, within the Year.	14 Number of Births, Female, within the Year.	15 Number of Deaths, Male, within the Year.	16 Number of Deaths, Female, within the Year.	17 Number of Deaf and Dumb Persons.	18 Number of Lunatics or Insane.	19 Number of Idiots.
			Detroit Chippewas.																				
1.	CHIPPEWAS.	1.	Chippewas of Swan Creek and Black River..	Michigan ...	.....	208																	
		2.	Chippewas of Swan Cr. and Black R., west.	Ind. Ter. W.	10	23	4	5	6	8	.....	.....	.....	.....	.....	.....	3	2	4	3			
			Saganaws.																				
		3.	Saganaws of Pewanuganink....................	Michigan ...	101	102	19	27	24	28	4	.....	2	.....	.....	3	...	1	1	2			
		4.	Saganaws of Natoway's Band.................	Michigan ...	46	200	54	59	40	40	7	1	.....	2	3	1	1	1	2	4	4		
		5.	Band of Ogakaning.............................	Michigan ...	102																		
		6.	Atawa's Band, Saganaws	Michigan ...	115																		
		7.	Band of Wabeseepenukaning..................	Michigan ...	23	94																	
			Lake Huron, South.																				
		8.	Chippewas of Thunder Bay.....................	Michigan ...	14	77	19	31	13	13	1	.....	1	10	8	.....	...	3	...	...	1		
			Chippewas of Grand Traverse Bay.																				
		9.	Eshquagonabee's Band	Michigan ...	24	127	49	20	27	28	3	.....	.....	.....	.....	.....	5	...	2				
		10.	Akosa's Band	Michigan ...	40	166	44	30	40	42	10	.....	.....	.....	.....	1	5	1	4	4			
		11.	Kagwatosa's Band................................	Michigan ...	19	81	22	19	17	21	2	.....	.....	.....	.....	.....	...	1	1	...	...	...	1
		12.	Band at Nagondagon	Michigan ...	15	46	5	8	16	14	3												
		13.	Band at Muhzheguning.........................	Michigan ...	18	70	20	14	16	19	1	.....	.....	.....	.....	.....	...	...	...	2			
			Chippewas of Michillimackinac.																				
		14.	Indians of the Beaver Islands	Michigan ...	7	35	12	9	7	7	.....	.....	.....	.....	.....	.....	1	3					
		15.	Manistee River, north	Michigan ...	13	73	23	21	12	16	1	.....	.....	.....	.....	.....	1	...	...	1			
		16.	Shabegoshing, or Carp River	Michigan ...	13	75	28	23	11	13	.....	.....	.....	.....	.....	1	...	3	1				
		17.	Ance's Band, Oak Point.........................	Michigan ...	23	92	24	25	21	21	1	.....	.....	1	1	4	11	3					
		18.	Point St. Ignace..................................	Michigan ...	8	42	9	18	7	7	1												
		19.	Chenos, Huron, north...........................	Michigan ...	15	45	16	5	11	13	.....	.....	.....	.....	.....	.....	1	...	1				
			Straits of St. Mary, &c.																				
		20.	Drummond Island Band.........................	Michigan ...	20	103	25	21	23	32	2	.....	.....	.....	.....	1	1	4	2	...	1		
		21.	Primary Band of Sault Ste. Marie	Michigan ...	58	195	31	29	54	78	3	.....	.....	.....	.. ..	.....	2	...	4	1	1		
		22.	Waishkee's Bay, foot of Lake Superior.......	Michigan ...	25	66	14	11	19	22	.....	.....	.....	.....	.....	.....	...	1	3	2			
		23.	Tacquimenon River	Michigan ...	12	51	9	9	15	18	.....	.....	.....	1	.....	.....	1	...	1				
		24.	Grand Island of Lake Superior.................	Michigan ...	13	57	9	7	21	20	.....	.....	.....	.....	.....	.....	1	...	...	...	1		
					734	2028	436	391	400	460	39	1	3	14	12	11	33	23	26	19	8		1

Number of Tribes phical Sub-Gr	Main Tribes, or Geographical Sub-Groups.	Number of Bands, o recognised Divisions	Names of the Divisions of the Group, as known to the Laws, or recognised by Treaties or Usage.	Geographical Position.	Number of Persons born Blind.	Number of Heads of Families who subsist their Families by the Chase.	Number of Heads of Families who subsist their Families by Agriculture.	Number of Males who are employed as Interpreters or Translators.	Number of Females who are employed as Interpreters or Translators.	Number of Persons occupied as Blacksmiths or Assistants.	Number of Persons occupied as Carpenters.	Number of Persons occupied as Shoemakers.	Number of Persons occupied as Wheelwrights.	Number of Persons occupied as Silversmiths.	Number of Persons occupied as Tailors.	Number of Persons engaged as Principals or Clerks in Trade.	Number of Persons who have studied or practise any of the Learned Professions.	Number of Male Children who attend School.	Number of Female Children who attend School.	Number of Children who can speak the English Language.	Number of Males who are engaged as Teachers.
			Detroit Chippewas.																		
		1.	Chippewas of Swan Creek and Black River..	Michigan ...	...	32	32	.....	.....	.....	...	...	.....	...	...	.....	.....	22	20		
		2.	Chippewas of Swan Cr. and Black R., west.	Ind. Ter. W.	...	.....	10	.....	.....	.....	...	...	.....	...	...	.....	.....	1	1	5	
			Saganaws.																		
		3.	Saganaws of Pewanuganink	Michigan ...	...	.....	.....	.....	.....	.....	...	...	.....	...	...	.....	.....	17	18		
		4.	Saganaws of Natoway's Band	Michigan ...																	
		5.	Band of Ogakaning	Michigan ...	...	.....	.....	.....	.....	.....	...	...	.....	...	...	.....	.....	20	18		
		6.	Atawa's Band, Saganaws	Michigan ...																	
		7.	Band of Wabeseepenukaning	Michigan ...																	
			Lake Huron, South.																		
		8.	Chippewas of Thunder Bay	Michigan ...	2	14	14														
			Chippewas of Grand Traverse Bay.																		
		9.	Eshquagonabee's Band	Michigan ...	...	26	37	.....	.....	2	1	...	.....	...	1	1	.....	87	10	5	
		10.	Akosa's Band	Michigan ...	...	11	8	.....	.....	.....	...	...	.....	...	1	1					
1.	CHIPPEWAS.	11.	Kagwatosa's Band	Michigan ...																	
		12.	Band at Nagondagon	Michigan ...																	
		13.	Band at Muhzheguning	Michigan ...																	
			Chippewas of Michillimackinac.																		
		14.	Indians of the Beaver Islands	Michigan ...	...	.....	.....	.....	.....	.....	...	...	.....	...	4						
		15.	Manistee River, north	Michigan ...	...	.....	17	.....	.....	.....	2	...	.....	...	1						
		16.	Shabegoshing, or Carp River	Michigan ...	...	10	3														
		17.	Ance's Band, Oak Point	Michigan ...	...	10	13	1	1	.....	1	...	.....	...	...	.....	.....	.....	.....	1	
		18.	Point St. Ignace	Michigan ...	...	.....	8														
		19.	Chenos, Huron, north	Michigan ...	...	10	5														
			Straits of St. Mary, &c.																		
		20.	Drummond Island Band	Michigan ...	1	18															
		21.	Primary Band of Sault Ste. Marie	Michigan ...	2	11	.....	.....	.....	.....	...	...	.....	...	...	.....	.....	2	1	5	
		22.	Waishkee's Bay, foot of Lake Superior	Michigan ...	...	6	10	1	.....	.....	...	...	.....	...	...	.....	.....	14	10	24	
		23.	Tacquimenon River	Michigan ...	...	12	.....	.....	.....	.....	...	...	.....	...	...	.....	.....	.....	.....	10	
		24.	Grand Island of Lake Superior	Michigan ...	1	21	.....	.....	.....	.....	...	...	.....	...	...	.....	.....	2	4	7	
					6	181	157	2	1	2	4				7	2		115	82	57	

II. ALGONQUIN GROUP.

Number of Tribes or Geographical Sub-Groups.	Main Tribes, or Geographical Sub-Groups.	Number of Bands, or recognised Divisions.	Names of the Divisions of the Group, as known to the Laws, or recognised by Treaties or Usage.	Geographical Position.	54 Number of Bushels of Oats raised.	55 Number of Bushels of Beans raised.	56 Number of Bushels of Peas raised.	57 Number of Bushels of Buckwheat raised.	58 Number of Bushels of Turnips raised.	59 Pounds of Flax raised.	60 Pounds of Hemp raised.	61 Pounds of Cotton picked.	62 Number of Fruit Trees of all Kinds.	63 Number of Melons of all Kinds raised.	64 Pounds of Maple Sugar made.	65 Pounds of Cheese made.	66 Pounds of Butter made.	67 Pounds of Honey, wild or domestic.	68 Number of Beef-Cattle killed or sold.	69 Estimated Value of the Agricultural and Horticultural Products of the Family during the Year.	70 Number of Horses.	71 Number of Mules.	72 Number of Oxen.	73 Number of Milch Cows.
			Detroit Chippewas.																					
		1.	Chippewas of Swan Creek and Black River	Michigan											1,100						68			
		2.	Chippewas of Swan Cr. and Black R., west	Ind. Ter. W.		15			50				150	276	60		70		4	$395	15		4	4
			Saganaws.																					
		3.	Saganaws of Pewanuganink	Michigan		2									2,315			200			16		2	
		4.	Saganaws of Natoway's Band	Michigan		10								12	2,400				4	$223	120		2	4
		5.	Band of Ogakaning	Michigan											1,000						4		4	2
		6.	Atawa's Band, Saganaws	Michigan											1,200								4	4
		7.	Band of Wabeseepenukaning	Michigan											3,500						5		2	1
			Lake Huron, South.																					
		8.	Chippewas of Thunder Bay	Michigan											8,000						3			
			Chippewas of Grand Traverse Bay.																					
		9.	Eshquagonabee's Band	Michigan		30							1	3,000	8,800					$80	13			
		10.	Akosa's Band	Michigan		19							30	2,800	15,580					$655	12			
1.	CHIPPEWAS.	11.	Kagwatosa's Band	Michigan		19								3,900	15,050					$100	17			
		12.	Band at Nagondagon	Michigan		9								600	2,150									
		13.	Band at Muhzheguning	Michigan										1,250							4			
			Chippewas of Michillimackinac.																					
		14.	Indians of the Beaver Islands	Michigan		80			20					100	20,000					$300				
		15.	Manistee River, north	Michigan		50	5		50					300	23,000					$300				2
		16.	Shabegoshing, or Carp River	Michigan												3500								
		17.	Ance's Band, Oak Point	Michigan											16,000						7			1
		18.	Point St. Ignace	Michigan	20				100						1,300						2			
		19.	Chenos, Huron, north	Michigan											12,000									
			Straits of St. Mary, &c.																					
		20.	Drummond Island Band	Michigan					250					500	7,000					$12				
		21.	Primary Band of Sault Ste. Marie	Michigan										300	17,000				1	$27	1		2	1
		22.	Waishkee's Bay, foot of Lake Superior	Michigan											7,600				3	$75			3	9
		23.	Tacquimenon River	Michigan					300					200	6,000					$71	2		3	1
		24.	Grand Island of Lake Superior	Michigan											3,500					$25				
					20	234	5		770				181	13,238	174,555	3500	70	200	12	$2263	289		26	29

CENSUS RETURNS OF THE INDIAN TRIBES OF THE UNITED STATES.

II. ALGONQUIN GROUP.

Number of Tribes or Geographical Sub-Groups.	Main Tribes, or Geographical Sub-Groups.	Number of Bands, or recognised Divisions.	Names of the Divisions of the Group, as known to the Laws, or recognised by Treaties or Usage.	Geographical Position.	37 Number of Females who are engaged as Teachers	38 Number of Male Sabbath-School Scholars.	39 Number of Female Sabbath-School Scholars.	40 Number of Males who can read and write.	41 Number of Females who can read and write.	42 Number of Males who have studied Vocal Music.	43 Number of Females who have studied Vocal Music.	44 Number of Females who can spin, knit, or weave.	45 Number of Pounds of Flax spun.	46 Number of Yards of Homespun wove.	47 Number of Pairs of Stockings knit.	48 Number of Females who can do Seamstress' or Mantuamakers' Work.	49 Number of Garments of all Kinds, made during the Year.	50 Number of Acres of Land cultivated.	51 Number of Bushels of Corn raised.	52 Number of Bushels of Wheat raised.	53 Number of Bushels of Potatoes raised.
			Detroit Chippewas.																		
1.	CHIPPEWAS.	1.	Chippewas of Swan Creek and Black River..	Michigan ...	...	...	...	11	...	.....	.....	...	...	...	...	.....	......	80	1,000	45	1
		2.	Chippewas of Swan Cr. and Black R., west.	Ind. Ter. W.	...	...	...	...	...	.....	.....	...	...	...	...	.....	......	25	500	...	
			Saganaws.																		
		3.	Saganaws of Pewanuganink....................	Michigan ...	...	...	...	...	...	.....	.....	16	...	...	18	32	......	......	766	...	
		4.	Saganaws of Natoway's Band..................	Michigan ...	...	...	...	...	...	.....	.....	...	...	...	...	1	750	45	100	...	
		5.	Band of Ogakaning..............................	Michigan ...	...	...	...	...	...	.....	.....	...	...	...	...	.....	......	......	150		
		6.	Atawa's Band, Saganaws	Michigan ...	...	...	...	...	...	.....	.....	...	...	...	...	.....	......	30	300	...	3
		7.	Band of Wabeseepenukaning..................	Michigan ...	...	...	...	...	...	.....	.....	...	...	...	...	.....	......	30	228		
			Lake Huron, South.																		
		8.	Chippewas of Thunder Bay.....................	Michigan ...	...	...	...	2	...	.....	.....	...	...	...	...	.....	......	40	200	...	7
			Chippewas of Grand Traverse Bay.																		
		9.	Eshquagonabee's Band	Michigan ...	...	22	27	10	2	9	5	13	...	...	...	2	101	150	539	...	8
		10.	Akosa's Band	Michigan ...	...	...	...	6	...	.....	.....	...	...	...	6	6	204	350	909	...	1,1
		11.	Kagwatosa's Band................................	Michigan ...	...	...	...	...	...	.....	.....	...	...	...	...	2	102	100	619	...	8
		12.	Band at Nagondagon............................	Michigan ...	...	...	...	...	...	.....	.....	...	...	...	...	.....	......	......	79	...	
		13.	Band at Muhzheguning.........................	Michigan ...	...	...	...	...	...	.....	.....	...	...	...	...	.....	......	......	301	...	1
			Chippewas of Michillimackinac.																		
		14.	Indians of the Beaver Islands	Michigan ...	...	...	...	14	2	.....	.....	...	...	...	...	.....	......	120	810	...	1,0
		15.	Manistee River, north	Michigan ...	...	8	9	20	12	.....	.....	...	...	...	...	.....	......	230	6,000	...	8
		16.	Shabegoshing, or Carp River	Michigan ...																	
		17.	Anee's Band, Oak Point.........................	Michigan ...	...	...	...	2	1												
		18.	Point St. Ignace..................................	Michigan ...	...	...	...	3	2	.....	.....	...	...	...	...	.....	......	......	....	...	8
		19.	Chenos, Huron, north............................	Michigan ...	...	...	...	1	...	.....	.....	...	...	...	...	.....	......	......	....	...	1
			Straits of St. Mary, &c.																		
		20.	Drummond Island Band.........................	Michigan ...	...	...	...	...	...	.....	.....	...	...	...	...	.....	25	20	200	...	2
		21.	Primary Band of Sault Ste. Marie...........	Michigan ...	...	2	1	...	...	.....	.....	1	...	...	...	35	100	10	....	...	2,0
		22.	Waishkee's Bay, foot of Lake Superior......	Michigan ...	...	14	10	2	2	.....	.....	4	...	...	...	.....	......	8	....	...	1,7
		23.	Tacquimenon River	Michigan ...	...	...	...	...	...	.....	.....	...	...	...	...	.....	140	4	....	...	1,5
		24.	Grand Island of Lake Superior................	Michigan ...	...	4	2	1	2	.....	.....	...	...	...	...	.....	61	5	....	...	60
						50	40	73	22	9	5	21			21	78	1183	1247	18,701	45	13,25

CENSUS RETURNS OF THE INDIAN TRIBES OF THE UNITED STATES.

II. ALGONQUIN GROUP.

Number of Tribes or Geographical Sub-Groups.	Main Tribes, or Geographical Sub-Groups.	Number of Bands, or recognised Divisions.	Names of the Divisions of the Group, as known to the Laws, or recognised by Treaties or Usage.	Geographical Position.	74 Other Neat Cattle.	75 Number of Sheep.	76 Number of Fleeces sheared.	77 Number of Hogs.	78 Number of Ploughs.	79 Number of Carts.	80 Number of Log-Chains.	81 Number of Crowbars.	82 Number of Hoes.	83 Number of Spades and Shovels.	84 Number of Axes, broad and narrow.	85 Number of Drays, Harrows, and Drills.	86 Number of Buggies and Pleasure-Wagons.	87 Number of Saddles and Bridles.	88 Estimated Value of Agricultural Implements.	89 Average Value of the "Plue" or "Skin," in the technical Sense of the Trade.	90 Number of "Skins" made during the Year.	91 Total estimated Value of the Year's Hunt to the Family.
			Detroit Chippewas.																			
		1.	Chippewas of Swan Creek and Black River..	Michigan ...	...	...	...	51														
		2.	Chippewas of Swan Cr. and Black R., west.	Ind. Ter. W.	7	...	...	15	5	...	3	...	14	1	12	...	....	14	$121	$40	80	$6.00
			Saganaws.																			
		3.	Saganaws of Pewanuganink......................	Michigan ...	1	...	...	18	1	...	4	...	25	...	55	1	....	8	$11			
		4.	Saganaws of Natoway's Band....................	Michigan ...	13	24	24	10	4	...	4	...	50	...	50	...	....	43	$35	....	410	$560.00
		5.	Band of Ogakaning..............................	Michigan ...																		
		6.	Atawa's Band, Saganaws	Michigan ...	1																	
		7.	Band of Wabeseepenukaning....................	Michigan ...																		
			Lake Huron, South.																			
		8.	Chippewas of Thunder Bay......................	Michigan ...	...	...	...	...	...	...	...	...	40	15	50	...	....	3	....	$60	.....	$43.00
			Chippewas of Grand Traverse Bay.																			
		9.	Eshquagonabee's Band	Michigan ...	...	...	...	60	...	...	...	...	51	9	63	...	....	11	....	....	.....	$25.00
		10.	Akosa's Band	Michigan ...																		
1.	CHIPPEWAS.	11.	Kagwatosa's Band................................	Michigan ...																		
		12.	Band at Nagondagon	Michigan ...																		
		13.	Band at Muhzheguning..........................	Michigan ...																		
			Chippewas of Michillimackinac.																			
		14.	Indians of the Beaver Islands	Michigan ...	...	...	...	30	...	...	...	...	50	1	200	...	....	....	$800			
		15.	Manistee River, north	Michigan ...	...	...	...	80	...	...	...	...	...	2	240	...	....	....	$700			
		16.	Shabegoshing, or Carp River	Michigan ...	...	...	...	...	...	...	...	...	6	...	70	...	....	....	....	$600	.....	$40.00
		17.	Ance's Band, Oak Point..........................	Michigan ...	...	...	...	...	...	...	...	...	...	...	70	...	....	10	....	$500	.....	$13.00
		18.	Point St. Ignace..................................	Michigan ...	...	...	...	...	...	...	...	...	10	3	26	1	....	2				
		19.	Chenos, Huron, north............................	Michigan ...	...	...	...	...	...	...	...	...	30	...	45	...	....	....	....	$250	.....	$25.00
			Straits of St. Mary, &c.																			
		20.	Drummond Island Band..........................	Michigan ...	...	...	...	2	...	...	...	...	30	...	50	....	....	....	$70	$2	680	$75.50
		21.	Primary Band of Sault Ste. Marie	Michigan ...	2	...	...	...	...	...	...	...	150	12	200	....	....	....	$400	$2	300	$60.00
		22.	**Waishkee's Bay, foot of Lake Superior.......**	Michigan ...	...	...	...	...	...	...	...	...	35	...	54	....	....	....	$60	$2	144	$46.00
		23.	**Tacquimenon River ....**	Michigan ...	...	...	...	...	...	...	...	...	100	...	100	....	....	....	$120	$2	200	$33.00
		24.	**Grand Island of Lake Superior.................**	Michigan ...	...	...	...	...	...	...	...	...	80	...	47	....	....	....	$5	$2	100	$10.00
					24	24	24	266	10		11		621	43	1332	2		91	$2,322	$1,460	1914	$936.50

II. ALGONQUIN GROUP.

Number of Tribes or Geographical Sub-Groups.	Main Tribes, or Geographical Sub-Groups.	Number of Bands, or recognised Divisions.	Names of the Divisions of the Group, as known to the Laws, or recognised by Treaties or Usage.	Geographical Position.	92 Amount of Cash Annuity per Capita received from the Government.	93 Number of Male Professors of the Christian Religion.	94 Number of Female Professors of the Christian Religion.	95 Number of Persons who still adhere to their native Religion.	96 Number of Members of Temperance Societies.	97 Number of Persons who are employed as Catechists, Exhorters, or Aids in the Dissemination of the Gospel in their Tribe.	98 Total estimated Fiscal Means and Resources of the Family, employed in their Subsistence, derived during the Year from Nos. 69, 91, and the aggregate of 92.	99 Average Fiscal Means of each Person in the Tribe.	100 Number of Clans, if the Tribe be organized in Clans.	101 Number of recognised Chiefs of the First Class (Civil).	102 Number of recognised Chiefs of the Second Class (Civil).	103 Number of War-Chiefs.	104 Number of Warriors fit to take the Field.	105 Number of the native Priesthood who officiate as Public Prophets.
			Detroit Chippewas.															
		1.	Chippewas of Swan Creek and Black River..	Michigan ...	....	81	......	......	......	....	....	.....	.....	2				
		2.	Chippewas of Swan Cr. and Black R., west.	Ind. Ter. W.	$13.00	......	......	5	......	....	$75.50	.....	1	1	1			
			Saganaws.															
		3.	Saganaws of Pewanuganink	Michigan ...	....	29	32	......	......	2								
		4.	Saganaws of Natoway's Band	Michigan ...	$2.04	8	14	80	32	1								
		5.	Band of Ogakaning	Michigan ...	....	51												
		6.	Atawa's Band, Saganaws	Michigan ...	....	32												
		7.	Band of Wabeseepenukaning	Michigan ...	....	30	......	45										
			Lake Huron, South.															
		8.	Chippewas of Thunder Bay	Michigan ...	....	6	4	67										
			Chippewas of Grand Traverse Bay.															
		9.	Eshquagonabee's Band	Michigan ...	....	32	40	54	......	....	....	.....	1	1	1			
		10.	Akosa's Band	Michigan ...	....	......	......	......	......	....	....	.....	1	1	1			
1.	CHIPPEWAS.	11.	Kagwatosa's Band	Michigan ...	....	......	......	......	......	....	....	.....	1	1	1			
		12.	Band at Nagondagon	Michigan ...	....	......	......	......	......	....	....	.....	1	1	1			
		13.	Band at Muhzheguning	Michigan ...	....	......	......	......	......	....	....	.....	1	1	1			
			Chippewas of Michillimackinac.															
		14.	Indians of the Beaver Islands	Michigan ...	....	......	......	18	7	1	....	.....	1	1	1	...	7	
		15.	Manistee River, north	Michigan ...	....	......	......	......	72	4	....	.....	1	1	1	...	20	
		16.	Shabegoshing, or Carp River	Michigan ...														
		17.	Ance's Band, Oak Point	Michigan ...	....	......	......	......	......	1								
		18.	Point St. Ignace	Michigan ...	....	......	......	......	......	1								
		19.	Chenos, Huron, north	Michigan ...														
			Straits of St. Mary, &c.															
		20.	Drummond Island Band	Michigan ...	$4.00	9	13	79	......	....	$91.50	.....	.....	1	1	2	16	
		21.	Primary Band of Sault Ste. Marie	Michigan ...	$4.00	50	70	70	2	....	$91.00	.....	.....	1	1	2	70	
		22.	Waishkee's Bay, foot of Lake Superior	Michigan ...	$4.00	12	14	42	21	....	$121.00	.....	.....	1	1	2	16	
		23.	Tacquimenon River	Michigan ...	$4.00	35	14	2	49	....	$108.00	.....	.....	1	1	2	10	
		24.	Grand Island of Lake Superior	Michigan ...	$4.00	25	25	7	50	....	$39.00	.....	.....	1	1	2	15	
						400	226	469	233	10	$526.00		8	15	12	10	154	

II. ALGONQUIN GROUP.

Number of Tribes or Geographical Sub-Groups.	Main Tribes, or Geographical Sub-Groups.	Number of Bands, or recognised Divisions.	Names of the Divisions of the Group, as known to the Laws, or recognised by Treaties or Usage.	Geographical Position.	122 Number of Carpenters.	123 Number of other Mechanics.	124 Number of Teachers.	125 Number of Missionaries supported in Part under Treaty Provisions.	126 Number of Persons under special or limited Employment in the Tribe, during the Year.	127 Number of Council-Houses.	128 Number of Mission-Houses or Establishments.	129 Number of School-Houses.	130 Number of Churches.	131 Estimated Value of all Public Buildings.	132 Number of Saw-Mills.	133 Number of Grist-Mills.	134 Number of Carding-Machines.	135 Number of Threshing-Machines.	136 Number of Fanning-Mills.	137 Number of Cotton-Gins	138 Number of Salt-Works.	139 Number of Looms in the Tribe.	140 Number of Spinning-Wheels.
			Detroit Chippewas.																				
		1.	Chippewas of Swan Creek and Black River..	Michigan ...	...	...	2	.....	......	...	1	1	1	$1000									
		2.	Chippewas of Swan Cr. and Black R., west.	Ind. Ter. W.																			
			Saganaws.																				
		3.	Saganaws of Pewanuganink....................	Michigan ...																			
		4.	Saganaws of Natoway's Band..................	Michigan ...																			
		5.	Band of Ogakaning..............................	Michigan ...																			
		6.	Atawa's Band, Saganaws........................	Michigan ...	...	...	...	.....	......	...	.....	1											
		7.	Band of Wabeseepenukaning...................	Michigan ...																			
			Lake Huron, South.																				
		8.	Chippewas of Thunder Bay.......................	Michigan ...																			
			Chippewas of Grand Traverse Bay.																				
		9.	Eshquagonabee's Band	Michigan ...																			
		10.	Akosa's Band	Michigan ...	...	...	1	1															
1.	CHIPPEWAS.	11.	Kagwatosa's Band..................................	Michigan ...																			
		12.	Band at Nagondagon	Michigan ...																			
		13.	Band at Muhzheguning...........................	Michigan ...																			
			Chippewas of Michillimackinac.																				
		14.	Indians of the Beaver Islands	Michigan ...																			
		15.	Manistee River, north	Michigan ...	...	...	1																
		16.	Shabegoshing, or Carp River	Michigan ...																			
		17.	Ance's Band, Oak Point..........................	Michigan ...																			
		18.	Point St. Ignace....................................	Michigan ...																			
		19.	Chenos, Huron, north.............................	Michigan ...																			
			Straits of St. Mary, &c.																				
		20.	Drummond Island Band	Michigan ...																			
		21.	Primary Band of Sault Ste. Marie.............	Michigan ...																			
		22.	Waishkee's Bay, foot of Lake Superior.......	Michigan ...	1	...	...	3	......	...	4	4	4	$6000									
		23.	Tacquimenon River	Michigan ...																			
		24.	Grand Island of Lake Superior..................	Michigan ...																			
					1		4	4			5	6	5	$7000									

CENSUS RETURNS OF THE INDIAN TRIBES OF THE UNITED STATES.

II. ALGONQUIN GROUP.

Number of Tribes or Geographical Sub-Groups.	Main Tribes, or Geographical Sub-Groups.	Number of Bands, or recognised Divisions.	Names of the Divisions of the Group, as known to the Laws, or recognised by Treaties or Usage.	Geographical Position.	106 Amount of Annuities paid in Coin.	107 Amount of Annuities paid in Goods or Merchandise.	108 Amount of Annuities paid in Provisions.	109 Amount of Annuities paid in Tobacco.	110 Amount of Annuities paid in Salt.	111 Sum set apart by the Tribe for Educational Purposes.	112 Sum expended for Cattle, Stock, and Agricultural Implements.	113 Sum expended for Iron, Steel, and Coal.	114 Sum expended by Government for Postage, Stationery, Rent, and other Contingencies.	115 Aggregate Sum expended by Government for Agents, and all other Persons employed by Authority.	116 Number of United States Agents.	117 Number of Sub-Agents.	118 Number of Messengers or Clerks.	119 Number of Interpreters or Translators.	120 Number of Blacksmiths and Assistants.	121 Number of Farmers and Assistants.
			Detroit Chippewas.																	
		1.	Chippewas of Swan Creek and Black River	Michigan																
		2.	Chippewas of Swan Cr. and Black R., west	Ind. Ter. W.	$200															
			Saganaws.																	
		3.	Saganaws of Pewanuganink	Michigan																
		4.	Saganaws of Natoway's Band	Michigan																
		5.	Band of Ogakaning	Michigan																
		6.	Atawa's Band, Saganaws	Michigan																
		7.	Band of Wabeseepenukaning	Michigan																
			Lake Huron, South.																	
		8.	Chippewas of Thunder Bay	Michigan																
			Chippewas of Grand Traverse Bay.																	
		9.	Eshquagonabee's Band	Michigan																
		10.	Akosa's Band	Michigan	......	.....	.....	.....	.....	....	.....	......	......	....	...	...	...	1		
1.	CHIPPEWAS.	11.	Kagwatosa's Band	Michigan																
		12.	Band at Nagondagon	Michigan																
		13.	Band at Muhzheguning	Michigan																
			Chippewas of Michillimackinac.																	
		14.	Indians of the Beaver Islands	Michigan																
		15.	Manistee River, north	Michigan																
		16.	Shabegoshing, or Carp River	Michigan																
		17.	Ance's Band, Oak Point	Michigan																
		18.	Point St. Ignace	Michigan																
		19.	Chenos, Huron, north	Michigan																
			Straits of St. Mary, &c.																	
		20.	Drummond Island Band	Michigan																
		21.	Primary Band of Sault Ste. Marie	Michigan																
		22.	Waishkee's Bay, foot of Lake Superior	Michigan	......	.....	.....	.....	.....	$5000	.....	$400	$50	$3710	...	1	...	1	4	1
		23.	Tacquimenon River	Michigan																
		24.	Grand Island of Lake Superior	Michigan																
					$200					$5000		$400	$50	$3710		1		2	4	1

CENSUS RETURNS OF THE INDIAN TRIBES OF THE UNITED STATES.

II. ALGONQUIN GROUP.

Number of Tribes or Geographical Sub-Groups.	Main Tribes, or Geographical Sub-Groups.	Number of Bands, or recognised Divisions.	Names of the Divisions of the Group, as known to the Laws, or recognised by Treaties or Usage.	Geographical Position.	141 Number of Horse-Mills.	142 Number of Public Ferries.	143 Number of Fisheries.	144 Number of Mines worked or leased.	145 Number of Kettles used in the Manufacture of Maple Sugar.	146 Number of Printing-Presses in the Nation.	147 Number of licensed Traders under Congress Laws.	148 Estimated Number of unlicensed Traders for Parts of the Tribe still within State Limits.	149 Number of Clerks and Interpreters employed in the Trade.	150 Number of Boatmen, Runners, or *Engagés* employed by the Principals.	151 Aggregate Number of "Skins," reduced to the Trade Standard of Value.	152 Gross Amount of Capital invested in the Trade in 1847.	153 Total Number of Hunters and Warriors.	154 Total Number of Agriculturists.	155 Total Number who profess Medical Skill.	156 Total Number, of all Classes, who officiate as native Priests or Jugglers.	157 Total Number, of all Classes, who have been educated.
			Detroit Chippewas.																		
		1.	Chippewas of Swan Creek and Black River	Michigan ...																	
		2.	Chippewas of Swan Cr. and Black R., west	Ind. Ter. W.	...	...	...	...	6	...	...	...	...	...	...	$40	5	10	...	...	5
			Saganaws.																		
		3.	Saganaws of Pewanuganink	Michigan ...																	
		4.	Saganaws of Natoway's Band	Michigan ...																	
		5.	Band of Ogakaning	Michigan ...																	
		6.	Atawa's Band, Saganaws	Michigan ...																	
		7.	Band of Wabeseepenukaning	Michigan ...																	
			Lake Huron, South.																		
		8.	Chippewas of Thunder Bay	Michigan ...																	
			Chippewas of Grand Traverse Bay.																		
		9.	Eshquagonabee's Band	Michigan ...																	
		10.	Akosa's Band	Michigan ...																	
1.	CHIPPEWAS.	11.	Kagwatosa's Band	Michigan ...																	
		12.	Band at Nagondagon	Michigan ...																	
		13.	Band at Muhzheguning	Michigan ...																	
			Chippewas of Michillimackinac.																		
		14.	Indians of the Beaver Islands	Michigan ...	...	...	...	...	...	...	...	...	...	...	...	...	...	35			
		15.	Manistee River, north	Michigan ...	...	...	...	...	...	...	...	...	...	...	...	...	...	72	...	...	72
		16.	Shabegoshing, or Carp River	Michigan ...																	
		17.	Ance's Band, Oak Point	Michigan ...																	
		18.	Point St. Ignace	Michigan ...																	
		19.	Chenos, Huron, north	Michigan ...																	
			Straits of St. Mary, &c.																		
		20.	Drummond Island Band	Michigan ...	...	...	7	...	200	...	...	...	...	...	...	$1360	18	...	8	1	
		21.	Primary Band of Sault Ste. Marie	Michigan ...	...	...	2	...	500	...	...	...	...	...	...	$600	81	...	...	2	
		22.	Waishkee's Bay, foot of Lake Superior	Michigan ...	...	...	...	...	211	...	...	5	...	...	...	$288	22	...	...	...	1
		23.	Tacquimenon River	Michigan ...	...	...	2	...	150	...	...	...	...	...	...	$400	24	...	1		
		24.	Grand Island of Lake Superior	Michigan ...	...	...	1	...	600	...	...	...	...	...	...	$200	36	...	4		
							12		1667							$2888	186	117	13	3	78

II. ALGONQUIN GROUP.

Number of Tribes or Geographical Sub-Groups.	Main Tribes, or Geographical Sub-Groups.	Number of Bands, or recognised Divisions.	Names of the Divisions of the Group, as known to the Laws, or recognised by Treaties or Usage.	Geographical Position.	158 Number of Persons who can read the Scriptures.	159 Total Number of Church Members, of all Christian Denominations.	160 Total Number who are of temperate Habits.	161 Total Number who may be ranked as Inebriates.	162 Total Number of Indians living in the Tribe, but not Members of it, and not entitled to Annuities.	163 Number of colored Slaves, or Persons of African Descent.	164 Number of Captives taken or adopted from other Tribes.	165 Number of Persons who live in a State of Polygamy.	166 Total Mass of the Tribe, fixed or roving, to be acted on by Civilization and Christianity.	167 Average Number of the Indian Family.	168 Number of Births over Deaths, during the Year.	169 Number of Deaths over Births, during the Year.	170 Average Number of Children by each Female, taking those between 16 and 60 Years of Age.	171 Value of Bead Work.	172 Amount received for Premises leased.	173 Number of Barrels of Fish sold.	174 Amount received for Fish.
			Detroit Chippewas.																		
		1.	Chippewas of Swan Creek and Black River..	Michigan ...																	
		2.	Chippewas of Swan Cr. and Black R., west.	Ind. Ter. W.	5	2	8	6													
			Saganaws.																		
		3.	Saganaws of Pewanuganink	Michigan ...																	
		4.	Saganaws of Natoway's Band	Michigan ...																	
		5.	Band of Ogakaning	Michigan ...																	
		6.	Atawa's Band, Saganaws	Michigan ...																	
		7.	Band of Wabeseepenukaning	Michigan ...																	
			Lake Huron, South.																		
		8.	Chippewas of Thunder Bay	Michigan ...																	
			Chippewas of Grand Traverse Bay.																		
		9.	Eshquagonabee's Band	Michigan ...																	
		10.	Akosa's Band	Michigan ...																	
1.	CHIPPEWAS.	11.	Kagwatosa's Band	Michigan ...																	
		12.	Band at Nagondagon	Michigan ...																	
		13.	Band at Muhzheguning	Michigan ...																	
			Chippewas of Michillimackinac.																		
		14.	Indians of the Beaver Islands	Michigan ...	4	35	33	2													
		15.	Manistee River, north	Michigan ...	24	72	72														
		16.	Shabegoshing, or Carp River	Michigan ...																	
		17.	Ance's Band, Oak Point	Michigan ...																	
		18.	Point St. Ignace	Michigan ...																	
		19.	Chenos, Huron, north	Michigan ...																	
			Straits of St. Mary, &c.																		
		20.	Drummond Island Band	Michigan ...	...	22	77	4	......	.....	.....	.....	79	...	...	...	......	...	...	190	$590
		21.	Primary Band of Sault Ste. Marie	Michigan ...	...	120	60	12	......	.....	.....	1	72	...	...	...	......	...	...	100	$400
		22.	Waishkee's Bay, foot of Lake Superior	Michigan ...	3	26	21	1	......	.....	.....	.....	46	...	...	...	......	...	...	25	$112
		23.	Tacquimenon River	Michigan ...	...	47	49	3	......	.....	.....	1	3	...	...	...	......	...	...	200	$600
		24.	Grand Island of Lake Superior	Michigan ...	3	50	50	7	......	.....	.....	.....	7	...	...	...	......	...	...	25	$125
					39	374	370	35				2	207							540	$1827

APPENDIX
Chapter I
Item 5

Thirteenth Census of the U.S.: 1910 – Indian Population

[This portion of the this census includes Native Americans living on Grand Island in 1910 with a total of 29 male and 21 females, 38 were born in Michigan with others coming from Canada and elsewhere in the United States. The instructions for "this modified form of the general schedule" as they reflect the prejudice of the general population towards the Native Americans with categories bordering on discrimination, e.g., "whether now living in polygamy, if living in polygamy, and whether the wives are sisters and living in civilized or aboriginal dwelling".]

INSTRUCTIONS FOR FILING THIS SCHEDULE.

This modified form of the general schedule for population is to be used in making the enumeration of Indians, both those on reservations and those living in family groups outside of reservations.

Detached Indians living either in white or negro families outside of reservation should be enumerated on the general population schedule (Form 8 – 1882) as members of the families in which they are found; but detached whites or negroes living in Indian families should be enumerated on this special Indian schedule as members of the Indian families in which they are found. In other words, every family composed mainly of Indians should be reported *entirely* on this special schedule, and every family composed mainly of persons not Indian should be reported *entirely* on the general population schedule.

Column 7. Age at last birthday – Some difficulty may be met in ascertaining the exact ages of Indians, as they frequently reckon their ages from notable events occurring in the history of the respective tribes.

Columns 18 and 19. Occupation – If the Indian is wholly self supporting, enter his or her occupation in columns 18 and 19 in accordance with the general instructions of returning occupations. If the Indian – man, woman, or child – has no occupation and is wholly dependent on the Government for support, write "Ration Indian" in column 18. If the Indian is partly self-supporting and partly dependent upon the Government, write the occupation in columns 18 and 19, and then the letter "R" (for ration).

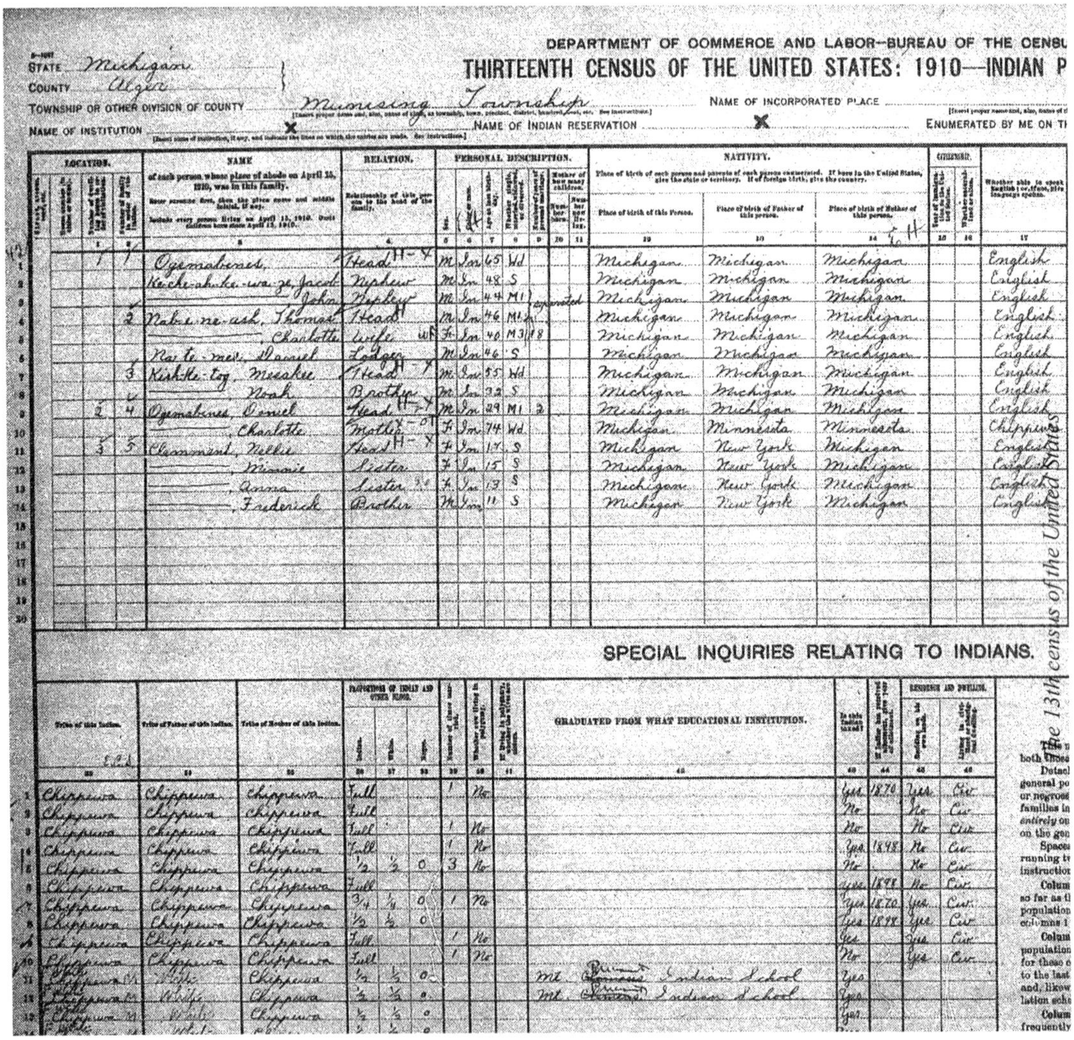

DEPARTMENT OF COMMERCE AND LABOR—BUREAU OF THE CENSU[S]

THIRTEENTH CENSUS OF THE UNITED STATES: 1910—INDIAN P[OPULATION]

STATE Michigan
COUNTY Alger
TOWNSHIP OR OTHER DIVISION OF COUNTY Munising Township
NAME OF INSTITUTION X
NAME OF INCORPORATED PLACE
NAME OF INDIAN RESERVATION X
ENUMERATED BY ME ON TH[E]

Dwelling no.	Family no.	Name	Relation	Sex	Color or race	Age	Marital condition	Years of present marriage	Mother of how many children: Number born	Number now living	Place of birth of this person	Place of birth of father	Place of birth of mother	Whether able to speak English; or, if not, give language spoken
1	1	Ogemabenes,	Head H-X	M	In	65	Wd				Michigan	Michigan	Michigan	English
		Ke-che-ah-ke-wa-ze, Jacob	Nephew	M	In	48	S				Michigan	Michigan	Michigan	English
		John	Nephew	M	In	44	M1	separated			Michigan	Michigan	Michigan	English
	2	Nabe-ne-ash, Thomas	Head H	M	In	46	M1	2			Michigan	Michigan	Michigan	English
		Charlotte	Wife	F	In	40	M3	18			Michigan	Michigan	Michigan	English
		Na-te-mer, Daniel	Lodger	M	In	46	S				Michigan	Michigan	Michigan	English
	3	Kish-ke-tog, Mesakee	Head H-X	M	In	55	Wd				Michigan	Michigan	Michigan	English
		Noah	Brother	M	In	32	S				Michigan	Michigan	Michigan	English
2	4	Ogemabenes, Daniel	Head H-X	M	In	29	M1	2			Michigan	Michigan	Michigan	English
		Charlotte	Mother X-O	F	In	74	Wd				Michigan	Minnesota	Minnesota	Chippewa
3	5	Clemment, Nellie	Head H-X	F	In	17	S				Michigan	New York	Michigan	English
		Minnie	Sister	F	In	15	S				Michigan	New York	Michigan	English
		Anna	Sister	F	In	13	S				Michigan	New York	Michigan	English
		Frederick	Brother	M	In	11	S				Michigan	New York	Michigan	English

SPECIAL INQUIRIES RELATING TO INDIANS.

	Tribe of this Indian	Tribe of Father of this Indian	Tribe of Mother of this Indian	Indian	White	Negro	Number of times married	Whether now living in polygamy	Graduated from what educational institution	Is this Indian taxed?	Year of allotment	Residing on his own land	Living in civilized or aboriginal dwelling
1	Chippewa	Chippewa	Chippewa	Full			1	No		Yes	1870	Yes	Civ.
2	Chippewa	Chippewa	Chippewa	Full						No		No	Civ.
3	Chippewa	Chippewa	Chippewa	Full			1	No		No		No	Civ.
4	Chippewa	Chippewa	Chippewa	Full			1	No		Yes	1848	No	Civ.
5	Chippewa	Chippewa	Chippewa	1/2	1/2	0	3	No		No		No	Civ.
6	Chippewa	Chippewa	Chippewa	Full						Yes	1898	No	Civ.
7	Chippewa	Chippewa	Chippewa	3/4	1/4	0	1	No		Yes	1870	Yes	Civ.
8	Chippewa	Chippewa	Chippewa	1/2	1/2	0				Yes	1898	Yes	Civ.
9	Chippewa	Chippewa	Chippewa	Full			1	No		Yes		Yes	Civ.
10	Chippewa	Chippewa	Chippewa	Full			1	No		No		Yes	Civ.
11	Chippewa	White	Chippewa	1/2	1/2	0			Mt. [illegible] Indian School	Yes			
12	Chippewa	White	Chippewa	1/2	1/2	0			Mt. [illegible] Indian School	Yes			
13	Chippewa	White	Chippewa	1/2	1/2	0				Yes			

1910 Census of Indian Population
The omitted right hand colums are the same as the census that does not include Native Americans.

APPENDIX
Chapter I
Item 6

Eighteenth Annual Report
of the
BUREAU OF AMERICAN ETHNOLOGY
to the
Secretary of the Smithsonian Institution
1896 - 97

by
J. W. POWELL
Director

NATIVE AMERICAN LAND CESSIONS IN THE UNITED STATES
By Charles C Royce

INTRODUCTION
By Cyrus Thomas

RIGHT TO THE SOIL DEPENDENT ON DISCOVERY

Among the various problems forced on European nations by the discovery of America was that of determining their respective rights in regard to the territory of the newly discovered continent. The fact that the country was inhabited by and in possession of a native population does not appear to have been taken into consideration in the resolution of this problem.

Each of the great nations of Europe was eager to appropriate to itself so much of the new continent as it could acquire. Its extent afforded an ample field for the ambition and enterprise of all, and the character, low culture-status, and religious beliefs of the aborigines afforded an apology for considering them a people over whom the superior genius of Europe might rightfully claim an ascendancy. The sovereigns of the Old World therefore found no difficulty in convincing themselves that they made ample compensation to the natives by bestowing on them the benefits of civilization and Christianity in exchange for control over them and their country. However, as they were all in pursuit of the same object, it became necessary, in order to avoid conflicting settlements and consequent war with one another, to establish a principle which all would acknowledge as the law by which the right, as between themselves, to the acquisition of territory on this continent, should be determined. This principle was, that discovery of lands gave title therein to the government by whose subjects or by whose authority such discovery was made, against all other European or civilized governments, which title might be consummated by possession.

In the establishment of these relations, the rights of the original inhabitants were, in no instance, entirely disregarded, but were necessarily, to a considerable extent, impaired. The (Native Americans) were admitted to be the rightful occupants of the soil, with a legal as well as just claim to retain possession of it (or rather so much as was necessary for their use), and to use it according to their own discretion; but their rights to complete sovereignty, no independent nations, were necessarily diminished, and their power to dispose of the soil at their own will, to whomsoever they pleased, was denied by the original fundamental principle that discovery gave exclusive title to those who made it.

While the different nations of Europe respected the right of the natives as occupants, they asserted the ultimate dominion to be in themselves; and claimed and exercised, as a consequence of this ultimate dominion, a power to grant the soil, while yet in possession of the natives. These grants have been understood by all to convey a title to the grantees, subject only to the Indian right of occupancy.

The history of America, from its discovery to the present day (about 1780), proves, we think, the universal recognition of these principles.

But the tribes of Indians inhabiting this country were fierce savages, whose occupation was war, and whose subsistence was drawn chiefly from the forest. To leave them in possession of their country was to leave the country a wilderness; to govern them as a distinct people was impossible, because they were as brave and as high spirited as they were fierce, and were ready to repel by arms every attempt on their independence.

What was the inevitable consequence of this state of things? The Europeans were under the necessity either of abandoning the country, and relinquishing their pompous claims to it, or of enforcing those claims by the sword, and by the adoption of principles adapted to the condition of a people with whom it was impossible to mix, and who could not be governed as a distinct society, or of remaining in their neighborhood and exposing themselves and their families to the perpetual hazard of being massacred.

Frequent and bloody wars, in which the whites were not always the aggressors, unavoidably ensued. European policy, numbers and skill, prevailed. As the white population advanced, that of the Indians necessarily receded. The country in the immediate neighborhood of agriculturists became unfit for them. The game fled into thicker and more unbroken forests, and the Indians followed. The soil, to which the crown originally claimed title, being no longer occupied by its ancient inhabitants, was parceled out according to the will of the sovereign power, and taken possession of by persons who claimed immediately from the crown, or immediately, through its grantees or deputies.

That law which regulates, and ought to regulate in general, the relations between the conqueror and conquered, was incapable of application to a people under such circumstances. The resort to some new and different rule, better adapted to the actual state of things, was unavoidable. Every rule which can be suggested will be found to be attended with great difficulty.

It has never been contended that the Indian title amounted to nothing. Their right of possession has never been questioned. The claim of government extends to the complete ultimate title, charged with this right of possession, and to the exclusive power of acquiring that right. The object of the crown was to settle the sea-coast of America; and when a portion of it was settled, it would have been as unwise as ungracious to expel them from the habitations because they had obtained the Indian title otherwise than through the agency of government. The very grant of a charter is an assertion of the title of the crown, and its words convey the same idea. The country granted is said to be "our island called Rhode Island;" and the charter contains an actual grant of the soil, as well as of the powers of government.

As early as September 22, 1783, while yet operating under the Articles of Confederation, the following "proclamation" was ordered by Congress.

The United States Congress have the sole power of regulating the trade, and managing all affairs with the Indians not members of any of the States. It is essential to the welfare of the United States that all cause of quarrel or complaint between them and the United States or any of them should be removed and prevented; therefore...prohibit and forbid all persons from making settlements on lands inhabited or claimed by Indians, without the limits or jurisdiction of any particular State, and from purchasing or receiving any gift or cession of such lands or claims without the express authority and direction of the United States Congress. By act of Congress updated in 1796 and 1804, entitled "An act to regulate trade and intercourse with the Indian tribes," the same principle was enacted into law.

These acts and treaties indicate, and in fact form, steps in the policy of the United States in its dealing with the Indians in reference to their lands. The object at present in referring to them is only to show the theory of the Government in regard to the Indian title.

It is clear, therefore, that although the United States has always conceded to the Native Americans right of occupancy to such lands as they were in possession of, yet they have always held the theory of the European powers, and claimed that the absolute right to the soil was in the Government.

However, as will be seen when allusion is made to the policy of the individual nations in their dealings with the Indians, there was some difference in regard to the extent of their right or title.

It would seem, in fact, that the United States proceeded on the theory that *all* the land was held by natives.

The right of occupancy in the Indians, until voluntarily relinquished or extinguished by justifiable conquest, being conceded, it became necessary on the part of the Government to adopt some policy to extinguish their right to such territory as was not necessary for their actual use.

As a natural corollary of this theory arose the question, With whom shall the Government treat? The Indians having no general government or regular political organization, but consisting of numerous independent tribes in a state of savagery, the usual policy of civilized nations in a case of conquest could not be adopted. As their claims were those of tribes or communities, and not individuals, it followed as a matter of necessity that the only policy which the Government could adopt was to recognize them as distinct political communities or nations.

Nevertheless, if we study carefully all the circumstances, and the pressing necessities of the Republic in its early days, we are likely to be convinced that it was not the part of wisdom then to hamper the struggles for national life with theoretic lines or legal technicalities, which stood in the way of practical progress. Humanity is an element which should attend every step of governmental as well as of individual progress, but political theories must be broadened, restricted, or varied in accordance with new and imperative necessities which arise.

It is doubtless true that the recognition of the Indian tribes as distinct nationalities, with which the Government could enter into solemn treaties, was a legal fiction which should be superseded by a more correct policy when possible. But necessity often makes laws, and in this instance forced the Government to what was, it its early days, probably the best possible policy.

That the Indians of the United States have been and are still considered wards of the Government must be conceded. It also must be admitted that, as a general rule of law, wards can not divest themselves of their title to land except through the decree of court or some properly authorized power. But in the case of the Indians the Government is both guardian and court, and as there is no higher authority to which application can be made, its decision must be final, otherwise no transfer of title would be possible.

However, it is not our object at present to theorize as to what should or might have been done, but to state that was done in this respect, and thus to show on what policy the various territorial cessions and reservations mentioned in the present work are based.

The correct theory on this subject appears to be so clearly set fourth by John Quincy Adams in his oration at the anniversary of the Sons of the Pilgrims, December 22, 1802, that his words are quoted, as follows:

> There are moralists who have questioned the right of Europeans to intrude upon the possessions of the aborigines in any case and under any limitations whatsoever. But have they maturely considered the whole subject? The Indian right of possession itself stands, with regard to the greatest part of the country, upon a questionable foundation. Their cultivated fields, their constructed habitations, a space of ample sufficiency for their subsistence, and whatever they had annexed to themselves by personal labor, was undoubtedly by the laws of nature theirs.

In order to show the correctness of the views expressed by Adams in the above quotation, and the absurdity of admitting the Indians' claim to the absolute right of the soil of the whole country, some comparisons are here introduced. These are simple comparisons between the Indian population and the extent of territory claimed by them.

Examining these estimates at the different dates, we see that the average, in round numbers, is 315,000. Now, assuming this to be a correct estimate, and allowing five persons to a family, this would give 63,000 as the whole number of Indian families in the United States. Assuming the area of the United States, exclusive of Alaska, to be 3,025,000 square miles, this would give to each Indian family a manor of 48 square miles, or 30,720 acres. Now, supposing, for further illustration, that the families were distributed uniformly over the whole territory, the state of Rhode Island, which now supports a population of 346,506 persons, or 69,101 families (allowing five persons to a family), would be apportioned among 26 Indian families; the state of Delaware would be allotted to but 43, and the whole state of New York, which now supports more than a million families, would be assigned to 1,025 lordly savages.

It is apparent, therefore, that the requirements of the human race and the march of civilization could not permit such an apportionment of the soil of the American Continent as this, even were the estimates trebled. That a population whose territorial needs would be amply supplied by the area embraced in the single state of Illinois should, on the score of being the first occupants of the country, be allowed the exclusive use of the whole territory of the United States is inconsistent with any true theory of natural rights. Moreover, it is not required by humanity, religion, nor any principle of human rights. This must be conceded. But what is the necessary consequence of such concession?

APPENDIX
Chapter I
Item 7

U.P. NATIVE AMERICAN Census by County, 1860 – 1900
SOURCE: Federal Census 1900

County	1860	1870	*1880	**1890	1900	2000 U.P. Population	2000 Native American Population	% Native American Population
Alger	-----	-----	-----	78	62	9,860	>325	3.3
Baraga	-----	-----	0	287	488	8,740	>1,024	12
Chippewa	244	734	788	441	449	38,400	>5,125	13.3
Delta	-----	15	158	217	249	38,500	>847	2.2
Dickinson	-----	-----	-----	-----	54	27,400	292	0.5
Gogebic	-----	-----	-----	1	58	17,400	>384	2.2
Houghton	279	73	24	10	18	36,000	180	0.5
Iron	-----	-----	-----	2	1	13,000	130	1.0
Isle Royale	-----	-----	0	-----	-----	-----	-----	---
Keewenaw	-----	3	0	0	0	2,300	23	0.1
Luce	-----	-----	-----	1	0	7,000	385	5.5
Mackinac	907	285	254	227	275	12,000	>1,704	14.2
Marquette	27	72	87	56	61	64,000	960	1.5
Menominee	-----	-----	121	129	101	25,000	>575	2.3
Ontonagon	0	19	74	59	27	7,800	>78	1.0
Schoolcraft	22	-----	134	43	32	8,900	>543	6.1
TOTAL	**1,479**	**1,701**	**1,640**	**1,551**	**1,875**	**317,000**	**12,600**	**0.4**

** Schoolcraft County split from Chippewa County in 1871.*
*** Alger County split from Schoolcraft County in 1885.*
> County includes reservation lands

APPENDIX
Chapter I
Item 8

NATIVE AMERICAN POPULATION 1600 – 2005

YEAR	U.S.A. (Native Americans)	MICHIGAN (Native Americans)	UPPER PENINSULA (Ojibway)	ALGER COUNTY (Grand Island Band)
1600		(6-8,000)		
1778	3.9 mil	(15,000)	(600)	
1812	5 mil	(9.000)	(3,000)	
1820	(470,000)	(8.800)	(820)	(95)*
1832	12.8 mil			(71)
1840s	13 mil	(32,000)	1,359 (1,096)	
1850s	23 mill	(31,000)	5,745 (400)	
1874	39 mil (300,000)	1 mil	60,000 (1,700)	(150)
1890	63 mil (59,000)** (250,000)***		175,000 (1,551)	1,240 (78)
1900	76 mil (237,000	2.5 mil	252,000 (1,875)	5,870 (62)
1910	92 mil (265,000)	2.5 mil	325,000	7,700 (28)
1950	150 mil (343,000)	6.4 mil	302,000	
2005	296 mil (2.4 mil)	10 mil	317,000 (13,400)	8,900 (357)

* 200 in Alger County Centennial book
** 1870-1890 excludes reservations
*** 1890-1990 includes reservations

U.S. Ojibwe census in 2005 totaled approximately 56,000.

APPENDIX
Chapter I
Item 9

Distribution of Sault Ste. Marie Tribe of Chippewa Indians – 2009
In U.S., Michigan, and Upper Peninsula

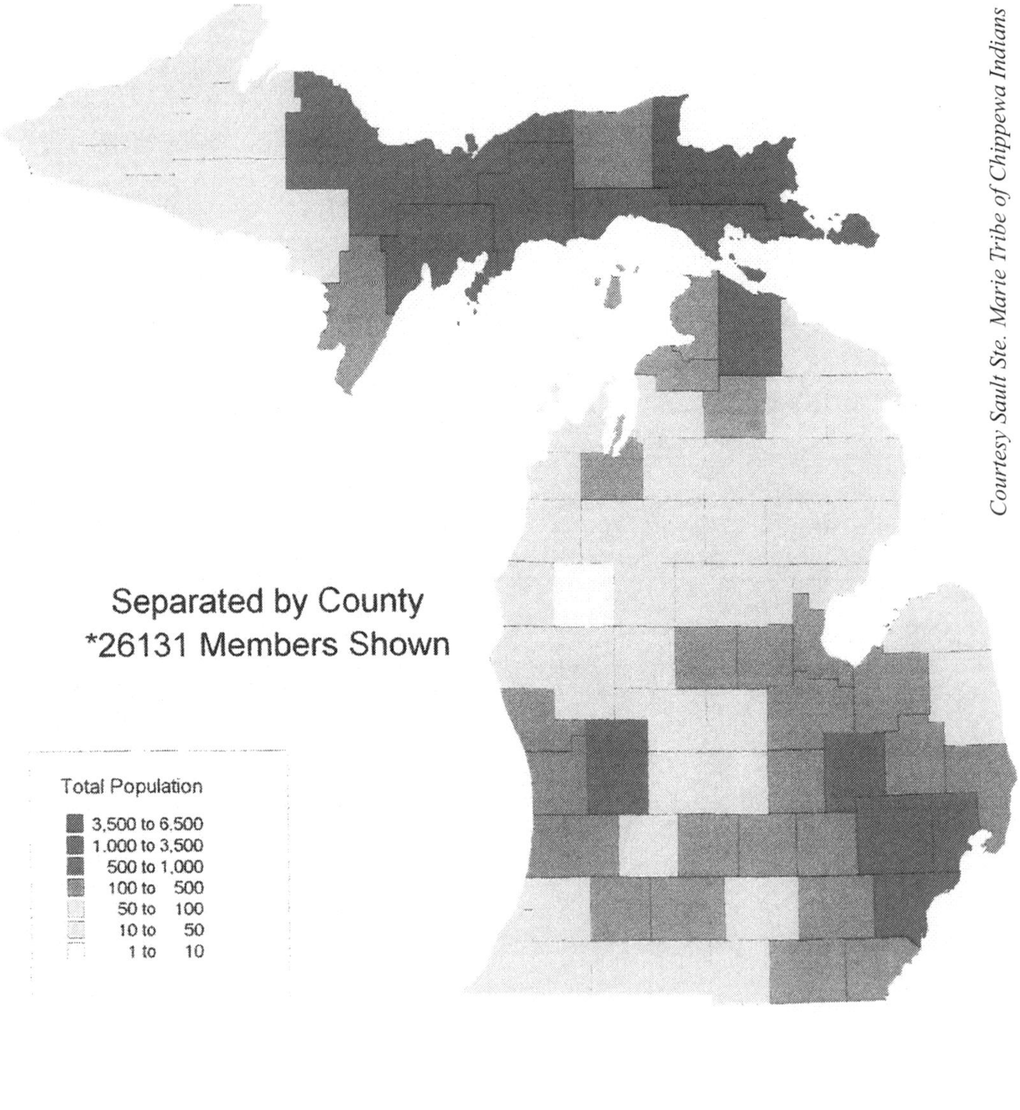

Courtesy Sault Ste. Marie Tribe of Chippewa Indians

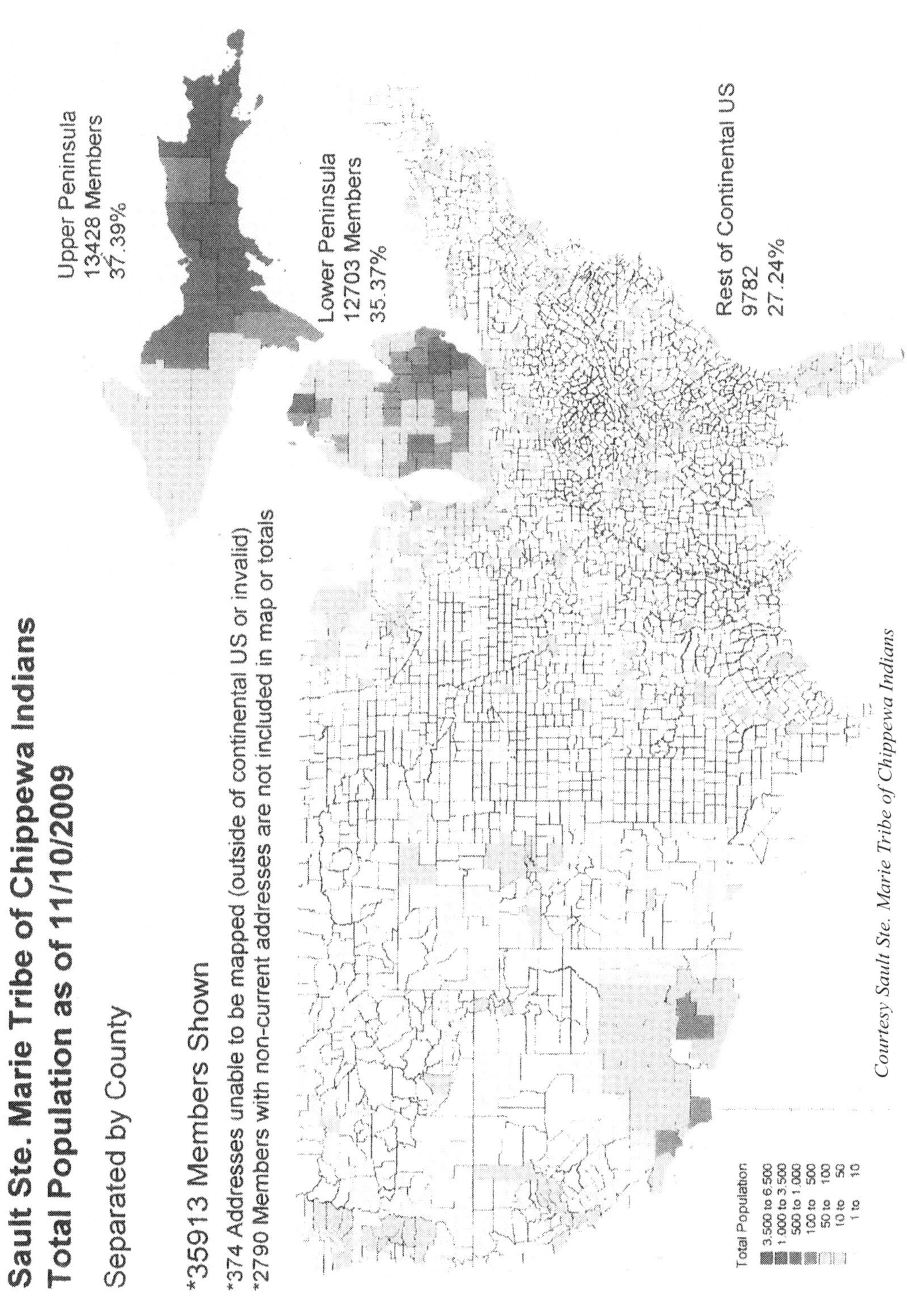

Courtesy Sault Ste. Marie Tribe of Chippewa Indians

APPENDIX
Chapter I
Item 10

Ojibwa Migration Map circa 1820

This native map, drawn on birch bark (which accounts for its shape) shows the migration legend of the Ojibwe, from the creation of their people (on the right) to their home in the upper mid-west (on the left). The left and central portions of the map show Lake Huron, Lake Superior and Red Lake in Minnesota. The right side of the map shows spiritual realities of the Ojibwe origins with important spirit guides symbolized along the route. The map is a sophisticated synthesis of spiritual and physical geography, revealing the vital importance of making maps in the context of your life and belief systems.

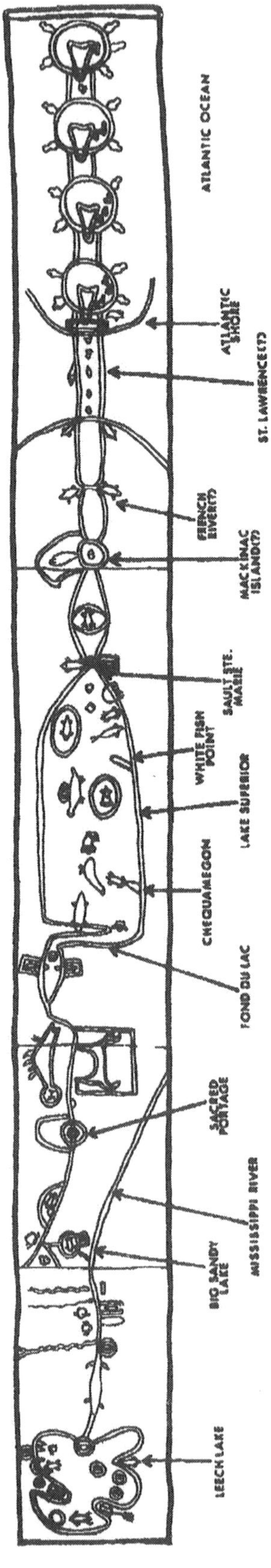

Courtesy University of Toronto Press, Glenbow Museum, Calgary

APPENDIX
Chapter I
Item 11

Dates of interest for Grand Island Chippewa as excerpted from *A Guide to Indians of U.P. 1621 – 1900*, by Russ Magnaghi 1984:

1792 Grand Island's summer residence of small band of Chippewas. They cultivate corn, potatoes, and pumpkins and live in cabins on south of Grand Island. They also fish with spears and torches at night for trout and whitefish.

1819 Thirteen Chippewas from Grand Island travel westward and are overwhelmed by a larger party of Sioux Indians.

1820 Indians of Grand Island were friendly to the Cass party inviting them to dance and make speeches. The band was very poor lacking both clothing and provisions. Some had just returned from a war expedition with the Sioux.

1830 A different observer describes the Grand Island tribe as consisting of 35 males, 24 females, and 12 warriors. They are healthy and well-clothed. They fish in the bay for herring, trout, and whitefish with spears and gill nets. They hunt red deer in the region between Lake Superior and Lake Michigan. Twenty of the Chippewa residing on Grand Island are from the Marquette area where they live most of the year. Others winter on Lake Michigan or Sault Ste. Marie.

1834 Indians on Grand Island demand compensation from the Federal Government for wood cut by a light ship crew.

1834 Grand Island Chippewa visited by George Copway, a young Chippewa who has become a missionary.

1846 Agreement between Jackson Mining Company and Marji Gesick, chief the Chippewa tribe, for leading them to iron ore in Negaunee. He was given 0.12 interest in the mine.

APPENDIX
Chapter I
Item 12

Dates of interest from Cleland's Rites of Conquest

- Prior to 1696, the French fur trading system was based on licenses granted from France through the Sault Ste. Marie headquarters. These licenses ended because of a glut of furs due to a change in fashions. The Native Americans of the Soo were called "Saulteurs".
- The Treaty of Washington of 1836 designated 13 million acres in northwest Lower Peninsula and eastern Upper Peninsula made up of fourteen reservations at a price of $1.5 million.
- In the second half of the 1800's Native Americans primarily cut timbers for mines and railroads and cord wood for lake steamers. They also produced maple syrup and became commercial fishermen.
- By 1900, Native Americans were attending government-funded boarding schools frequently far off the reservations. Although many of these were operated by churches, the children were treated poorly. This was a result of a U.S. government policy to "Americanize" the Native Americans
- In 1850, Native Americans were offered Michigan citizenship if they renounced their tribal status. Native Americans had become "true refugees with no help and none on the horizon".
- In 1853, Native Americans were given full rights of citizenship. Federal recognition of the Soo tribe was not until the late 1970's and therefore became eligible for assistance through the Bureau of Indian Affairs on a "government to government basis". In the early 20th century, there was a migration to the cities with 80% of Native Americans living in cities at the current time.
- The Michigan Soo tribe was established in 1975 with a federal grant from HUD.
- By the late 20th century, Native Americans were using the legal system to win back fishing rights granted to them in the Treaty of 1836.
- By 1990, the Michigan Native American population was 62,000 (0.6% of Michigan population) with 80% in the Lower Peninsula and 20% in the Upper Peninsula. Michigan Native American birth rate remains high at approximately four times average.

APPENDIX
Chapter II
MICHIGAN TIMBERLAND – 2007

State of Michigan –

36.4 million acres (56,000 square miles) total
19.3 million acres of forest land – 95% suitable for timbering
8.4 million (56%) individual ownership
6.6 million (36%) government (federal, state, local)
4 million acres for game and wildlife
2.1 million – corporate wood users
1.5 million (8%) forest industry
642,000 acres – tribal Native American land; other federal

U.P. –

10,250,000 acres (16,000 square miles)
Timberland increased 7% since1980.

300 miles east to west and 121 miles north to south (1 sq. miles/640 acres) Upper Peninsula - 1/3 acreage of Michigan and 3% of the population

Federal forest –
Hiawatha – east and central

State forests - 450,000 acres (upper state forest)
420,000 acres – Escanaba State Forest
1,010,000 acres – Superior State forest
Forest Industry timberland ownership – 8% of timberland
650,000 acres – Mead Paper Company
450,000 acres – International Paper Company
500,000 acres – Forestland Group – NE CCI forest; Benson forest; Shelter Bay forest.
2006 – 271,000 acres conservation easement to The Nature Conservancy

Alger County

587,000 acres (1918 sq. miles) Population: 9,862

Onota Township

61,000 acres (98 sq. miles) Population: 310

Tyoga Lumber Company – 1906 7,000 acres

Tyoga Historical Pathway – 200 acres – 2008

APPENDIX
Chapter IV
Item 1

Alger County Deed
Liber 18 p 20
L. L. Bailey to Tyoga Lumber Co
March 23, 1905

L. L. Bailey
Carrie J. Baile
Geo. M. Spalding
Florence W. Spalding
Chas H. Roberts
Mary S. Roberts
J. H. Gardner
Mary N. Gardner
Thos. M. Archer
Mary G. Archer
Frank Hammund
Amy E. Hammund

Tyoga, A Bit of Pennsylvania in Michigan, by Michael R. Zuidema (page 229) explains the sequence of ownership of the Tyoga Lumber company prior to March 23, 1905.

APPENDIX
Chapter IV
Item 2

A town called 'Tyoga'

PATHWAY PLANS PAVED ROAD TO U.P.'s PAST
Escanaba Daily Press
October, 1979

By RICK RUDDEN, Staff Writer

ESCABABA – Mike Zuidema has been trying to complete a vivid picture of pioneering life in the Upper Peninsula. He's been at it for the past 18 months.

Although it is not being painted on paper or canvas, this particular "picture" is very slowly unfolding in the Laughing Whitefish State Forest Campground in Alger County.

Brushes and paints aren't this artist's tools. Instead, Zuidema is using old foundations, eye-witness accounts and library and tax records in pursuit of his task.

Zuidema, Department of Natural Resources area manager in Escanaba, is attempting to piece together the times of a 70-plus-year-old village that has been muddled by both time and memory.

Once he has gathered all the information he can, Zuidema will incorporate it into a combination historical-interpretive pathway which will run through 1.5 miles of the campground.

Before Zuidema began investigating the history of the village in the spring of 1978, it was just one of hundreds of once-existing places in the U.P. that are not known by people living even in their close proximity.

They are not known because the general stores, railroads, post offices, barns, logging camps and schools that made up these villages are all gone and forgotten.

There is not much left of the buildings that made up what now could be nicknamed "Zuidema's village."

Remnants of some old foundations, which are all that is left of the community, can be found…but only by expending some effort. Through a nearby clump of thick brush and trees, another large and strange foundation is found.

Zuidema explains that this foundation is all that is left of an old steam engine that once powered the huge band saw in a lumber mill. The foundation is 12 feet wide, 11 feet long and over six feet high, and Zuidema speaks of the giant piece of mortared rock as if it is the prized find of the old village.

The foundation for the steam engine is almost totally covered with moss, grass, brush and trees. If it had not been for efforts made by Zuidema, it soon would probably have been completely reclaimed by nature and forever lost in history.

MEMORIES AND REMAINS...

Zuidema found that the remains of the village and memories of it have much in common:

Foundations can still be found. With more digging, more foundations and more memories could probably be recovered. But no matter how much exploring is done, the physical remains of the village and its memories could never be totally retrieved.

Time does that to man-made buildings and man-made memories.

Like a detective solving a crime, Zuidema is attempting to locate eyewitness accounts of the village by people who once lived there.

His witnesses are in their 80s and 90s and must try to recall events that took place over 70 years ago.

Zuidema said the time lapse gives for sketchy information. Yet he is lucky to have any first-hand information at all.

"There is a definite problem in trying to figure out history that is 80 years old," he said. "Practically nobody is living that actually lived or worked there."

In talking with these sources he has come up with some differences in dates and in stories. "How do you weed out what's right and what's wrong? he asks.

TIOGA OR TYOGA?

The answer to the question is not all that cut and dried.

"On the old tax records, it says 'Tioga'; The guy who owed and operated the mill, George Burtis, came from Tioga, Penn. So it doesn't make sense for a guy who came from Tioga, Penn., to name his mill the Tyoga Mill. You would think he would use the same spelling," Zuidema said, explaining another side to the village spelling bee.

No matter how it was spelled, a very real village was there, and it "was quite a complex," Zuidema said.

"The Detroit, Mackinac and Marquette Railroad had a spur off the main line which ran in Tyoga to pick up lumber," he said, adding that "to build a spur all the way in there and to build this big mill, it was a big time operation."

ONLY LASTED FIVE YEARS

Zuidema said he would make "a wild guess that 100 people or so had to make up the small community confined to just a few acres."

The information he received from the Ford Museum concerns his pet project: the old steam engine foundation.

He went to the museum while on vacation to see if he could compare the huge steam engine which had once been the heartbeat of Tyoga to the steam engines at the museum.

He had recorded the bolt placement of the foundation of the Tyoga steam engine and compared it to the bolt placement of the steam engines in the museum in an attempt to find what specific type of steam engine had been used in Tyoga.

Despite this effort, Zuidema struck out. Even after traveling downstate, he still had no indication of what type of steam engine had been in Tyoga.

But earlier this year Zuidema found a steam engine in Nahma, in the eastern portion of Delta County. Its bolt placement is very similar to the bolt placement of the foundation of the Tyoga steam engine. But Zuidema was still confused because the two were not identical. He then wrote John Bowditch, curator of the Power and Shop Machinery Division of the Henry Ford Museum, for some answers.

Bowditch told Zuidema that the engine in Nahma and the engine used in Tyoga were probably the same, the only difference being that one was for right-hand operation and the other was for left-hand operation.

Bowditch told Zuidema that "There is virtually no doubt in my mind that your engine is the same as the one at Nahma, most side-crank engines were available either right or left-handed.

NEED TO TIE UP LOOSE ENDS

"I have been just gradually digging into the history of this area I'd say a year ago last spring, and now all I've got is a couple loose ends to clear up before its all completed," Zuidema said.

He first considered building a trail at the Laughing Whitefish Campground "just to give the campers there something to do." He later began investigating the history of Tyoga last year after learning that years before there had been an old sawmill near the campground.

"I was thinking possibly we could have an interpretive trail, because at that time I had no idea that I could try and make it an historical pathway," he explained.

He found out about the sawmill when "one day I was working out there and I stopped and talked to a guy who told me it (the campground) had been a site of an old sawmill, so he got me interested in it and I started checking it out."

Zuidema is planning to create a historical hiking trail in the campground that would possibly be made up of interpretive posts that would explain the area and help create a feeling of the small community during that era.

He has tentatively planned 13 different stops along the pathway, which he hopes will be completed by the beginning of next year's camping season.

The foundation of the steam engine will be able to be seen easily by hikers from across the Laughing Whitefish River once the DNR clears out some brush and trees.

From this stop across the river from the old steam engine foundation, Zuidema plans on going into great detail into the history of the sawmill and the surrounding community.

Although he has been gathering information on the project for the last year and one-half, as far as DNR officials are concerned, it is a low priority project that Zuidema can work on only as he has time.

"I'm not on this full-time, they did tell me I could keep working on it but that it's low priority," he explained. "I haven't been spending too much time on it, I do most of my work at home at night while watching television."

When the project is completed, an almost forgotten, but very real, piece of Upper Peninsula history will have been preserved.

APPENDIX
Chapter IV
Item 3

Logging in Days Gone By…

TYOGA, A BIT OF PENNSYLVANIA IN MICHIGAN

By Michael R. Zuidema

The Timber Producer, February, 1982

The Upper Peninsula is full of long forgotten villages, logging camps and homesteads which flourished around the turn of the century. Scores of towns and camps were created as the lumbermen began to harvest the virgin forests of this land of Hiawatha. Most of these settlements thrived, waned and then disappeared as the great forests were cut and the mills moved on to other locations. Areas where our ancestors had lived were now taken over by Mother Nature who proceeded to fill in the openings with brush and trees. Today, only the observant eye can detect evidence of the activity that once took place.

Information about these sites varies. Sometimes there are no records and the area's past identity is lost. In many cases, information was passed down from parent to child as the elder often reminisced about the old days. Less frequently, someone is alive today whose memory encompasses events which happened there.

In the case of Tyoga, a typical Upper Peninsula sawmill and logging town of the turn of the century, two gentlemen were found who actually worked there. Ninety-six year old Walter Alexander of Marquette worked the winter of 1906 at Tyoga loading sleighs with white pine logs with the contract logging crew of Dan McEachern; and the late Russell Boogren, of Munising, who taught nine students at Tyoga the school year of 1908, about a year after the mill was sold and dismantled. From their account, together with newspaper articles, labor records, corporation records, township and Alger County Historical Society records and other information from descendents of people who lived there, an accurate history of Tyoga gradually was pieced together.

I had been in charge of the Laughing Whitefish State Forest Campground just across the river from the Tyoga town site, for over ten years before becoming aware of the existence of the old logging and sawmill community. Having already developed the "Ninga Aki" (Mother Earth in the Algonquian language) self-guided nature pathway at the Portage Bay State Forest Campground in the Garden Peninsula, I saw the potential for a similar hiking trail at the Laughing Whitefish. While working on the trail layout one day, I met an elderly local resident. After talking for awhile, the conversation lead to Tyoga. From this meeting my interest was aroused and over two years of research began. Tyoga once again was beginning to come to life.

As the virgin forests were being cut in the northeast and Canada, interest grew in the vast white pine, hemlock and hardwood forests of the Lake States. It was common for investors from other parts of the country to put money into Michigan. In 1902, a group of businessmen from Wellsboro, Pennsylvania, including Leonard Harrison, Fred W. Siemens and Horace B. Packer, purchased 1,200 acres, which included the eventual town site of Tyoga, from Daniel Powell of Marquette, the namesake of the present day Powell Township of Marquette County.

Leonard Harrison, one of the most successful lumbermen from Tioga County, Pennsylvania was undoubtedly the leader of this Michigan venture having, by far, more experience in the lumber business than either Siemens or Packer. Fred W. Siemens was primarily in the furniture business in Wellsboro as well as being a director and president of the First National Bank. Horace B. Packer was a lawyer who was active in politics. Packer was a member of the Pennsylvania State Senate in the early 1890's and was in Congress for two terms during the Spanish-American War. Besides Michigan, the three invested in timberlands in Pennsylvania, New Hampshire, West Virginia, and Newfoundland.

At the same time this group was beginning to build a sawmill and logging town in Michigan, they were active in a similar venture in West Virginia. The three were large stockholders in the Tioga Lumber Company of Tioga, West Virginia; Harrison was secretary of the corporation.

Just as the three men brought the name of their home county to West Virginia, they brought the same name to Michigan. Obviously wanting to keep the two companies separated for tax purposes, a modified spelling of the name Tioga was chosen – Tyoga. Oddly enough, the modified version of the spelling was also used at home. Harrison was a charter member of the Tyoga Country Club in Wellsboro. The club still exists today.

As the town of Tyoga began to take shape, loggers were busy in the woods felling logs for the mill to cut. Railway crews were busy installing a spur into Tyoga from the Duluth, South Shore and Atlantic Railway, mainline between Onota and Deerton, formerly the Detroit, Mackinac and Marquette Railway. On March 7, 1905, the Tyoga Lumber Company was incorporated. A few days later Harrison, Siemens, and Packer sold the corporation for $41,200, a considerable sum of money in those days, to another group of investors from Pennsylvania.

Noted for their remarkable business judgment, the 17,700 acres of virgin yellow poplar and white oak at their Tioga Lumber Company in West Virginia must have seemed a more promising investment to Harrison, Siemens, and Packer. The expertise of Leonard Harrison and company in setting up a sawmill town and logging crews was apparent. Three million board feet of logs equally divided between pine and hardwood had already been cut in the woods, and the mill and town were well under construction.

The new owners, however, lacked the lumbering experience of the first owners. New President L.L. Bailey, a banker, was President of the Tioga County Savings and Trust Company, George M. Spalding, Secretary-Treasurer of the Corporation was engaged in several businesses including a drugstore, coal and building supply company and later was the Trust Officer for the Tioga County Savings and Trust Company. One of the stock holders, Thomas M. Archer of Crooked Creek, Pennsylvania, was a lumberman. It is known that he looked after a logging operation in northern Lycoming County, Pennsylvania; but whether he helped run Tyoga is not known.

The new owners must have relied heavily on the men hired by Harrison to run the Tyoga operation. It is believed that many of the key men that built and logged at Tyoga were brought in from Pennsylvania by Harrison. One of the key figures at Tyoga was a local man, Dan McEachern, who was accompanied by his two sons, Neil and Danny, plus four women cooks, and a crew of 30 men, were contracted to cut pine on the company's land about two miles from Tyoga. The logging crew operated by the company cut mostly hemlock and hardwood.

By February of 1906, the two crews had two million board feet of logs on the rollways at the mill. Sleigh hauling and cutting of another three million feet of hemlock, pine, and hardwood logs was in progress. At this time, the *Munising Republican* reported: "The new mill is ready to begin operations as soon as the belting arrives. The machinery has been tested, and fires have been alive under the boilers for the past two weeks. The company expects to install a railroad during the coming summer, and after this year will do away with team hauling from the woods. A complete railway with logging equipment will be ordered this summer, and the tracks laid from the mill to the lumber camps."

To power the band saw which could cut about 35,000 board feet of hardwood or 50,000 board feet of pine and hemlock per 10 hour work day, the Tyoga lumbermen installed a 130 horsepower Atlas steam engine. It took a total of 40 men to operate the mill. Slabwood from the sawmill burned to produce steam power. The fieldstone steam engine foundation, which was 11 x 12 feet, and stood 6-1/2 feet high, is still intact at the mill site today.

With the population of Tyoga growing, the need for postal service became apparent. On February 15, 1906, John W. Bailey became Tyoga's first Postmaster.

In May of 1906, the Tyoga owners purchased an additional 1,000 acres of land to bring their total holdings to over 7,000 acres, most of which was covered with virgin timber. Their last land acquisition alone included an estimated 60 million board feet of hemlock, pine, and hardwoods which would be sufficient volume for several year's sawing. With the demand for hemlock at the unprecedented levels, the future of Tyoga looked bright.

Tyoga Lumber Company was busy logging, operating their sawmill and constructing a new mill to make shingles and railway ties during the spring of 1906. Newspaper reports indicated this new mill would be started as soon as the machinery could be installed and operations would be carried on throughout the year. No other records have been found indicating the activation of this smaller mill, if it were started up, the additional labor to run it would have easily put the total population of Tyoga to well over 150 people.

Working six days a week, ten hours per day for $30.00 per month including room and board, the lumberjacks pulled their misery whips (cross-cut saws), loaded sleighs and drove their teams of horses to the mill. Although some of the loggers lived in homes with their families or stayed at one of the boarding houses, most of the loggers stayed in the bunk house provided by the company. Walter Alexander recalls that if your straw-filled bed was not soft enough, you simply raided the horse barn for more straw. Compared to the bedbug ridden bunk houses, the comforts of sleeping at one of the boarding houses was well worth the extra expense. Some of the lumberjacks who slept in the bunk houses claimed that they could only get to sleep if they turned their underwear inside out and hurried to fall asleep before the lice got inside again.

In contrast to the lumberjacks $1.25 per day, the workers in the Tyoga mill received a higher wage of $1.75 per day, however, had to pay for their own room and board.

Tyoga seemed to be full of activity and was prospering. A long hard week's work was followed by a dance in the school room every Saturday night. Men from other logging camps two or three miles upstream on the Laughing Fish River, as it was known in those days, were frequent visitors on weekends for both dancing and trading at the company's general store. Trains ran in and out of Tyoga picking up manufactured lumber and the company still had millions of feet of standing timber on their 7,000 acres.

From this point on, a series of events occurred that may have led to the second sale of Tyoga.

On July 20, 1906, the *Munising News* reported that Game Warden E. B. Eddy of the Soo visited the Tyoga Lumber Company and found a state law being violated by the company. Sawdust and other refuse was being dumped in the Laughing Fish River. No arrest was made, but the company promised to burn their refuse in the future.

Then came the winter of 1906-1907. On February 8,1907, the *Munising New* reported: "If much more of the "Beautiful" (snow) falls, 'tis altogether likely several camps in the vicinity of Munising will be obliged to suspend operations for the winter." Other reports noted that camp laborers were scarce that winter despite the fact high wages were being paid. Felling crews were finding it extremely difficult to walk from tree to tree and the horses weren't doing much better skidding, and the winter was only at the half-way mark.

With the owners in distant Pennsylvania, There must have been considerable pressure on the company's manager. Whether or not the company was making money is not known, but it may have been that the winter of 1906-1907 was too much and the men in Pennsylvania weren't looking forward to another Lake Superior shoreline winter. Near the end of the winter on March 18, 1907, Tyoga Lumber Company was sold to the Cleveland Cliffs Iron Company.

The Tyoga holdings consolidated well with C.C.I. land in the area. Although most of the Tyoga property was within two to three miles of the mill, some of it extended out over six miles, and the Tyoga stockholders undoubtedly made a good profit.

Walter Alexander in the meantime was hired by the railroad. During the summer of 1907, Alexander recalls making the run into the mill to pick up lumber and the mill was still operating. C.C.I must have dismantled the mill that fall after the winter's inventory of logs were sawed. When Russell Boogren taught there in 1908, the mill was gone.

C.C.I. began seeking the luxuries a larger town had to offer their workers, preferably with a more centralized location for their mill in relation to other C.C.I. lands in Alger County; also wanted bay facilities and improved railway connections. Therefore the Tyoga mill moved to Munising. The nine-foot diameter flywheel of the Atlas steam engine served C.C.I. well until August of 1926 when a fire destroyed the Munising mill.

Whatever the reason for the sale and subsequent dismantling of Tyoga, memories of the little village will be long lasting. Sometimes, if you listen closely, you can still hear the broom swinging and Mrs. Nestor Koski screaming as she chased a black bear out of her kitchen; you can almost hear the arguing and sometimes the fighting of the lumberjacks at the Tyoga Saturday night dances as they competed for each other over the next dance with the beautiful Mrs. Ross Chamberlain, and the music of the nearly five foot long horn Mrs. Dan McEachern blew to announce to the hungry loggers their vittles were hot and ready to be served.

Michigan became a state in 1837. This was followed by the Homestead Act of 1862. Some wealthy men gave money to others who were to go out and file a claim on a section of land. A section was 160 acres. The filing fee cost $26.00. That man then was required to build a cabin with at least two windows in it, and to live there for no less than five years. The lumber rights went to the man who loaned him the money.

The lumber barons bought up all the land they could. The way of life shifted for everyone, Native Americans as well as Europeans, as progress, purchase and providence began to unfold. Logging camps began to pop up all across the U. P. The cutting of the big pines began in earnest. Saw mills grew up overnight. Some small towns had as many as five or six mills within the city limits. Everything changed for everyone.

The dream of Tyoga is still alive. Now we visit it and follow a well-marked trail that "shows and tells" its history. Many of descendants of the people who spent time there in its heyday still live in the area. Most are hardy, friendly, hard-working, willing to spin stories and spill them into listening ears. In this book you will learn about some of them, and the way they lived. It will be worth your while. You may even recognize them should you dig into your family's past. Where did you come from long ago?

Courtesy Michigan DNR

DNR crew constructing north bridge over Laughing Whitefish River, including Terry Porpour, DNR (Ret.)

APPENDIX
Chapter IV
Item 4
Joe Wagner
Map of Tyoga Historical Pathway with names of signs and their GPS coordinates

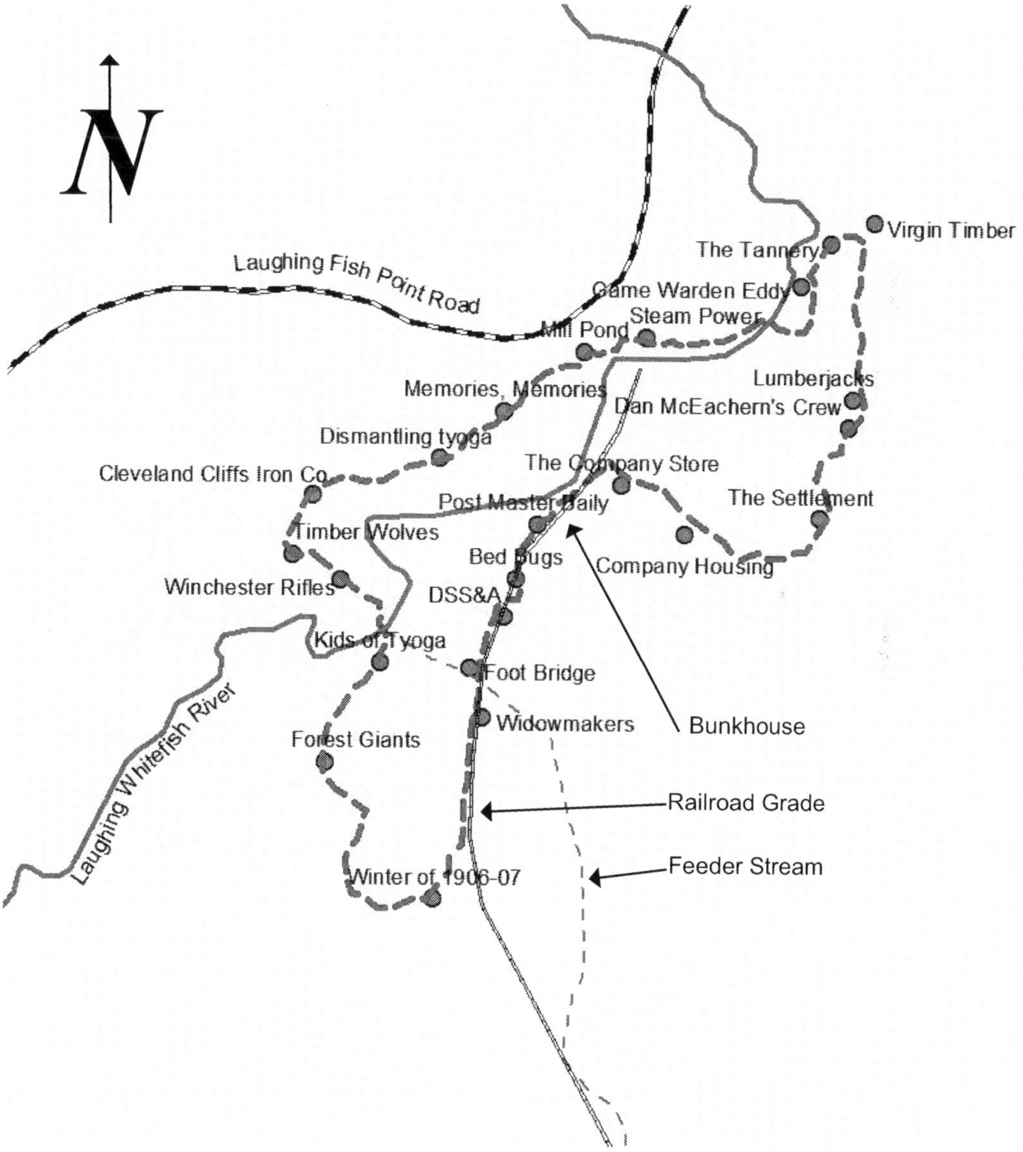

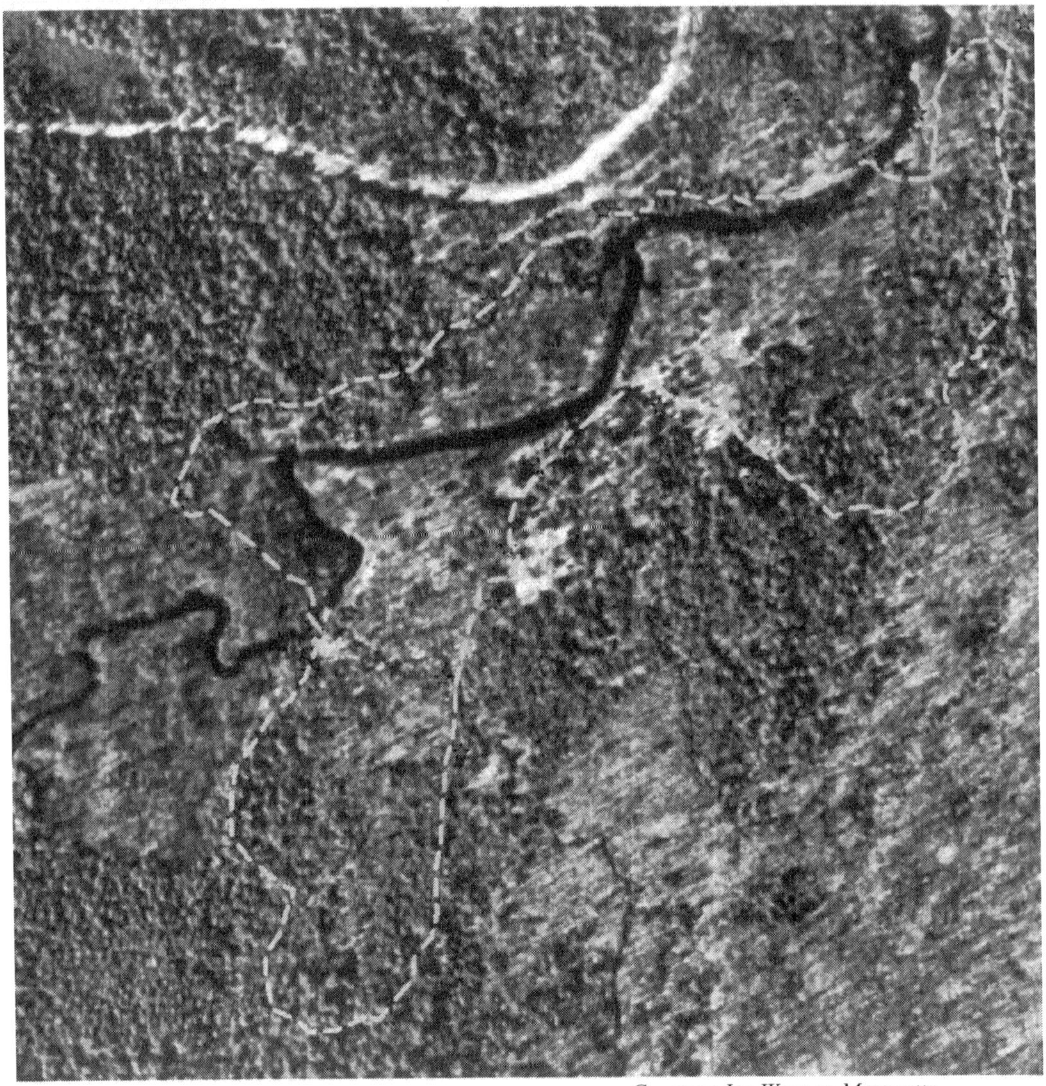

Courtesy Joe Wagner, Marquette

Aerial Photo with dotted line marking the Tyoga Pathway
- - - - - - Trail
▬▬▬ River

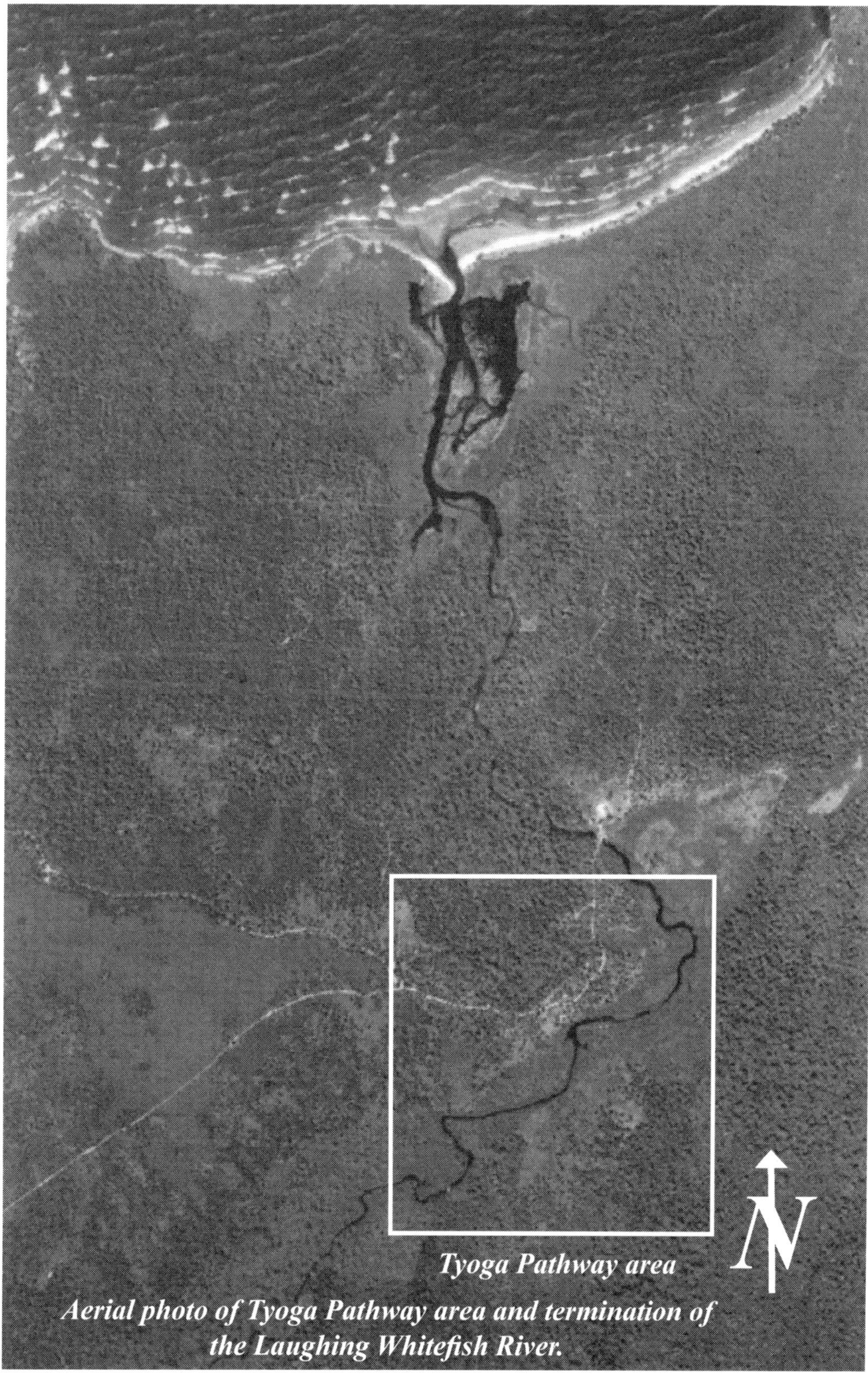

Courtesy Michigan DNR, 1941

Aerial photo of Tyoga Pathway area and termination of the Laughing Whitefish River.

APPENDIX
Chapter IV
Item 5

Narrative on the signs of the Tyoga Historical Pathway and their GPS Coordinates

TRAILHEAD PARKING AREA

As the virgin forests were being cut in the northeast and Canada, interest grew in the vast white pine, hemlock and hardwood forests of the Lake states. In 1902, a group of businessmen from Wellsboro, Pennsylvania, led by Leonard Harrison, purchased the eventual town site of Tyoga and 1200 adjacent acres. When the Tyoga Lumber Company was incorporated on March 7, 1905, three million board feet of logs had been cuts in the woods and Tyoga, a sawmill and town site, was well under construction. A few days later, Harrison and his associates sold the corporation for $41,200, a considerable sum of money in those days, to another group of Pennsylvania investors.

Very little information was known about Tyoga until recently when the personal accounts of Walter Alexander of Marquette and Russell Boogren of Munising, now both deceased, were pieced together along with newspaper articles, labor records, corporation records, township and county historical records and other personal accounts, constructing an accurate historical record which will keep Tyoga alive, despite mother nature's attempt to reclaim its identity.

Relive some of Tyoga's history as you go back in time and walk the Tyoga Pathway.

* * * * * * * * * * *

Waypoint description from signs by Mike Mike Zuidema

In italics are additional remarks in 2004 by Mike Zuidema, Mike made the 63 mile trip to Tyoga from his home in Escabana countless times to work on the pathway.

GPS Waypoints are latitude/longitude – see map and aerial photos on 3 previous pages.

<u>MILL POND</u>

Waypoint 1

46.51325438/-87.02671582

The Laughing Fish River, as it was known then, was dammed downstream to create a mill pond in the area you are now viewing. The winter's cut of logs was rolled into the pond and then hauled up the jack ladder onto a raised deck in the mill from which the logs were rolled into a carriage which fed the bandsaw. It took 40 men to operate the Tyoga mill. The mill was two-storied. The logs were sawed on the first floor and the saw filing room was located directly above the saw. Saws were changed through a trap door in the saw filing room.

* * * * * * * * * * *

If you look down along the riverbank closely, you can see some evidence of the mill pond dam.

STEAM POWER

Waypoint 2

46.51268735/-87.02778208

The Tyoga mill was powered by an Atlas steam engine which was mounted on the foundation just across the river from this point. The 130 horsepower steam engine pushed the bandsaw blade through 50,000 board feet of pine and hemlock each 10 hour work day. The millworkers were paid $1.75 per day, out of which they paid for their room and board at the large boarding house or one of the private homes.

* * * * * * * * * *

The cleft in the foundation is where the flywheel went through. 30 feet to the left is the three foot high masonry base for a steam boiler, fueled with sawdust and chips.

Following the left bank of the Laughing Whitefish River, the campground was to the left. At the time of Tyoga, a number of small frame houses were in this location (see Photos on pages 71-74).

To build the north footbridge, timbers here were hauled in right from the campground with a little tractor or something. For the second (south) bridge across the Laughing Whitefish River we dumped all the timbers and all the planking into the river at the County Road bridge, took off our shoes and took out our wallets and made probably the last log drive ever on the Laughing Whitefish River.

Just east of the north footbridge is a bridge designed and constructed in 2005 by volunteers because of deep beaver sloughs on the original trail.

GAME WARDEN EDDY
Waypoint 3
46.51224654/-87.02864232

The DNR at work in 1906? July 20, 1906 Munising News headlines; "Time to call a halt" reported, "Deputy Game Warden Eddy, of the Soo, was in Alger and Marquette counties recently. At the plant of the Tyoga Lumber Company, near Deerton, Mr. Eddy found the state law violated by the company, which was depositing its sawdust and other refuse in the stream on which it is located. The management agreed to burn the refuse in the future, so no arrests were made. The state game warden's department proposes to see that the mill company observes the law."

* * * * * * * * * * *

I see a blue confidence marker up ahead on that big tag alder. This trail is a jungle right now. This has got to be herbicided and cleaned up with a weed whacker. I've done it both ways (weed whacker versus herbicide). Blue marks are called confidence makers. They are there to let people know, "Hey, you're still on the trail. Cool it. You don't have to turn back." This particular tree, a cherry, would be a good place to have a confidence maker. You spray the tree by rotating the nozzle quickly to create a dot the size of a pop can.

OOPS! Look down here, we've got a big Bruno right here (bear tracks). Pretty fresh, too. I brought along baggies for poop because we are doing DNA on scats that we think are mountain lion. This next set of tracks could be wolf or coyote.

But anyway, this particular spot was probably never harvested at the time of Tyoga logging and I would say that this area probably could be considered a small virgin forest. These hemlocks are all 300 years old, at least.

The triangular blue confidence markers are by the Michigan DNR. The trail is listed on Page 458 of the *Trail Atlas of Michigan, 2nd edition.* The triangular blue confidence markers are scattered along the trail.

THE TANNERY
Waypoint 4
46.51203356/-87.02972618

During the spring and early summer, hemlock trees were felled and the bark hand peeled and stacked into piles which were eventually skidded to Tyoga and shipped by rail to the tannery in Munising. Tannic acid was extracted from the bark at the tannery and used in processing green native hides into leather.

VIRGIN TIMBER
Waypoint 5
46.51194748/-87.03025969

In May of 1906, the Tyoga owners purchased an additional 1000 acres of land to bring their holdings to over 7000 acres, most of which was covered with virgin timber. Their last land acquisition alone included an estimated 60 million board feet of hemlock, pine and hardwoods, which would be sufficient volume for several year's sawing. With the demand for hemlock at unprecedented levels, the future of Tyoga looked bright.

LUMBERJACKS
Waypoint 6
46.51188319/-87.03038776

The Tyoga lumberjacks were mainly Finns, Englishmen, and French-Canadians. They were a hardy, tough lot, capable of gargantuan labors, and were filled to the brim with the zest of living. Two separate crews cut the virgin timber surrounding Tyoga, a contract crew led by Irishman Dan McEachern and a company crew. McEachern's crew, which cut only white pine, consisted of thirty-two lumberjacks accompanied by four women cooks. The company crew was larger and cut mainly hemlock and hardwoods.

DAN McEACHERN'S CREW
Waypoint 7
46.51133191/-87.03064525

Courtesy Mike Mike Zuidema

Dan McEachern logging crew from Blaney Park - Tyoga – 1906
***Front row** included Duncan Sinclair, Allen MacAuley with baby Laverne, Mrs. Allen MacAuley, Mrs. Dan McEachern with 5 ft horn, Kate McEachern, Mrs. England, and Neal McEachern. **Second and third rows** included Fred England, Ed Musselman, Charles Mattison, Ernie Peterson, and August Lagman. **Fourth and fifth rows** included Dan McEachern Jr., August Carlson, and Dan McEachern (third from right in top row).*

We're coming to the Dan McEachern Crew sign. This picture was probably taken just after the turn of the century and is in the neighborhood of 90% accurate as to those actually working at Tyoga. The picture of this crew was actually taken at Blaney Park on Highway 2. A relative of Dan, Holland McEachern, lived in my neighborhood, less than a mile away. He is the great grandson of Dan McEachern. He told me about the picture hanging up in one of those stores at Blaney Park. The guy at Blaney Park let me take it and make copies of it. In those days, a lot of lumberjacks were also farmers. They farmed near Manistique and Gulliver and other nearby areas. In the winter, they worked in the woods. Dan was the boss and he had his set crew. I'm sure from time to time he would get some new guys so that is why I'm saying this sign is probably 90% authentic. I was able to get many of the names of the men, women, and baby in the picture from the store owner at Blaney Park. At that time, some were still living. Tyoga Lumber Company hired a second company crew.

THE SETTLEMENT
Waypoint 8
46.51109906/-87.02997822

You have been walking down the main road from Tyoga into country a former resident described as an "awful wilderness". Proceeding down the pathway one enters the cleared Tyoga townsite which mother nature is gradually reclaiming to brush and trees. Besides the sawmill, the structures within this opening included private homes, a boarding house, company sleeping camp, company cook shanty and store, horse barns and the blacksmith shop. The cook shanty, a central building in any logging camp, was large enough to feed forty men. Potatoes, beans, pork, bread, molasses, strong tea and coffee could be found on the table in some form three times a day.

* * * * * * * * * *

There is definitely a bear population back here. If they don't maintain the grassy areas and you're walking through thick brush and taller grasses, bear would not see you as quickly as if you had cleared it out and could see several yards away. I don't think anybody has ever been hurt here to my knowledge but with poor trail maintenance you could have a big problem.

There is moss on this trail growing through the grassy opening. This is good because the moss helped to define the trail.

COMPANY HOUSING
Waypoint 9
46.51069698/-87.02944136

1907 photos and historical accounts indicate six to ten wood frame homes built by the company existed uphill across the river from this marker. The house with the porch was most likely occupied by the company superintendent (see Page 73). Nearby were log cabins built by families who came to work in the woods and mill. Some of these families boarded workers for 25 cents a day, including laundry.

THE COMPANY STORE
Waypoint 10
46.51326335/-87.02670207

The Tyoga Lumber Company store, pictured here, stood a short distance straight out from this marker. Food, tobacco, clothing and various odds and ends were sold to Tyoga residents, as well as to lumberjacks from other logging camps several miles upstream. The log building to the left of the store is a cook shanty. This picture was taken from the other side of the store towards where you are now standing. The next portion of the pathway travels down the DSS & A Railroad spur which served Tyoga.

POST MASTER BAILEY
Waypoint 11
46.51339126/-87.02586941

As the population of Tyoga approached 150 residents, the need for postal service became apparent. On February 15, 1906, Jon W. Bailey became Tyoga's first Post Master. Lena Dilger was the settlement's first school teacher, followed by Russell Boogren in 1908. The school house still stands today in the village of Deerton. In 1911, after Tyoga was abandoned, the structure was hauled by horse team on sleigh runners in the winter via the railroad grade to its present location.

BED BUGS!
Waypoint 12
46.51360902/-87.02421659

Two hours before daylight, six days a week, the lumberjacks were awakened by the woods boss as he stuck his mug in the bunkhouse doorway and yelled, "Daylight in the swamp!" Compared to these bed bug ridden sleeping camps, the comforts at the boarding house were well worth the extra expense. Some lumberjacks claimed the sleeping camps so bad that the only way they could get to sleep was to turn their underwear inside out and quickly fall asleep before the lice got inside again. Contract logger Dan McEachern's camp, complete with cook shanty, horse barn and sleeping camp was on the west edge of Tyoga.

DULUTH SOUTH SHORE & ATLANTIC RAILWAY
Waypoint 13
46.51894787/-87.01785757

Better step off the railroad grade, as it is almost time for the Duluth, South Shore and Atlantic Railway to make its run into Tyoga to deliver supplies and pick up lumber and hewed cedar railroad ties. Although Tyoga residents most likely had neither the extra money nor the time, they could take one of the annual round trip excursions from Munising to Detroit, Buffalo, Cleveland, and Toledo via the DSS & A and D&C Navigation for the low rates of Detroit $6.50, Toledo $7.00, Cleveland $8.00 and Buffalo $8.50.

FOOTBRIDGE
Waypoint 14
**

This little bridge at the 6-foot wide creek is where I caught a rainbow trout with my hands. Technically it was illegal but this was the first and only rainbow trout I ever caught with my bare hands. I was told how to do it, by reaching down and coming up from behind them, and just stroking their belly real lightly and then work up toward their gills and clamp down. I picked a beautiful rainbow trout up out of the water here. I got the lesson from a violator. He told me how to do it, I tried it and it worked. Of course, I released it.

WIDOW MAKERS
Waypoint 15
46.51641897/-87.02565769

Life in a logging camp was rough. The work was dangerous and death or serious injury, in the form of a falling limb, an exploded boiler, or a wrecked train, was very common. Old newspapers are replete with accounts of such accidents.

Burial customs were quite casual during these early days, which accounts for the two or three unmarked graves which are reported to be located a short distance from this location. The men who are buried there died attempting to make a living in the surrounding forest.

* * * * * * * * * *

There are no marked Indian graves in Onota Township. This area was used by the Ojibwe for hunting and fishing. An occasional Indian may have died and been buried here but no evidence can be found.

THE WINTER OF 1906-1907
Waypoint 16
46.52125223/-87.02767421

Tyoga was full of activity and was seemingly prospering. A long hard week's work was followed by a dance in the school house every Saturday night. Men from logging camps upstream were frequent visitors on weekends for both dancing and trading at the company's general store. Then came the winter of 1906-1907 that must have been hard financially on the Tyoga owners. On February 8, 1907, the *Munising News* reported: "If much more of the "Beautiful" (snow) falls, 'tis altogether likely several camps in the vicinity of Munising will be obliged to suspend operations for the winter." Other reports noted that camp laborers were scarce that winter, despite the fact high wages were being paid. Felling crews were finding it extremely difficult to wallow from tree to tree and the horses were not doing much better skidding, and the winter was only at the half-way mark.

FOREST GIANTS
Waypoint 17
46.51668929/-87.02539424

Many of the white pines, the lumberjacks felled with their axes and misery whips (two man crosscut saws) were 4 to 5 feet in diameter. Then they were loaded onto sleighs and hauled by horse teams to the mill.

When the Pilgrims landed at Plymouth Rock, these trees would have been either seedlings or pole size; and they were in the prime of their life when the Liberty Bell rang out the birth of our new nation.

The tallest trees stood like columns in an immense cathedral, reaching heights of up to 150 feet tall. The cry of "tim-b-e-r" rang out through the woods and after each ringing call a forest giant would come crashing down with an earth-shaking jar.

THE KIDS OF TYOGA
Waypoint 18
46.51609149/-87.02570815

Brook trout fishing on the Laughing Fish River was excellent and 3-1/2 pounders were relatively common. Charlie and Arthur Risku, whose father hewed railroad ties in the woods with a broadax, were just young boys when they lived at Tyoga. The brothers used to go down by the dam with a wheelbarrow when the water was let out after the log drives from the upstream logging camps and pick up trout which were left high and dry. For recreation, the youths at Tyoga had the old swimming hole in the summer and in the winter their skating rink.

WINCHESTER RIFLES
Waypoint 19
46.13588744/-87.03344716

In the early days a man's rifle was his companion where or whenever he went for a walk in the woods. The two most common rifles used in the Tyoga area were the .44-.40 and .30-.30 Winchesters. These rifles were most commonly purchased at Spencer's Store in Marquette at $13.00 and $12.50 respectively. A box of 50 .44-.40 cartridges cost $1.00. Although serving venison was forbidden in the logging camps, it was an important part of the family menu. Deer licenses which allowed the shooting of 2 deer cost $1.50.

* * * * * * * * * * *

Again, fresh bear tracks in the mud. This should be cleared out so both the bear and the people have better visibility in case they are in the same vicinity. No one likes to be startled by an unexpected visitor. It is sort of a safety factor..

TIMBER WOLVES
Waypoint 20
46.51137675/-87.03806768

Prior to the logging of the virgin forests, the white-tail deer population was low because there was no browse available. But white-tail populations have the capability of exploding if habitat conditions are right. The removal of the forest giants that blocked the natural light meant hardwood saplings could begin their growth, and they make for excellent deer feed.

Market hunting began to be a problem and following his visit to Alger and Marquette counties, Game Warden Eddy reported the "wholesale slaughter of deer." Between the market hunters and the timber wolves, many expected deer populations to be reduced to pre-logging populations. On November 18, 1904, the *Munising News* reported "Deer are doomed - Pack of 50 wolves near Dorsey." (It is a few miles south of here). However, the market hunting was controlled and the wolves began to be exterminated allowing the deer population to remain high

THE CLEVELAND-CLIFFS IRON COMPANY
Waypoint 21
46.50395189/-87.04301074

Whether or not the Tyoga Lumber Company was turning a profit is not known, but the owners in distant Pennsylvania, not familiar with tough Lake Superior shoreline winters, most likely figured it was a good time to sell before having to go through another winter. On March 18, 1907, the Tyoga Lumber Company was sold to the Cleveland-Cliffs Iron Company. The Tyoga holdings consolidated well with C.C.I. land in the area. Although most of the Tyoga property was within two to three miles of the mill, some of it extended out over six miles, and C.C.I. owned most of the land between the parcels. It was a good acquisition for C.C.I. and the Tyoga stockholders undoubtedly made a good profit, selling the Tyoga corporation for $140,000.

DISMANTLING TYOGA
Waypoint 22
46.49497454/-87.04574383

During the summer of 1907, Walter Alexander, while working for the D.S.S. & A. Railway, made a run into the Tyoga to pick up lumber and the mill was still operating. C.C.I. dismantled the mill that fall after the inventory of logs was sawed. When Russell Boogren taught nine students there in 1908, the mill was gone and just a few families remained. Seeking the luxuries a larger town had to offer their workers, a more centralized location for their mill in relation to other C.C.I. lands in Alger County, bay facilities and improved railway connections, the Tyoga mill was moved to Munising. The nine-foot diameter flywheel and the Atlas steam engine served C.C.I. well until August of 1926 when a fire destroyed the Munising mill.

* * * * * * * * * *

D.S.S.& A. – Duluth South Shore and Atlantic – locally known as the "darn sure slow and abusive." Prior to 1887, it was the D.M. & M. (Detroit, Mackinac, and Marquette) and subsequently the Wisconsin Central.

MEMORIES, MEMORIES
Waypoint 23
46.49384600/-87.05121160

Memories of the little village of Tyoga will be long lasting. Somehow, if you listen closely, you can still hear the broom swinging and Mrs. Nestor Koski screaming as she chased a black bear out of her kitchen, the arguing and sometimes the fighting of lumberjacks at the Tyoga Saturday night dances as they competed for the next dance with the beautiful Mrs. Ross Chamberlain, or the music of the nearly five-foot long horn Mrs. Dan McEachern blew to announce to the hungry loggers their vittles were hot and ready to eat. Many similar stories of early Michigan logging history have been lost forever, but Tyoga and a part of its history will be known to future generations and will live on.

This is the last stop before returning to the start of the pathway. We hope that your hike through a portion of Michigan's logging history has been an educational and enjoyable experience.

APPENDIX
Chapter IV
Item 6

OLD LUMBER DAYS CAN BE RELIVED ON NEW ALGER COUNTY PATHWAY

By DAVE SCNEIDER
Mining Journal
8/12/87

TYOGA – The days of giant white pine and lumberjacks using real horse power instead of engines measured by horsepower can be relived near Deerton,

That drift into the past is accomplished by the recently opened Tyoga Historical Pathway, a 1.4 mile trail near the Laughing Whitefish River State Campground.

Coursing its way through what was once the bustling little community of "Tyoga", 20 information plaques placed along the trail will paint the picture of life in a lumbering town around the turn-of-the-century.

"There were a lot of old towns scattered around the Upper Peninsula that usually had lumber camps associated with them," said Mike Zuidema, Michigan Department of Natural Resources timber management specialist at Escanaba.

"Tyoga was a little sawmill town which was typical of others around the area."

Although Zuidema's current job has taken him out of the Alger County area, he carried the project through that he started 10 years ago.

"A lot of our state forest campgrounds have trails associated with them, usually nature trails," Zuidema said. "I thought this (Laughing Whitefish) campground is a little out of the way and doesn't have a beach or anything, and it needed an attraction – something for people to do."

It appears the trail will be popular with campers, as several walked up to where a bridge was being completed on the trail last week.

The work on the bridge and rest of the trail has been performed by a Michigan Youth Corps group out of Munising, using materials and equipment from the DNR.

After working 2-1/2 years researching the town of Tyoga, the project ended up in Limbo because of budgeting restrictions at the DNR.

But a resourceful Zuidema and a helpful Onota Township turned the project into reality. With technical assistance from Zuidema, the township board applied for and was awarded a $4,300 Michigan Equity Fund grant to complete the project.

Greeting walkers along the meandering path is a site that once sported a town housing 150 people, including wives and children, a school teacher and postmaster. A sawmill cut 50,000 board feet of pine and hemlock a day, and a mill pond soaked the wooden giants.

Not all the trees fell, however, as Zuidema said a stand of huge hemlock at least 300 years old somehow escaped the cross-cut saw along the trail.

The sawmill and surrounding timberlands were sold to Cleveland-Cliffs Iron Co. in 1907. The mill was moved to Munising during the fall of 1907, and operated by CCI there until burning down in 1926.

The mill site, tree stand and most of the rest of the attractions are along the section of the trail across the Laughing Whitefish River from the campground. Zuidema cautions visitors to be careful on the path, which travels through some rough country.

The pathway is on the first road east of the Laughing Whitefish bridge on M-28 leading north to the campground.

APPENDIX
Chapter IV
Item 7

AN INTERVIEW WITH FRED CANON
By Fay Swanberg

Fred Canon relates his experience with the timber industry in the late 19th century:

"Lumberjacks cut logs, pine, hardwood into logs, peeled bark and made short-stuff. Short-stuff was ties, pulp. Hardwood applies to beech, birch and maple. And pine was white pine. Very little jack pine logged for logs. Good for pulpwood. Swamper – 3 men in a crew. One man did the notching, measured the tree, if his saw breaks, he had to cut half of them, the sawyers cut the other half. 1st they cut the tree down, then limbed the top. Other than the timber fitter, the other 2 men of the team were the sawyers, used a cross cut saw. The other man in the team had to figure out the size of the logs to cut, 16 or 14 or whatever size seemed to work out best. This man called a timber fitter, he notched, measured. Swamper came along and cut the limbs off. Throw brush out of way so man with a team could come along and skid them out of there. It was also the swampers job to make a road the best way out of there. Had to have 2 or 3 swampers for a team. The skidway was at the sleigh road (logging road) or railroad, whatever it would be. The swamper would make a road from the skidway back. Timber fitter would have to figure, if a tree had a bend, whether he could get 10 or 12 foot logs." They tried to cut the trees as close to the ground as possible. Before Fred's time they had quite tall stumps left in the summer, but in his time they stopped logging when the snow began to get deep rather than leave the big stump there. They'd shut the camps down.

"They went into the camps in Sept. and cut until they could haul, sometime in Dec.; sometimes the last of Dec. before they could haul. Then they'd quit cutting all together and just hauled. They hauled what they had on skids, and they generally figured to have enough on skids to run until middle of Feb. or 1st of March. Never figured on hauling after March 1st. If the snow went down some springs, so they could cut, they'd cut a few if they were just a short haul to the river. The logs were all put in the river in those days. They decked these logs all over the woods. There might be from 5,000 to 40,000 in one of those decks. Had 3 to 5 million feet on decks in the winter in some of those camps, when they started to haul. Most generally had 3 million. When a camp he knew of had 5 million ft. the pine was awful thick.

"On the Escanaba and Manistique River that never freeze in the winter they had a skid just far enough back from the river so that the logs would roll right into the river. This would give them a chance to scale them and stamp them. They floated logs down this way all winter. On Escanaba River they used scow (batteau) to get out in the river. On our river here you jumped on a log if you wanted to go out in the river.

Courtesy Superior View Studios

Cross-haul system with chains to lift logs and cant hooks to direct them.

"They didn't try to jam them. If a log would float all the way down this would be fine, but they jammed themselves. They'd break the rollways loose to start them and they would go down. When the jam was right up to the rollway so there was no more space they took the crew, they had 2 men they called the breakers who were stationed at a bend in the river. The bends all had names and a crew of breakers at each bend. You might sit under a tree smoking your pipe all day, that is if the logs were going right down. They were on bends for a couple of miles. When they started to jam they'd have to jump right out there with his peevie and break that jam and then he'd have to get back into shore before they let go too much. He had a packsack on his back with his lunch, a No.2 tomato can with a bale on it either tied on his suspenders or tied to his pack to brew coffee and tea in. If the logsflowed smoothly all he had to do was sit there and eat his lunch and cook his tea. Sometimes that would be a pretty busy job and other times there would be days he could just sit there. When there'd be no more logs coming he'd start back up the river, whether it was quitting time or not, because he knew up there somewhere someone had a jam. They'd all chip in and break that jam loose, then he'd have to hurry back down to his own bend. You'd just keep going back up, joined by breakers until you found the jam. Next morning there was a mile or 2 miles of river that was clear. When there was a jam not too far away the water leveled raised. They didn't care if there was a jam down the river as long as there was a cleared space to roll logs in. They'd go back next morning and started rolling logs in again until there wasn't more space in the river. They had dams on those rivers. They'd open the gates on those. The flood would last only 6-8 hours and then the water would go down. When the dams were opened there would be no jams after they'd been opened up about 3 hours, then the logs started moving down the river if they weren't jammed solid."

The life in a lumber camp must have been colorful with big meals and lots of music, pipe smoking, and stories around the potbellied stove in the bunkhouse.

Fred continues: "Dried fruit in those days, came in 25# wooden boxes. Lots of pork & beans. In fall they'd put in potatoes, beggies, vegetables. Lots of pastry. That table would be just loaded down. Pie everybody. What didn't go at noon was put on for supper if they wanted it then. Those cooks all good bread bakers. Made with Hoppe's yeast, bought dry, water added. Always kept a big jug of it. Kept building it up, take out so much, put more in. Came in about 4# package. Some used yeast cakes and potato water, but the best was Hoppe's yeast. Big loaves. If a man wasn't a good cook, they didn't keep him. Sourdough pancakes every morning for breakfast. Lots of fried cakes and cookies (wonder if Hoppe's yeast was dry or in cakes since Fred said they'd break off so much, even though he agreed it was dry and water added.) No eggs until after 1900. Lots of meat, 3 times a day. Beef only 3-4 cents lb. Had salt pork and corned beef, which they called redhorse. Always a barrel of salt pork. In summer it was salt pork and ham, you didn't get very much -. In the spring it was all ham on the river. In later years they did furnish eggs on the river. They'd boil them hard. Only fish was from boys who'd go out on Sunday, lakes teeming with fish. In fall the men would love to go out on Sunday and bring in lots of fish for the cook. In those days when summer came it was awful hot, winters hard. Plenty of rain, very few years with dry spell. Year after when they'd cut the pine down. In the later 90's from 94 to 1900 get the forest fires, but not in early years." (Worst fire Fred remembered: 1896.)

Do's and Don'ts of Camp Life:

Drinking – none in camps. Strict. Moved into camp around September, stay the winter, start decking. Pile them with chain in middle. 2 men on each side to keep the log straight. Might bring back a bottle, this would be gone the 1st day. Put them in great big piles all over the woods. Start to haul to river maybe sometime before Christmas, about Dec. 1 he'd say.

Smoking: Safe to say there wasn't a man in camp who didn't use tobacco in some form. Smoke or chew. Nobody smoked cigarettes. No Canadian Frenchman smoked anything but pile. When Austrians (foreigners) came in was the 1st cigarette smoking. Some smoked clay, corn cobs, and some cheap cherry bowls. Some chewed tobacco. 1900 before cigarettes appeared.

Recreation: Horseshoe. A game going in the camp. Most always someone to play instrument. Violin, fiddle, mouth organ, accordion. Some pretty good singers, sang those French songs, always a card game going on. No poker seen, later on. Possible at night when Fred didn't see.

Read: Most couldn't. Used the English language awful rough. Awful hard to understand. Had to guess about part of what they were telling you.

Disposition: Easy to get along with. No fighting. All they knew of was work. Never go in trouble.

Build: Big, husky fellows.

Contortionist: String a logging chain between 2 pine trees, wasn't too tight since light decking chain. Hopped around and performed on that chain as well as any in a circus.

Religion: No. Used it awful rough if they did.

Women: Never heard of any of those old time lumberjacks getting in any trouble with a woman. Great, great, respect for a woman. Remarkable how they respected a woman if she happened to come into camp."

APPENDIX
Chapter IV
Item 8

AN INTERVIEW WITH CARL LINNA
By Bea Anderson
2007

Carl Linna, age 77, was interviewed in 2006 concerning his memories of the logging industry in the first part of the 20th century. Although Carl spent much of his later years driving a truck, he was a remarkable man, working the *New York Times* crossword puzzles **in ink**. Following our interview, he gave me a handwritten piece that is worthy of inclusion:

<u>The Old Homestead</u> – "What was unique about this particular area was that it contained trees that were not ordinarily present in the forest in general, that surrounded it. There were hickory trees, white oak, butternut, Lombardy poplar, apple and cherry trees in an orchard gone wild, and even wild grape vines. There was a long fence made of rocks that forms a boundary on one side that had once been a tilled field. One could imagine the backbreaking labor that had been necessary to tame in part what must have then been a wilderness. In many other parts of the woods I have come across what I call "broken dreams". The people who lived in these places, toiled, had families, good and bad times, mostly bad, then died and are forgotten. Traces of their presence are fast disappearing as nature reclaims what is her own. Sometimes it seems that humans are the great intruders. What was once beautiful in its natural and original state is destroyed by man and nature gets busy at once to repair the damage. After we are gone, little islands of our presence remain as scars on the landscape, and in time they, too, disappear."

<u>A Finnish Joke</u> – "Sometimes at lunchtime we would sit on stumps at the landing, where the wood was decked up for hauling, and tell long ridiculous jokes and stories. One of them was this:

Toivo and Eino were partners in the woods, they cut pulpwood the old-fashioned way, with a bucksaw. Toivo was a good worker but Eino was more of a schemer and did his best to arrange matters so that Toivo would get the bulk of the work. After a few years, Toivo caught onto what was happening and decided to dissolve the partnership. He bought a new chainsaw and came to work with it, telling Eino, I am going on my own, I don't want to be your partner anymore. Eino was very angry but held his temper in check, vowing to himself to find an opportunity to get even. The two still worked adjacent strips, side by side, 50 foot or so separating them from one another. One day Eino saw his chance to get even, and dumped over a big tree on Toivo, killing him instantly. Then he went to the lumber camp boss, reporting the incident as an accident. The boss said, Eino, since Toivo was your work partner for a long time, it would seem appropriate that you should be the one to tell the widow, telling her also that it had been an accident. So the next day was the funeral and Eino went to it. The widow took hold of Eino's hand and led him to the coffin to view the body. Eino took a quick look and said, Too bad about Toivo, have you decided what you are going to do with the chainsaw?"

Ethnic Lumberjack Rivalry – "A Finn, a Norwegian and a Swede all worked at the same lumber camp. Each thought he was a better lumberjack than the others. They often came to blows about who was best but couldn't seem to resolve the question. The boss tired of the endless quarreling and told the three they would, as a condition of their employment, have to submit to his decision as to who was the best. We will, the four of us, take a walk in the woods together, and a situation is sure to arise, said the boss, where each of you can demonstrate your skills, and I will decide the winner. So they walked along a trail and soon they saw a skunk that ran into a hollow log. Now, said the boss, this is going to be the contest. Whoever stays in the log the longest time with the skunk is the winner.

Me first, said the Norwegian and dived into the hollow log. He managed to stay 10 minutes then crawled out, coughing and choking.

I can beat that, said the Swede and stayed 15 minutes.

You guys ought to join the Girl Scouts, said the Finn as he dived into the log. He stayed an hour and a half, then the skunk crawled out."

APPENDIX
Chapter IV
Item 9

AVIAN SIGHTINGS OF THE
TYOGA HISTORICAL PATHWAY 2007
By Scott Hickman

A total of 77 species of birds have been sighted.

Waterfowl - Mallard, Common Merganser, Blue and Green-winged Teal, Wood Ducks

Fowl - Ruffed Grouse

Bitterns, Herons, Egrets, and Ibises - American Bittern, Great Blue Heron

Diurnal Raptors - Broad-winged Hawk

Rails, Cranes, and Relatives - Sandhill Crane

Shorebirds and Relatives - Wilson's snipe, American Woodcock, Ring-billed Gull

Owls - Barred Owl

Kingfishers - Belted Kingfisher

Woodpeckers - Yellow-bellied Sapsucker, Downy Woodpecker, Hairy Woodpecker, Black-backed woodpecker, Northern Flicker, Pileated Woodpecker

Songbirds - Olive-sided Flycatcher, Eastern Wood-Pewee,Yellow-bellied Flycatcher, Alder Flycatcher, Eastern Phoebe, Great Crested Flycatcher, Blue-headed Vireo, American Crow, Common Raven, Black-capped Chickadee, Red-breasted Nuthatch, White-breasted Nuthatch, Brown Creeper, Winter Wren, Sedge Wren, Golden-crowned Kinglet, Ruby-crowned Kinglet, Hermit Thrush, American Robin, Cedar Waxwing, Nashville Warbler, Northern Parula, Yellow Warbler, Chestnut-sided Warbler, Magnolia Warbler, Black-throated Blue Warbler, Yellow-rumped Warbler, Black-throated Green Warbler, Blackburnian Warbler, Black-and-White Warbler, American Redstart, Northern Waterthrush, Mourning Warbler, Common Yellowthroat, Canada Warbler, American Tree Sparrow, Clay-colored Sparrow, LeConte's Sparrow, Fox Sparrow, Song Sparrow, Lincoln's Sparrow, Swamp Sparrow, White-throated Sparrow, Harris's Sparrow, White-crowned Sparrow, Dark-eyed Junco, Snow Bunting, Rose-breasted Grosbeak, Indigo Bunting, Red-winged Blackbird, Purple Finch, American Goldfinch, Evening Grosbeak.

APPENDIX
Chapter IV
Item 10

TREES OF TYOGA
2007

HARDWOODS	**LATIN NAME**
"Paper" Birch (White)	*Betula papyrifera*
Black Cherry	*Prunus serotina*
Sugar Maple Acer	*Saccharum*
Red Maple Acer	*Rubrum*
Yellow Birch	*Betula alleganiensis*
American Beech	*Fagus grandifolia*
Quaking Aspen	*Populus tremuloides*
Common Choke Cherry	*Prunus virginiana*
Hornbeam	*Carpinus caroliniana*
Balsam Poplar	*Populus balsamifera*
White Ash	*Fraxinus americana*
Black Ash	*Fraxinus nigra*

SOFTWOODS	
Eastern Hemlock	*Tsuga canadensis*
Northern White Cedar	*Thuja arborvitae*
White Spruce	*Picea glauca*
Balsam Fir	*Abies balsamea*
Eastern White Pine	*Pinus strobus*
Black Spruce	*Picea mariana*

OTHER	
Tag Alder	*Alnus serrulata*
Nine Bark	*Physocarpus opulifolius*
Royal Fern	*Osmunda regalis*
Thimbleberry	*Rubus parviflorus*

Trees of the Tyoga Historical Pathway

This trail is a 1.6 mile loop, and is marked with both Michigan Department of Natural Resources blue confidence markers and signs that explain the history of the Tyoga Lumber Company (1905-1907). Numbered trees along the way have their name and species attached. This is a detailed list of those trees. Please do not harm the trees or other natural elements of the trail – leave what you find so others may enjoy the beauty. How many can you identify?

A guide to the trees on the trail

1) Ninebark (Physocarpus opulifolius)
The ninebark is a shrub with a maximum height of 30 feet. It has shiny red berries that mature from May to July and provide a source of nectar that attracts multiple species of birds.

2) Mountain Ash (Sorbus americana)
This small tree is less than 20 feet in height, with long, thin, pointed leaves. The fruit is a bright orange color and is favored by cedar waxwings and grouse. Native Americans used the berries as a source of Vitamin C to treat scurvy.

3) Paper Birch (Betula papyrifera)
These trees are noted for their smooth, white bark that easily peels off. When Compared to other trees, their lifespan is short, with a full height of 70-80' reached by 70 years. It attracts a variety of wildlife, from moose to ruffed grouse.

4) Black Cherry (Prunus serotina)
The mature version of this tree has distinctively fissured dark bark. The young trees may resemble young birches, but are typically distinguished by their shiny, lengthened leaves. Its fruit is astringent and bitter, but attracts numerous birds.

5) Sugar Maple (Acer saccharum)
Despite its status as a common tree, sugar maples are important to forests. It has a high tolerance for shade and grows in a variety of soils, allowing it to grow in many areas. It is also well known as the "maple syrup" tree.

6) Red Maple (Acer rubrum)
Found throughout the eastern United States, this tree is highly adaptable to many environments. Its leaves separate it from other maples through both variation, and their brilliant red coloring in autumn.

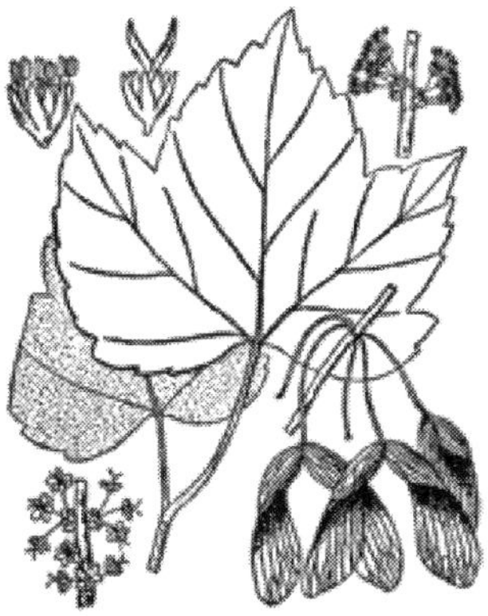

7) Eastern Hemlock (Tsuga canadensis)
A very long lived tree, this specimen is 250-300 years old. This tree prefers cool, humid climates. The bark of a mature tree is a deep brown and consists of furrowed ridges. It grows well in shaded or sun-filled environments. The tannin contained in the bark is known for its past use in tanning leather. Eastern Hemlock stands provide shelter for many animals.

8) Yellow Birch (Betula alleghaniensis)
The bark of this tree is a silvery-yellow. The leaves are similar to other birches - alternate with serrated edges. This tree is a food source for many moths and butterflies. The inner bark is noted for its wintergreen scent.

9) Northern White Cedar (Thuja arborvitae)
A coniferous tree, the cedar is a sweet smelling tree with brownish bark that peels easily. Its foliage is distinct for a spray-like configuration and its flat, scaly appearance. It grows well in swampy areas and is a favorite food of white-tailed deer.

10) White Birch (Betula papyrifera)

11) White Spruce (Picea glauca)
Another evergreen tree, the white spruce has a wide range - from Canada through the Midwest. It features a distinctive scent, gray-brown bark, and slightly drooping branches. Its branches are

often densely clustered and it prefers wet areas. It was used medicinally by Native Americans. However, the tree contains a neurotoxin and any ingestion may be harmful.

12) Yellow Birch (Betula alleghanesis)

13) Eastern Hemlock (Tsuga canadensis)

14) American Beech (Fagus grandifolia)
Covered in smooth, silver bark, the beech tree is shade-tolerant and produces a nut that is covered by a spiky husk. It is an important food source for many species, from porcupines to black bears. The leaves remain on the tree until spring.

15) Quaking Aspen (Populus tremuloides)
The name of this tree comes from the tendency of its rounded leaves to tremble at the smallest breeze,
producing a soft rustling sound. This happens because the stalk connecting the leaf to the stem is flattened. The aspen is a tall tree that grows quickly, and prefers cooler climates.

16) Choke Cherry (Prunus virginiana)
This species is known for its popularity with birds, despite the astringent quality of its fruit. It is a bushy tree, commonly less than 25 feet in height. The fruit is noted for its high level of antioxidants.

17) White Spruce (Picea glauca)

18) Balsam Fir (Abies balsamea)
Well known for its popularity as a Christmas tree, this is in part due to its strong fragrance and the springy quality of its needles. The bark of the balsam is a gray color and often has resin blisters.
It provides food and shelter for moose and a variety of other animals.

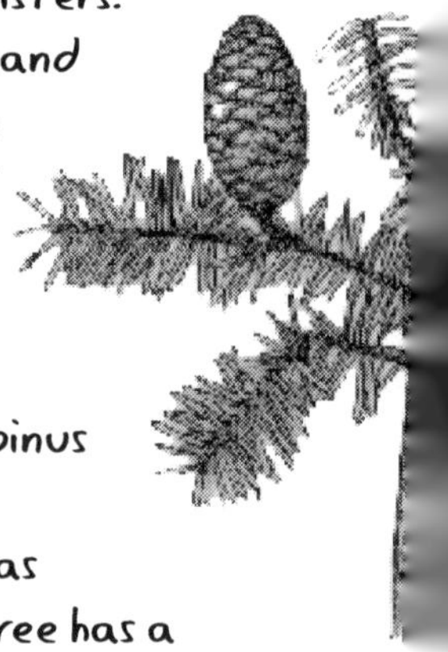

***19) American Hornbeam** (Carpinus caroliniana)
Commonly known as ironwood, this tree has a small, crooked stature but a wide range of tolerance for soil, climate, and precipitation. It grows well in shaded understories of forests and is a key food source for gray squirrels.

20) Quaking Aspen (Populus tremuloides)

21) Eastern White Pine (Pinus strobus)
White pines are distinguished by their groupings of five needles, whic[h] are soft and pliable. They are a lon[g] lived tree, easily reaching beyond 250 years. The cones mature in 3–5 year cycles and the

*Error: A mature tag alder was mistaken for ironwood

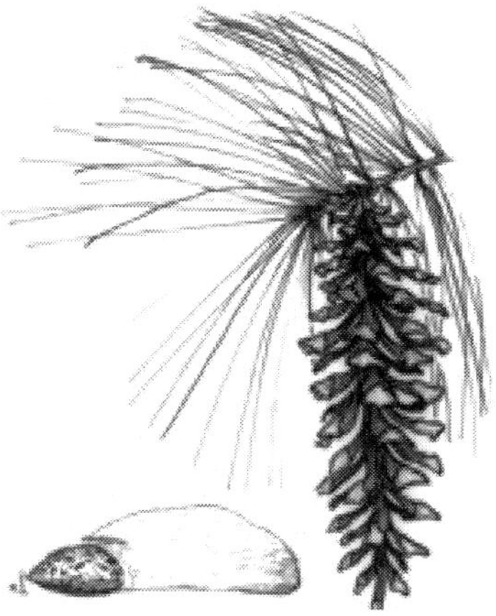

branches are known to be whorled and upswept. They remain a valuable source of commercial timber. (Although there are no red pines in Tyoga – the difference is that these trees have three brittle needles.)

22) Black Spruce (*Picea mariana*)
Native to North America, this evergreen grows slowly and is recognized for its drooping branches. Compared to the white spruce, it has shorter rigid needles and smaller cones. Forest fires commonly burn through black spruce stands, and the tree seeds quickly after a fire. Black spruce can also reproduce through what is known as layering. When the lower branches are compacted by snow or otherwise forced to the ground, they can take root and become new trees.

23) Balsam Poplar (*Populus balsamifera*)
These poplars have a rapid rate of growth and moderate lifespan. The name comes from its sticky strongly scented buds. It requires wet soils and has a shallow root system.

Brochure by Catherine Kimar

24) White Ash (*Fraxinus americana*)
This is a deciduous tree that can reach a maximum height of 125 feet. The bark is a light gray–brown color and the leaves are long, compound, and opposite. The wood of the white ash is preferred for usage in sports equipment and furniture because of its lightness, strength, and flexibility.

25) Black Ash (*Fraxinus nigra*)
The black ash is different from the white ash in that its leaves are narrower and do not have stalks. It prefers a wetter soil, and its bark is darker brown. The wood is well suited to basketry because of the separation of the growth rings – allowing the layers to be removed fairly easily.

26) Tag Alder (*Alnus rugosa*)
Tag alders are a member of the birch

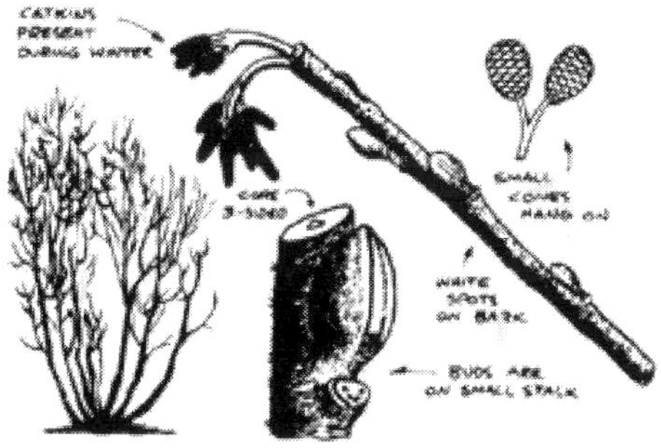

family and common to the Upper Peninsula. They grow in bottomlands, next to streams. The cones often stay on the twigs through the winter.

Tyoga is located on North Point Road, 2 miles north of M-28, just east of Laughing Whitefish River.

APPENDIX
Chapter V
Item 1

CHARLES HULSE
An archaeological evaluation of Rock Kilns and Finn Hollow
Michigan State University, 1977

"The survey of Finn Hollow and Rock Kilns (Gumerman, Hulse, and Polk 1977) and the later test excavations of portions of the site area indicate that the site is extremely valuable from both an archaeological and an interpretive standpoint. Despite the damage done by unthinking bottle hunters or amateur archaeologists, the site remains well preserved and is stratigraphically intact. Test excavations in portions of Rock Kilns indicate a deposition of cultural material which is rich in artifactual material and was virtually undisturbed for the last 80 years.

This site is historically and archaeologically valuable for many reasons. First, the Rock Kilns represents an industrially-oriented settlement which played a major role in the development of the Upper Peninsula of Michigan. Rock Kilns, however, is the only major charcoal kilns site in Michigan on Federally owned lands and forms a line in the chain of historic industrial sites in the Great Lakes area. Combined with interrelated studies of industries such as logging, mining, and transportation, a broad picture of development of Michigan can be gained through the preservation and study of Finn Hollow and Rock Kilns.

Another extremely important aspect of the site is its cultural or ethnic composition. The majority of the Upper Peninsula was settled by immigrants of Europe and Canada. However, few of the original settlements survive today. Rock Kilns and particularly Finn Hollow represent one of the few remaining sites where a pure cultural affiliation can be archaeologically examined. Finn Hollow, in particular, is an area where one cultural group settled, leaving behind in the soil traces of their beliefs, struggles and life ways. This is the only known site of this type in the Great Lakes area.

The vast amount of historical and documentary evidence on Rock Kilns is another reason for the site's historical and archaeological potential. With such valuable historical sources in combination with archaeological investigation, a total picture of the site and its role in the history and development of the Upper Peninsula can be obtained. Only recently have administrators and scholars recognized the potential for study and interpretation of such early industrial sites. For years similar sites in the eastern United States have enjoyed wide popularity and academic interest. Michigan's site, though presently neglected, comprises a resource equal to or surpassing those found in other parts of the country. Therefore, for many reasons, the sites in Michigan, and particularly Finn Hollow and Rock Kilns, should be protected and preserved for the future."

APPENDIX
Chapter V
Item 2

DAN O'ROURKE
Rock Kilns a case study of Michigan's Charcoal Industry
Michigan Tech University, 1997

The Rock Kilns site deteriorates a little more every year. The main agents of decay are illegal collectors and the environment. Collecting eliminates this information by removing artifacts from sites and destroying the archaeological record. Archaeological resources are nonrenewable.

The 1977 survey noted more than 100 recent pot holes from looters. The 1996 survey also found numerous recent looter pits. A partial remedy is to place signs at the site warning that unauthorized digging is illegal as many collectors are ignorant of this fact. The best solution is education making people aware of the ramifications of their actions through signs placed at the site or by public meetings held by Forest Service officials about collecting's adverse effects. Community involvement is essential to site preservation. There is no justification for destroying a site's potential just to get trinkets.

While collecting is destroying the archaeological record the environment is destroying the kilns. Freeze/thaw cycles and vegetation are damaging the masonry. Roots are slowly tearing apart the remaining structures. The erosion of mortar between stones presents the greatest threat to the kilns. Preserving the top of the remaining walls is crucial to prevent water entering the wall. Excavations showed that rubble from the upper wall is reinforcing the lower walls, and that the lower walls are in minimal danger of deterioration.

The first step should be removing all vegetation from the kilns with stabilization efforts to follow. This may weaken the wall further. The interior walls of the kilns contain a layer of waste product (creosote) from kiln operations which helps preserve the wall. I also recommend erecting a wooden support structure bracing the kilns interior wall to counteract the stress of gravity.

Consultation with a specialist in masonry restoration should accompany any preservation effort. Various laws and guidelines govern such undertakings, such as the Secretary of the Interior's guidelines for restoration. A professional trained in preservation may offer additional stabilization recommendations. State facilities and agencies such as the Michigan Historical Center's parks at Fayette and Fort Wilkins State Parks have restored kilns.

The complete document, *Rock Kilns, A Case Study of Michigan's Charcoal Industry*, is an excellent recounting of the charcoal and iron furnace industry of the last half of the 19th century.

APPENDIX
Chapter V
Item 3

Interview with Eino Nummela by Bea Anderson
August 16, 2006

Eino first saw Silver Lake at age six or seven which was about 1924. His dad and a friend went fishing in the lake. Eino was left on land so he explored all the buildings. All the doors were open and he could see some signs that looting had taken place. It looked like things were missing that should have been there.

Eino said there was about a five-mile clearing on the North Shore of the lake where the buildings were. The lodge and cabin area had a lovely lawn, many wildflowers, and lots of forget-me-nots.

The horse barn was at the extreme east end of the lake, almost where Silver Creek runs out. The lodge still had some furniture but it had been looted too. From there he saw two or three cabins and they were also looted.

Eino continued west and found the boat house at the northwest end of the lake. It was built over a slough so the boats could be moved by water directly to the lake. There was no wall on the south end of the boathouse. This open end provided easy access to the lake. It could only hold two small boats with motors. The shoreline was somewhat swampy but where the buildings were there was a hard bottom. He does not recall any sandy beach but many people swam in the lake so he didn't think the bottom was "mushy".

Eino does not remember a cattle barn and never saw buildings on the south rim of the canyon wall. A pipe was installed from the base of the falls one mile through Silver Lake to provide water for the lodge and other buildings. This pipe was placed some distance under the water to prevent freezing. He stated there was some type of sluice built that diverted the water flow to the spring below. This apparently created enough pressure to get the water to the lodge and other buildings. He also said the Penegor Lumber Co. built a platform halfway up the rim of the falls and created extra pressure. He doesn't remember exactly when that occurred.

After Nate Kaufman suddenly abandoned the lodge, etc., the Penegor Lumber Co. built a camp on the west end near the falls. Matt Pantti worked for the Penegors and got the pipes from the lake routed to the Penegor camp, men's camp, cooking camp, etc. Eino said he and other locals came into the area from the south side but the road was very poor. He used a model T at this time.

Eino said Nate had a road built from a little west of Onota Hill to his property. It was very overgrown when Eino was young but Penegor with his logging made a really good road and it was used for many years. Last time Eino tried to find the road it was very overgrown. In a previous time it had a steel gate with a large concrete foundation. The last time Eino looked for it he couldn't determine the exact site.

When Nate and family abruptly left with everything intact, he had not arranged for any caretakers so things disappeared overnight.

Matt Wittenberg and Mr. Spence both bought the lumber from the lodge and they built houses for themselves. Mr. Wittenberg built his house on the east end of the main road (M-67) in Chatham; and Mr. Spence built his house about 1 mile south on M-67 on the south side of the Bohemian Creek. Tom Niemi bought the horse barn and built a cattle barn at his farm.

Eino's family lived in Rumley from 1919 through 2000. Eino had hunted and fished in this area for about 50 years. Last time he was deer hunting near Silver Lake was about 1964 but all was so overgrown he couldn't find any foundations, etc. He was with his longtime friend, Frank Ritola. They found so many beaver dams and flooding at the east end by the Silver Creek that it was almost impossible to reach the site.

Eino said that many huge pike were caught in Silver Lake for many years. In the last number of years it is rumored that fine brook trout have been caught. Could they be offspring of the brook trout carried in a large washtub from the Slapneck Creek by Nate Kaufman? In a diary entry October 19, 1892 "Moccasin" Bill Oliver states that Nate Kaufman brought in 12 very nice brook trout and planted them in Silver Lake.

Eino W. Nummela was born on June 24th of 1916. He died at the Dominic Jaccobetti Home for Veterans on March 9, 2008.

"Moccasin" Bill Oliver made daily entries in his diary between 1874 and 1931.

APPENDIX
Chapter V
Item 4

Bea Anderson's Early Life in Alger County

My husband's family, the Parkkilas, were from Haapavesi, Finland. They left there to escape the Communist Bolsheviki. My father's name was Akseli Antinpoika Parkkila. His American name was Axel. He was ten years old when they landed at Laughing Fish Point in the winter of 1901. They walked south to the Halfway House near the north bridge crossing the Laughing Whitefish River. Here they spent the night. The next day they followed the river. When they found a suitable location, they built a log house and homesteaded there for several years. Part of the log house is still standing. The forest is gradually reclaiming it. The walls have been gradually caving in. I visited there in the late 1990s and took a few pictures of what remains.

When I was very young my mother told me of when she was a little girl. Her dad took her with him to the railroad depot in the town of Rock River where they picked up their year's supply of flour, sugar and coffee. They would order it from a store in Marquette and they would ship it to Rock River by train.

All of the people living in the area provided food for themselves during the growing seasons. They put up canned goods and come fall they packed the root cellars full of potatoes, carrots and rutabagas. They made sauerkraut from cabbages and stored it in large crocks. They raised their own chickens for both eggs and meat. Some raised cows, pigs and geese. Some traded goods with each other to make up the difference so they could all survive the winters. There were always opportunities to hunt deer, fish, and trap for furs also. The people were used to providing for themselves and helping out each other.

I was eight years old when I first saw the Rock Kilns. The year was 1932 and it had to be in middle to late August. I remember this clearly because my mother was pregnant with my sister, Sarah, who was born October 12th of that year. My dad knew the black berries were ripe so we made the trip to Rock River to do the picking. That was when I first saw the kilns. I was enthralled. They were still intact and looking up through the tops of them was a wonder to the eyes. Altogether we picked twenty-four quarts of berries, some from inside the kilns and the rest from around them.

My dad never told me anything about the kilns and I'm sure I never asked. I didn't realize what had taken place there some thirty years before. There was very little evidence of a town, or of a huge business related to logging and making charcoal for blast furnaces. We had been there that day for the serious business of only picking berries.

Years later I told my husband and our three boys about the experience. We stopped there numerous times and searched for the kilns but only found a few rocks, nothing like what I remembered from my childhood. We searched the south side of the Onota-AuTrain Road and couldn't see the kilns. What a disappointment! Finally I said to my dear husband, "I'm not looking south anymore." It took me a while to realize that the road we had been on (Route 28) hadn't been there when I was young. We turned around. The railroad grade was raised up higher than the road and it was to the north. We walked over to it and began looking northwest and that's when we saw the kilns. Oh, what a joy!

Now I would like to say to all you explorers of any age: Landmarks change. And so do the names of little towns. Don't give up. Keep looking! Find some old-timers still living in the area. Tell them what you're looking for. Ask them for their stories. Most of them would enjoy telling you what they know.

~ ~ ~ ~ ~

The following are accounts by various local people who knew and remembered some of these events. My mother, Sophia Johnson Parkkila, used to tell me about the teens in the area going dancing on an outdoor porch at the Silver Lake clubhouse. She described it as a porch with white railing which was over some part of the clubhouse below. My mom was born in 1901 so I imagine this was somewhere in the era of 1914 to 1920. To my regret I never asked more questions. In fact I think she said it was burned.

Courtesy Bea Anderson

This picture of the Parkkila Homestead taken in 1965 shows Elmer Watland inspecting the remains of the first home of the Parkkila family which was built in 1896 at a location south of Deerton on the Laughing Whitefish River.

~ ~ ~ ~ ~

My friend Marti Lappanen and I were responsible for maintaining the North Country Trail from the Rock River Road to west of the Rock River canyon just past the Laughing Whitefish canyon and north of its falls. We covered this area many times per year. On one of these long days we took a shortcut and to my amazement we actually saw this clubhouse overlooking the southern shore of Silver Lake almost exactly as described long ago. I took my brother and one of my cousins to see this because they had often heard a similar story. I've not been able to find anyone who knows if this building was built by Nate Kaufman or was this thing done by locals using lumber from the lodge, etc. Several local people knew this building.

~ ~ ~ ~ ~

Recently another old-timer told me that the camp I described was built in the early 50's by Maki, Henry Enyyard and other local men. He also said that when the Silver Lake property came into new hands, the locals burned down the cabin. So now I know that was not the camp my ancestors danced in and that there was a similar building way back in the early 1900's and this may have been a replica? Don Hakkola took me to this site and yes, it seemed like the same spot. He had some pictures (not too clear) of huge caves in this wall of the Rock River canyon. In fact some of the caves were so large that they were used as hunting camps for many years. Some hunters just put wood doors over the large openings. There was even a small access from one cave to another. To get into some, a step ladder was needed.

~ ~ ~ ~ ~

Scott Hosteller told me the Penegor Lumber Co. was by what we call the "pipe falls". He build a platform at the base of the falls so water would run over it and keep the food cool.

~ ~ ~ ~ ~

Now for some old-time rumors:

Number one, there was a narrow gauge Railway part of the way to Silver Lake. This much is true. Story is that a train robbery occurred. The train crew hung the robber but never got the money. Rumor was the coinage was dumped below the "pipe falls". And number two, men attending at the lodge used to taunt each other as to who could throw a silver dollar across the lake – apparently all are still in the lake.

Comments from Bea Anderson related to Silver Lake

Number one: One time I called a Penegor. I can't remember his first name. He lived down in Kentucky then. He told me lots of stuff but one thing really stuck in my head. He said that when he was ten years old he drove a logging truck from the lumber camp in the woods all the way to the railway siding with a full load of logs to boot. He even unloaded them. Who'd trust a kid with a job like that nowadays?

Number two: Long ago, Toivo, a very good and old friend of mine, was working near Silver Lake. There were a lot of mushy places there abouts. One time when he was using his horse to pull logs, they had to go through a lot of mush. The horse got himself good and stuck so Toivo had to unharness him, and then he tried and tried to coax the horse out. The horse couldn't get free, and the more he tried the more stuck he got. Finally Toivo went to where some other loggers were working and told them about his horse. They came and they too tried to help to get the horse out. Finally they were all pretty convinced it couldn't be done. That's when Toivo went and got his gun. Now when that horse saw the gun, or maybe got a whiff of it, who knows; but that's when he tried real hard. He struggled and jerked about and then somehow he got one leg out, and then he got another out. He jerked and jerked, until pretty soon he came all the way up and out of that mush, and then he snorted just like he was gonna say, 'Put the gun away.'

Number three: Another thing: I didn't like it when the new maps come out and you find out that someone changed the name of a lake or something. Now that's disgusting. Anyway that's the way I felt until one time when the name of Johnston Lake was changed. And I learned something else. It was Charlie Johnston who brought the idea of the gin pole to the area. It seems no one had seen one before. Although this concept had many applications elsewhere like raising the mast of the ship. He rigged one up that had a block and tackle and he used it for lifting logs. It was slick as could be. He showed the others how to use it. It made loading and unloading logs so much easier. That's the reason they changed the name of Johnston Lake to Ginpole Lake. Everybody was calling him "Ginpole Charlie" so they all knew who it was named for anyway.

APPENDIX
Chapter VI
Item 1

Congressional Act of 1856 ceding lands to State of Michigan and subsequently to the Chicago and Northwest Railway*

By an act of Congress approved June 3, 1856…certain lands were granted to the State of Michigan to aide in the construction of a railroad from Marquette to the Wisconsin state line near the mouth of the Menominee River. …The State of Michigan by an act of its legislature approved February 14, 1857, accepted said grant afterwards by an act of the legislature approved March 18, 1860, confirmed the said grant of lands as conferred by the railroad…to the Peninsular Railroad Company of Michigan, now the Chicago & Northwestern Railway Company by acts of consolidation dated October 21, 1864.

These maps of railroad lines of the U.P. over 143 years show our expansion and then construction of rail lines built to support the mining of copper and iron ore, along with the logging industry. By 1910 the copper rush of the Keewenaw had ended. Trucks replaced most of the rail lines for the timber industry. Rail and freighters continued to transport iron ore.

This transaction is a part of the history of many deeds in Onota Township.

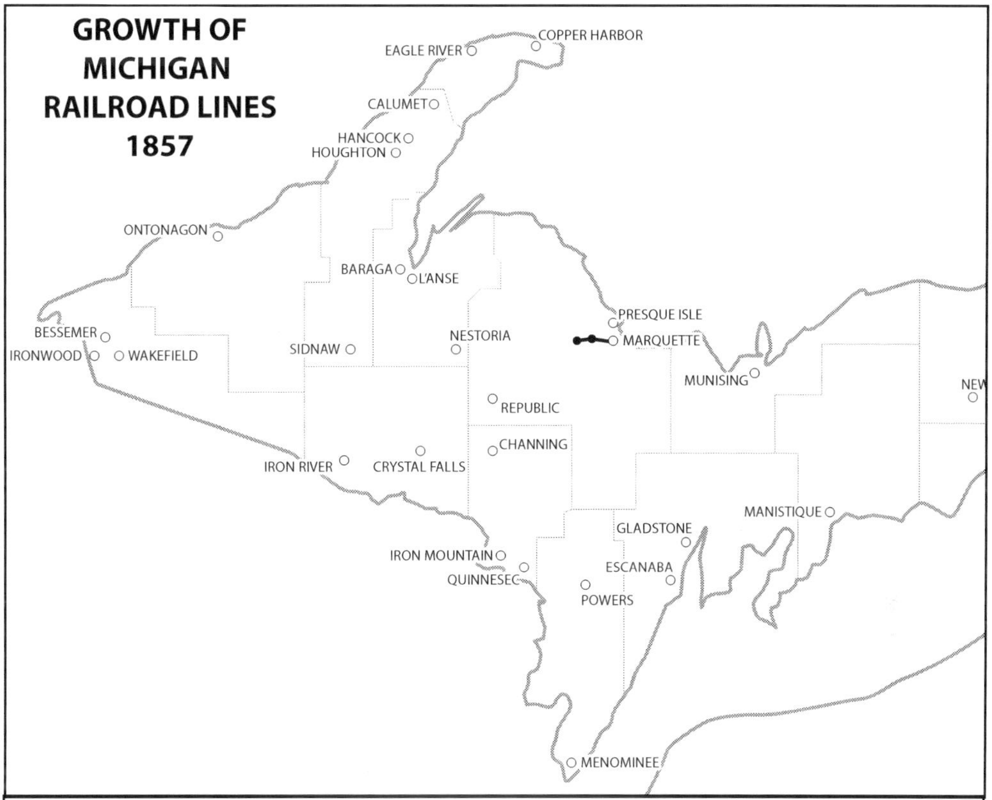

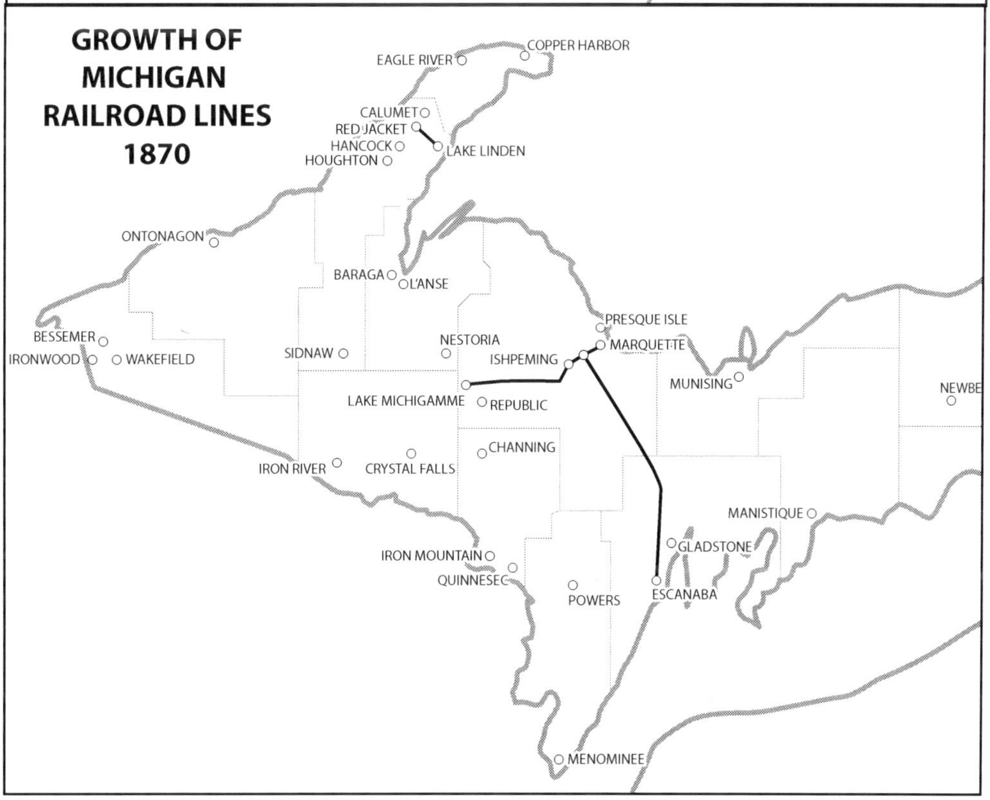

Information from http://www.michiganrailroads.com

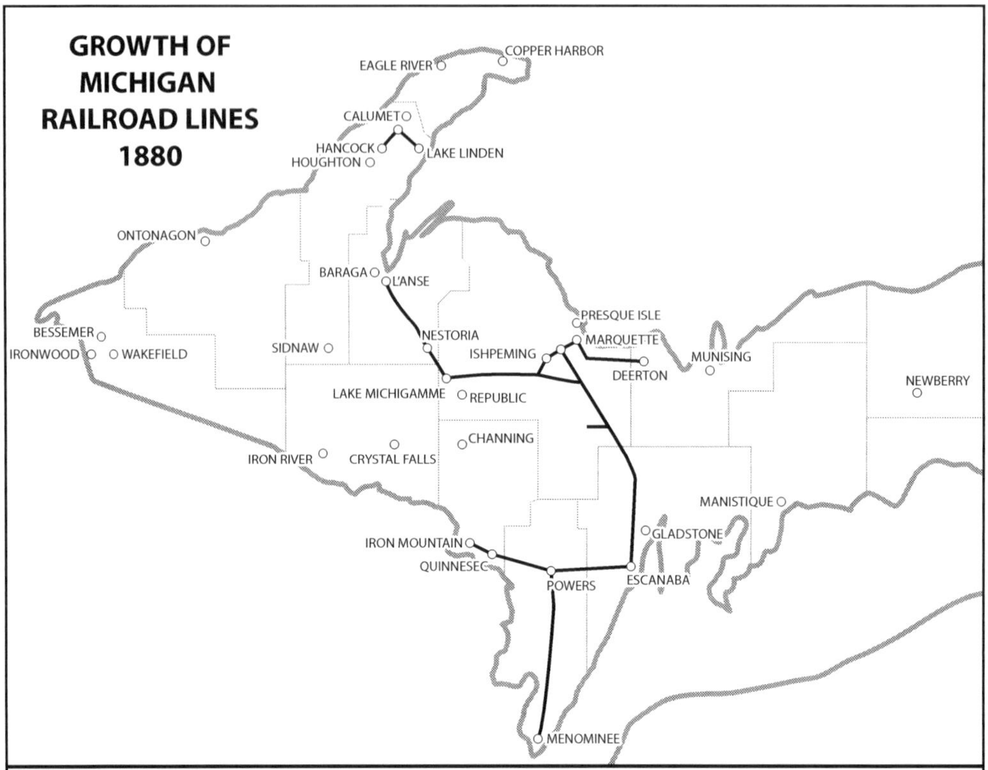

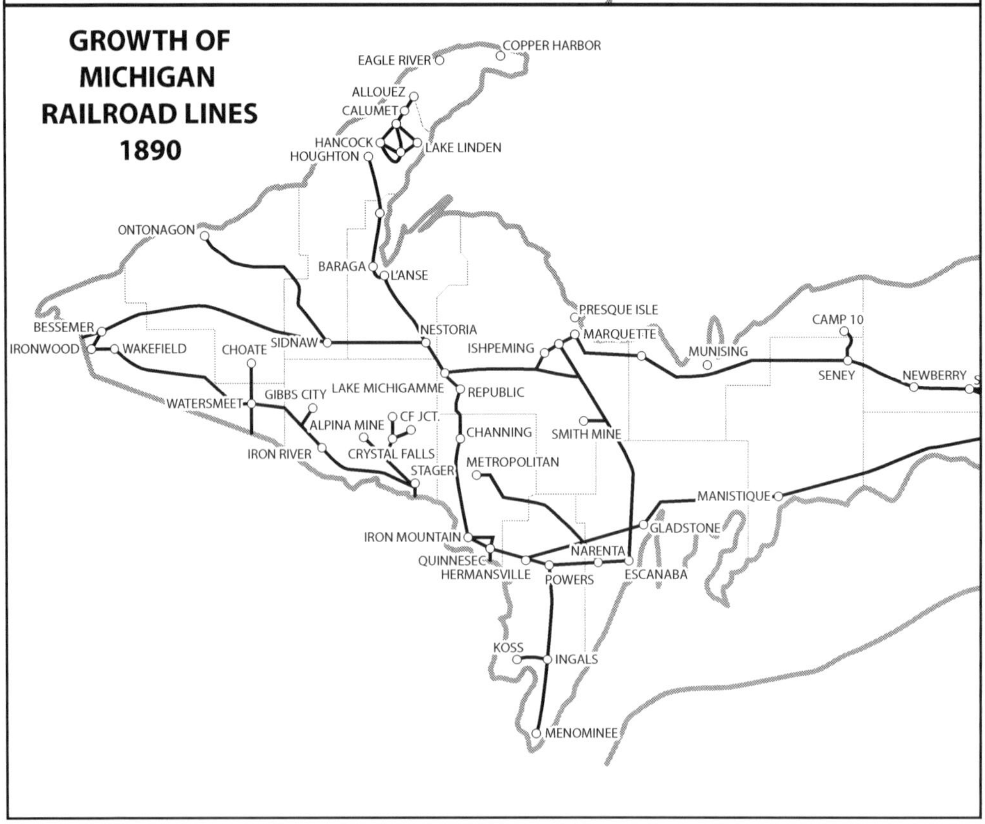

Information from http://www.michiganrailroads.com

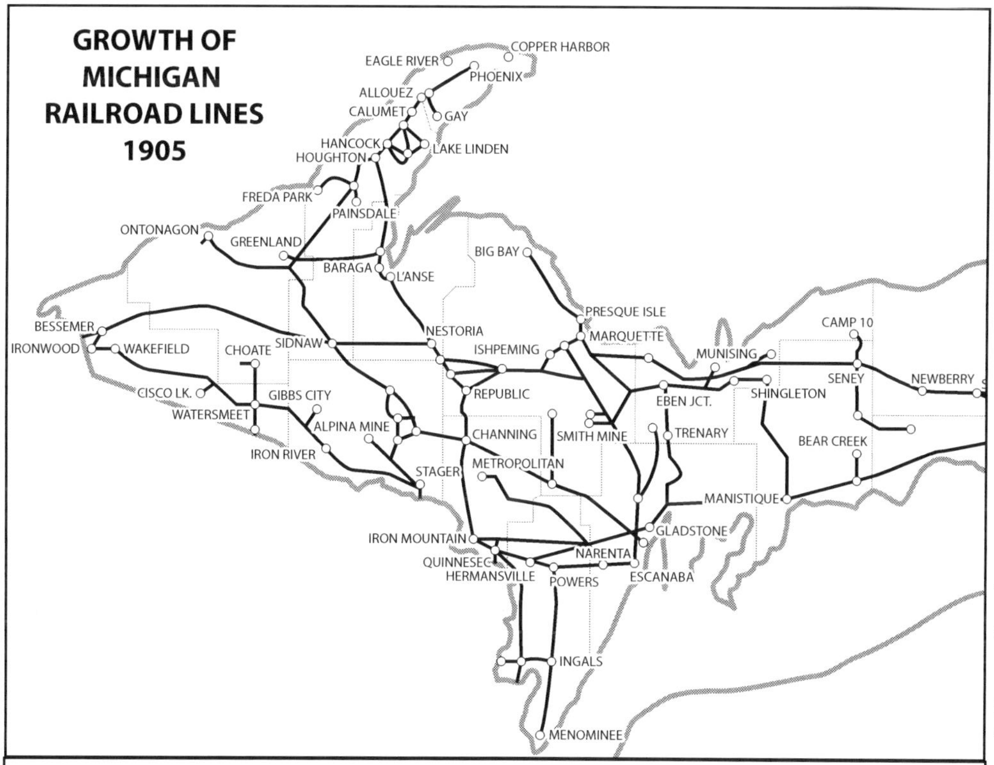

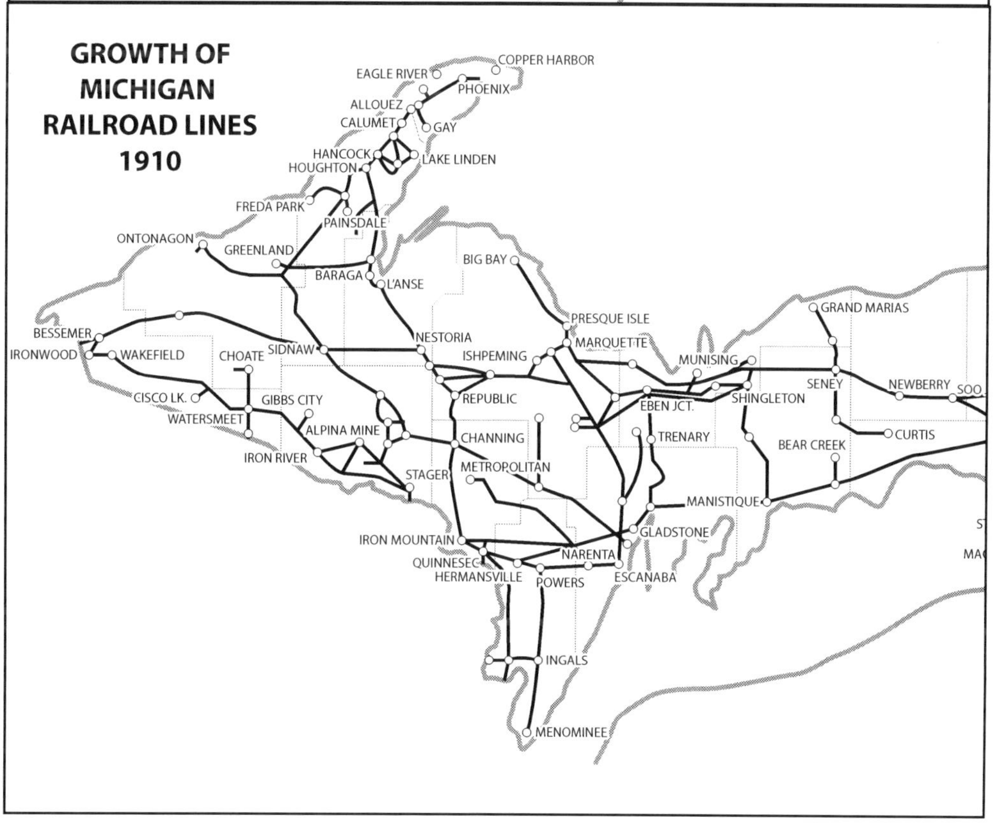

Information from http://www.michiganrailroads.com

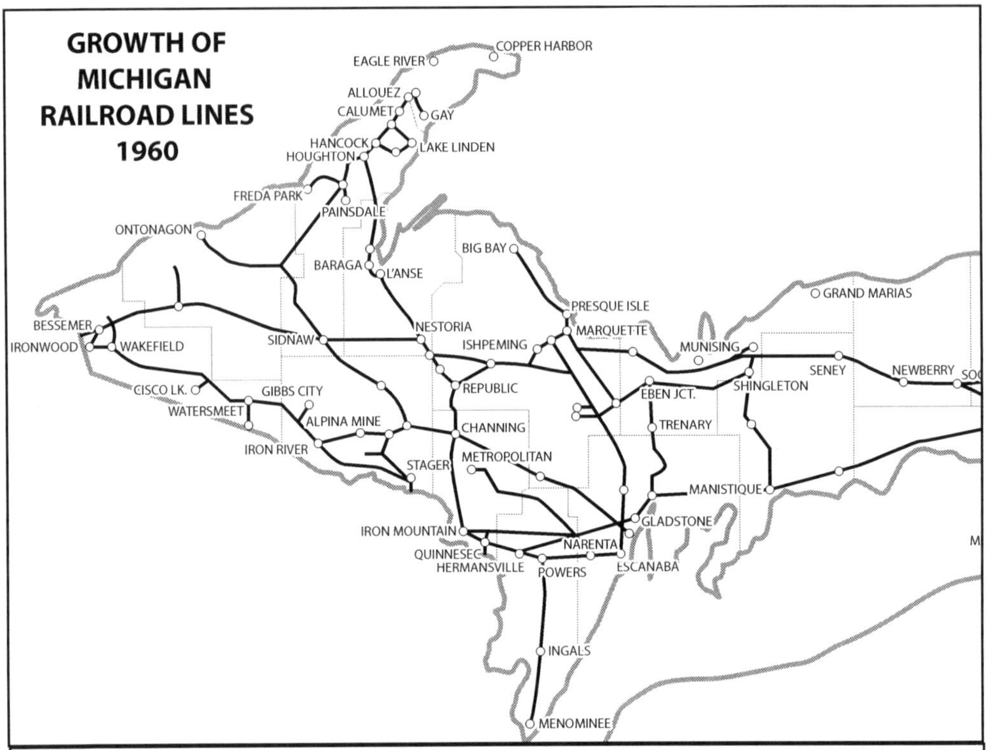

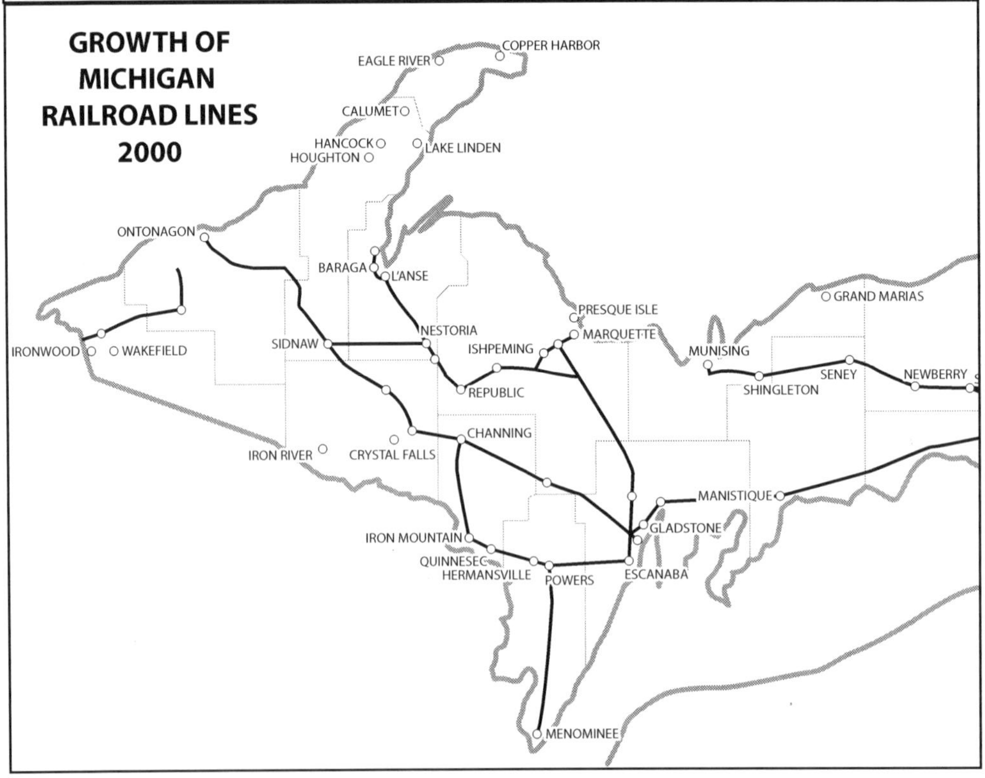

Information from http://www.michiganrailroads.com

APPENDIX
Chapter VI
Item 2

Letter from George Wagner to Michigan Governor J. J. Bagley*

Marquette, June 11, 1873

Geo. Wagner to Governor J. J. Bagley

Dear Sir:

At my last visit at Lansing about six weeks ago you promised me to recommend the completion of our contract on the S. St. M. & G. I. State Rd. extension, but I am now informed by Mr. Selden that nothing was done at that meeting & that at the last meeting of the Board of Control, you agreed to let us finnish our contract if we would build the /road from Chocolay to my Mill (near Tyoga) for $1,000 per mile from there at the rate of our contract.

To this your Honor: I am sorry to say that I have no more to lose and therefore be obliged to give up my share of the contract.

Mr. Wetmore informs me that some one from here reported to be willing to built the above section of road for one section Swamp Land per mile. If this is true, you may tell him that he may have my share in that contract with the greatest pleasure.

When I entered in the contract to built this road last fall it was not my intention to make a fortune out of the State, but only to get the road built as soon & as good as possible for it was so badly needed.

Showing the necessity of this road I will only state my own case:

On the first of May the provisions which I had laid in last fall at my Mill gave out, & the only way I could get anything down to save the people there from starvation, which consisted of twelve men, three women, & eight children, was by carring supplies to them by the Native Americans-, around the shore of Lake Superior. Two weeks since that time the Mill had to stop for the men was not able to work, for the want of fuel. Today the 11th day of June the ice is so close a shore that it is impossible for me to get a small boat through with supplies and the steamers do not stop there.

Under the present state of things I am obliged to lay up my mill as soon as my present logs are sawed for I have no Company to back me nor any means to built a road to lead me to the outside world.

---

*Alger County Historical Society archives

APPENDIX
Chapter VI
Item 3

WESLEY PERRON RAILROAD COLLECTION*

The Daily Mining Journal
Marquette, MI
September 25, 1905, Monday

DSS&A Ry.

Excursion Rates for Fishermen and Campers

Commencing Saturday, May 13, 1905, and continuing until further notice, the DSS&A will sell excursion tickets to the following point at the very low rates named:

	Mileage from Soo	Marquette	Negaunee/Ishpeming
Gordon		.50	1.00
Sand River	140	.50	1.00
Deerton	136	.75	1.25
Onota	134	.75	1.25
Rock River	129	.75	1.25
Au Train	125	.75	1.25
Munising Junction	113	.75	1.25
Wetmore	110	.75	1.25
Shingleton	105	1.00	1.50
Creighton	95	1.25	1.75

The Peter White Library in Marquette houses the Wesley E. Perron Railroad collection, an extensive chronological collection of articles from news media about railroads throughout the U.P. This has proved to be an excellent source for some of the information about the Tyoga Lumber Company.

---

*Peter White Library, Marquette

APPENDIX
Chapter VI
Item 4

WESLEY PERRON RAILROAD COLLECTION
Peter White Library

Michigan Railroad Commission, 1907

Duluth, South Shore & Atlantic Ry.

Road owned
In Mi. – 470.62 miles

Road operated
In Mi. – 474.16 miles

Cost of property - $46,342,418.12

Revenue from operation -	$ 3,161,052.56
Operating expenses -	2,462.033.44
Net deficit -	179,725.93

Traffic earnings passenger -	$ 1,173,805.32
" " freight -	1,961,999.32

Equipment estimated value - $ 1,380,113.00

Locomotives -	73
Passenger -	45
Bags & Exp -	17
Box -	1,266
Platform -	666
Ore -	856
Way cars -	36
Others -	96
Total cars & engines -	3,055

Stations
In Mi. – 45

Employees
In Mich. – 1,723

Baggagemen -	17
Brakemen -	109
Conductors -	62
Engineers -	73
Firemen -	72
Laborers -	485
Shopmen -	348
Yardmen -	35
Others -	522

Total tons moved one mile - 202,766,915

Bridges and trestles – 226
Grade crossings (at, over or under) – 45
Fencing – 599.25 miles
No. of ties put in in 1906 – 161,314

Accidents – 40
Total killed – 17
" injured – 36

Average daily compensation (inc. officers [those days are long gone]) – $2.09

APPENDIX
Chapter VII
Item 1

CCI STATISTICS*

Personnel and output between the spring purchase of the Tyoga Lumber Company by C.C.I. and the fall closure in 1907.

	April	May	June	July	August	Sept.
Foreman	1	1	1	2	2	1
Carpenter	5	4	2	1	2	1
Cook & Cookees						
Laborers	20	10	15	20	10	7
Bark Peelers						
Cedar Makers				3		
Inspector						
Bladesmith	1	1	1	1	1	
Night watchman	1	1	1	1		
Clerk	1	1	1	1	1	
Bookkeeper	1	1	1	1	1	1
Total Men	27	25	22	30	15	9
Horses		8	5	2	5	2
Total Cars Shipped/Month	10	5	14	73	41	7
Total Posts Cut			200			
Total Ties Cut			21			
Poles Cut			24			
Logs Skidded		150	50			

Shipped by Railroads

DSSA B&O
C&NW M&SE
M Ry
C&O

*Source: CCI Annual Report, fall 1907

APPENDIX
Chapter VII
Item 2

Tyoga Lumber Co to CCI

Liber 12 p 523
Alger County Courthouse – 3-25-05

The following people have their signatures on the deed:

L. L. Bailey	Carrie J. Bailey
Geo. M. Spalding	Florence W. Spalding
Chas H. Roberts	Mary S. Roberts
J. H. Gardner	Mary N. Gardner
Thos. M. Archer	Mary G. Archer
Frank Hammund	Amy E. Hammund

APPENDIX
Chapter VIII

ABBREVIATIONS
of sources for maps of Lake Superior and the Upper Peninsula

ACHS	Alger County Historical Society
CCI	Cleveland Cliffs Iron
DNR	Department of Natural Resources
DSS&A	Duluth, South Shore, and Atlantic
LS	Lake Superior
LSWP	Lake Superior Watershed Project
LWR	Laughing Whitefish River
MCHS	Marquette County Historical Society
MM&D	Marquette, Mackinac and Detroit Railroad
MSU	Michigan State University
NMU	Northern Michigan University
PWL	Peter White Library, Marquette
SV	Superior View, Marquette
THP	Tyoga Historical Pathway
TLC	Tyoga Lumber Company
TNC	The Nature Conservancy
U.P.	Upper Peninsula of Michigan
WFL	Wisconsin French Library
WHS	Wisconsin Historical Society

APPENDIX
Chapter VIII
Item 1

Excerpt from *The Pioneer Woodsman*
as Related to Lumbering in the Northwest
George Henry Warren - Minneapolis, 1914

Surveying and Selecting Government Timber Lands

The Continental Congress, through its committee appointed expressly for the work, inaugurated the present system of survey of the public lands in 1784. The system consists of parallel lines six miles apart running north and south, designated as "range lines"; also of other parallel lines, six miles apart running east and west, designated as "township lines". Any six miles square bounded by four of these lines constitutes a "township". The territory within these two range lines and two township lines is subdivided into "sections", each one mile square, by running five parallel lines north and south across the township, each one mile from its nearest parallel line, and, in like manner, by running five other parallel lines east and west across the township from the east range line to the west range line, each line one mile from its nearest parallel line. In this manner, the township is subdivided into thirty-six sections each one mile square. The four township corners are marked by posts, squared at the upper end, and marked on the four sides by the proper letters and figures cut into the four flat faces by "marking irons", each flat surface facing the township for which it is marked. (See page 145)

Midway between adjacent section corners is located a "quarter corner", on the line between the two adjacent sections. This is marked by a post blazed flat on opposite sides and marked "% S". There are also two "witness trees" or bearing trees marked ^ S. In addition, one tree in each of the four township corners is blazed (a smooth surface exposed by copping through the bark into the wood) on the side of the tree facing the stake, and the same letters and figures as are on the nearest face of the stake are marked thereon. These letters and figures give the number of the township, range and section touching that corner. In another blaze below the first, and near the ground, are marked the letters "B T", meaning "bearing tree". The surveyor writes in his field book the kind and diameter of tree, the distance and direction of each bearing tree from the corner post, and these notes of the surveyor are recorded in the United States land office at Washington. Even if the stake and three of the bearing trees should be destroyed, so that but one tree be left, with a copy of the notes, one could relocate the township corner. The section corners within the township are marked in a similar manner.

APPENDIX
Chapter VIII
Item 2

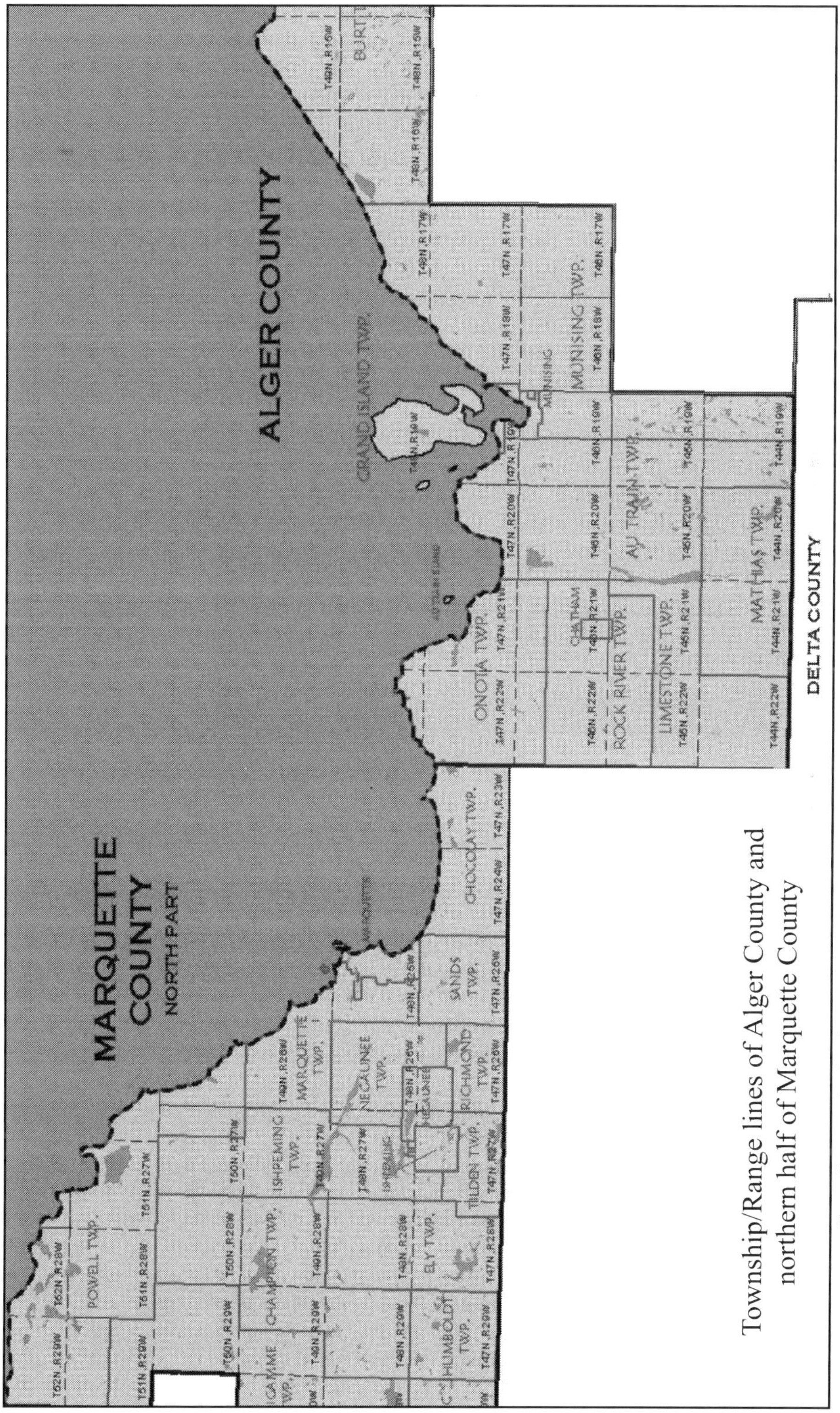

Township/Range lines of Alger County and northern half of Marquette County

Alger County plat book 2005

APPENDIX
Chapter VIII
Item 3

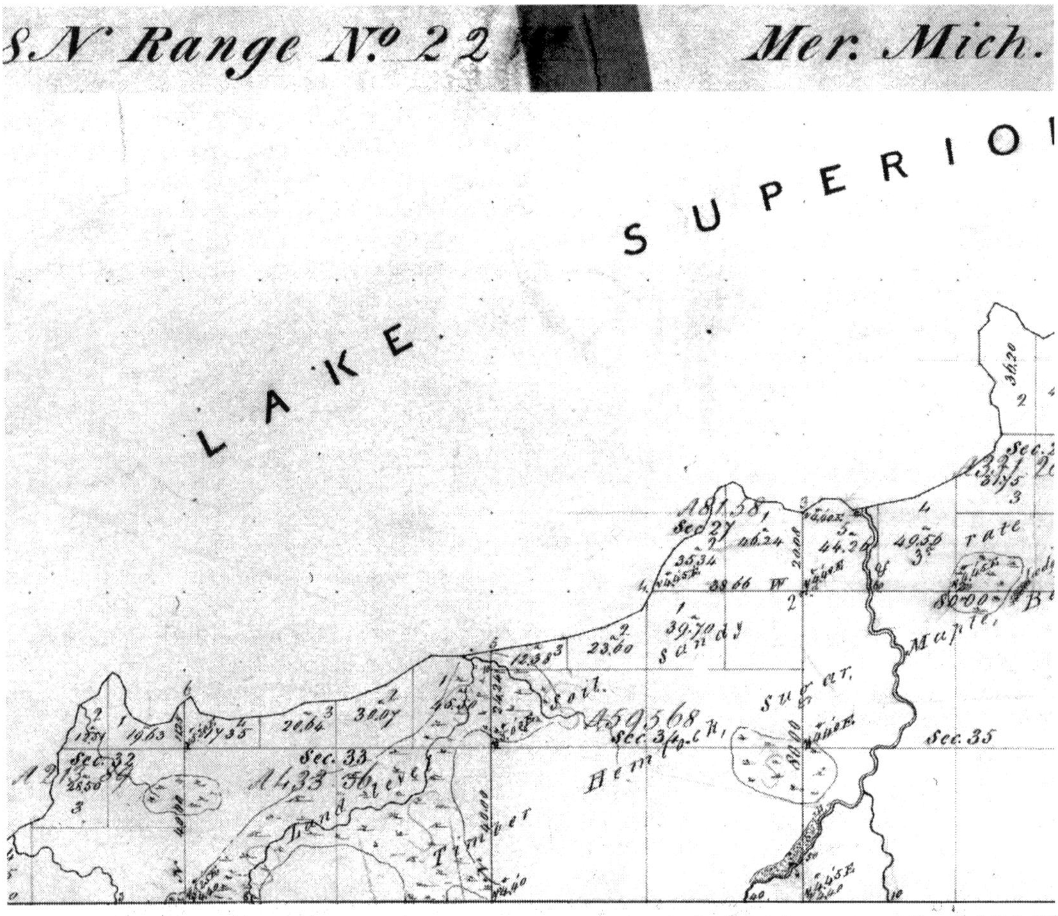

Total number of Acres 3.321.23

Surveys Designated.	By Whom Surveyed.	Date of Contract.	Amount of Surveys. M.	Chs	Lks	When Surveyed.	When C... the Sur...
Township lines.	Wm A. Burt	April 1st 1841	6	14	30	3d Quar 1841	
Subdivisions.	Alfred Millerd	Bond Dec. 18 1852	13	62	53	2nd Quar 1852	

Courtesy MSU Library

William A. Burt survey map of township 48 north and range 22 west, 1841
Tyoga settlement is located where map shows section 35.

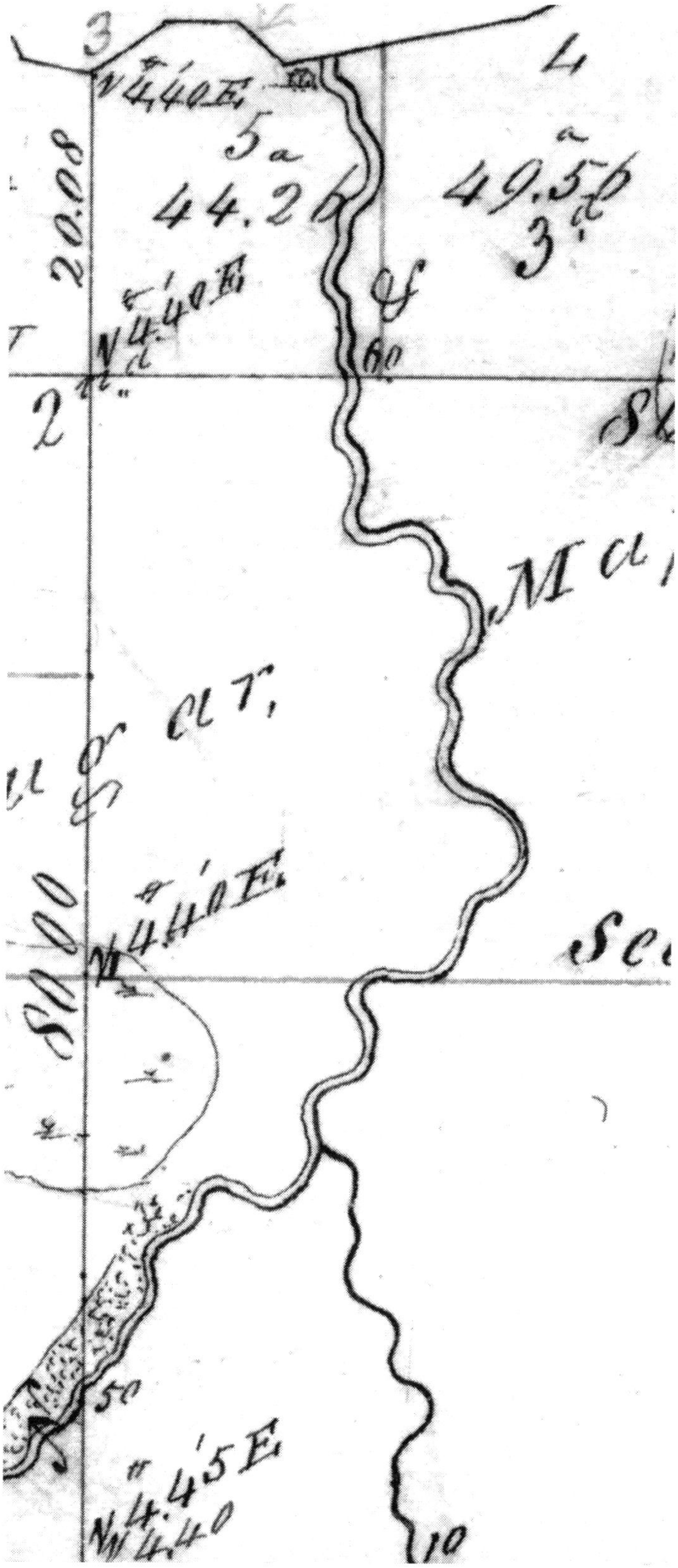

Courtesy MSU Library

This map is a magnified view of western half of Section 35, Burt 1841 surveyed feeder streams which joined the Laughing Whitefish where the southern Tyoga Historical Pathway bridge crosses Laughing Whitefish River.

APPENDIX
Chapter VIII
Item 4

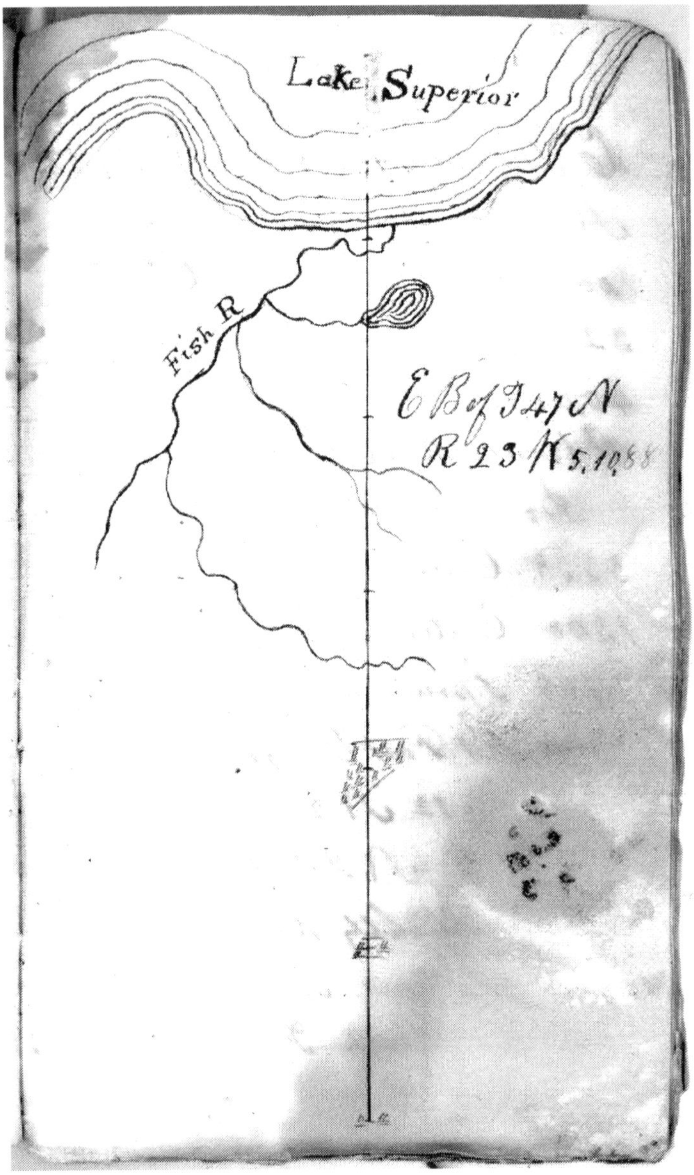

Courtesy J. M. Longyear Research Library

William A. Burt Field notebook drawing of the North/South line forming the east border of Township 47 North and Range 23 West. The Sand River labeled as Fish River

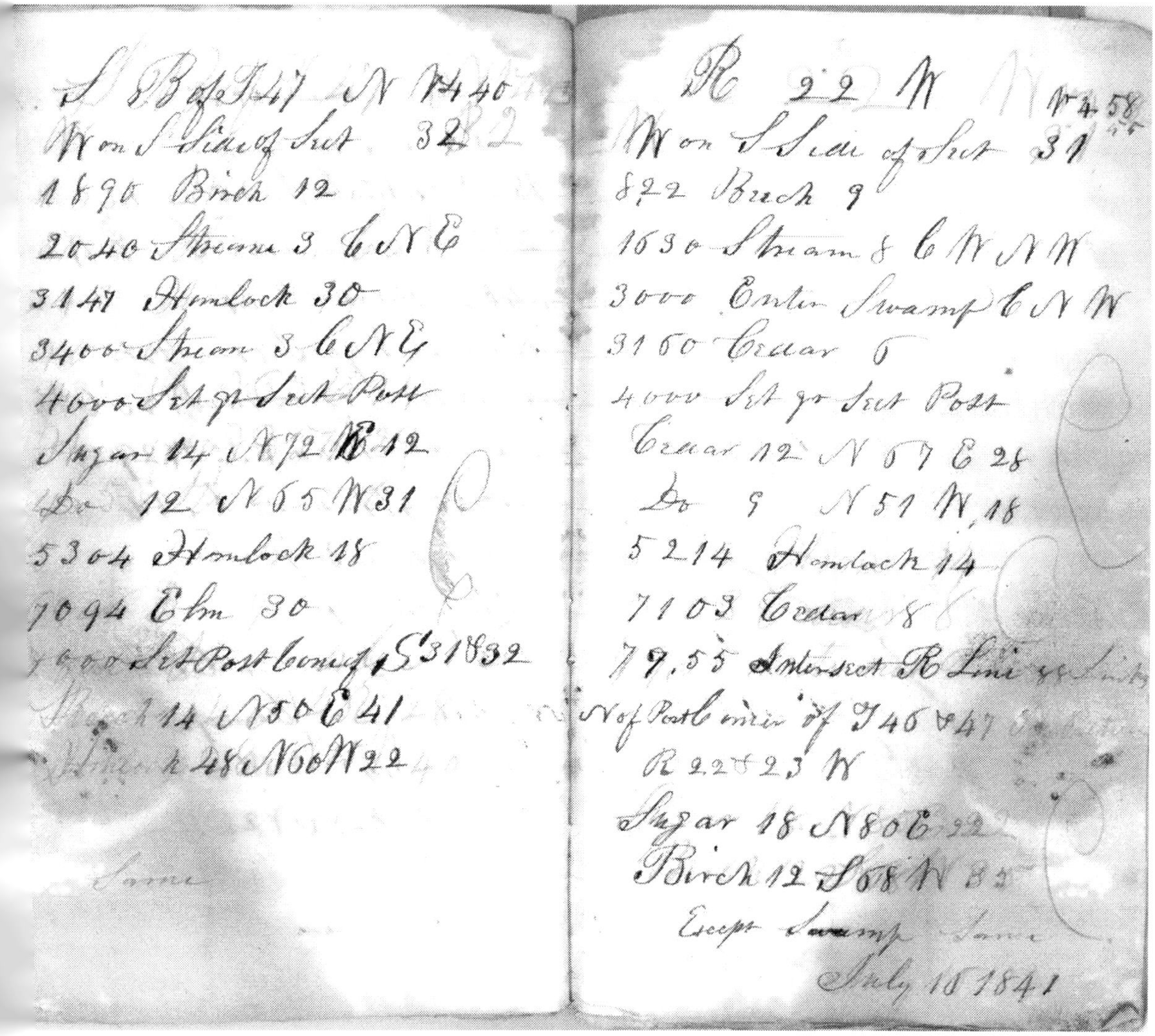

S Bet T 47 N R 40
W on S Side of Sect 32
1890 Birch 12
2040 Stream 3 C N E
3147 Hemlock 30
3400 Stream 3 C N E
4000 Set qr Sect Post
Sugar 14 N 72 E 12
Do 12 N 65 W 31
5304 Hemlock 18
7094 Elm 30
8000 Set Post Corner S 31 & 32
Beech 14 N 50 E 41
Hemlock 48 N 60 W 22
Same

R 22 W
W on S Side of Sect 31
822 Birch 9
1630 Stream 8 C N N W
3000 Enter Swamp C N W
3160 Cedar 6
4000 Set qr Sect Post
Cedar 12 N 67 E 28
Do 9 N 51 W 18
5214 Hemlock 14
7103 Cedar 8
79.55 Intersect R Line
N of Post corner of T 46 & 47
R 22 & 23 W
Sugar 18 N 80 E 22
Birch 12 S 08 W 85
Except Swamp Same
July 18 1841

Courtesy J. M. Longyear Research Library

The trees described in area on previous page include hemlock, cedar, spruce, beech, maple, tamarack, fir and elm.

APPENDIX
Chapter VIII
Item 5

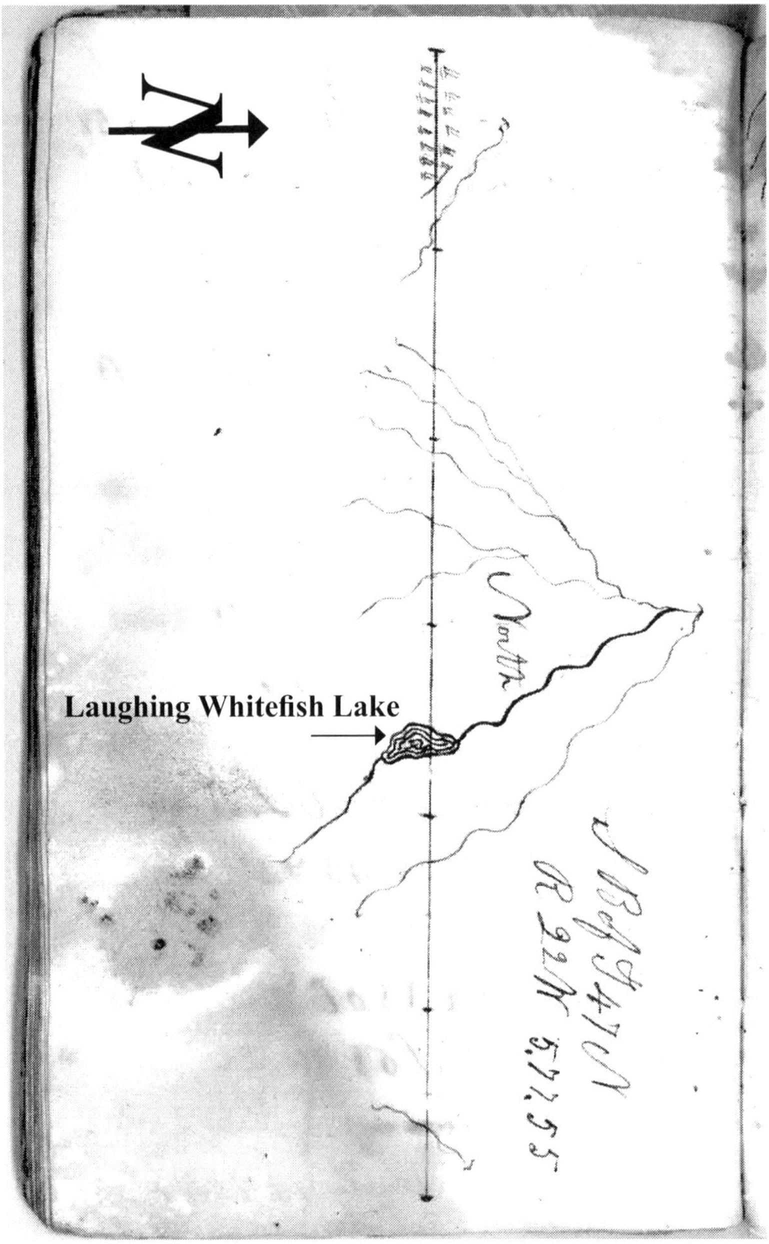

Burt's description of south boundary line of T47 North, Range 22 West
Tyoga settlement is located in northern part of T47-R22.

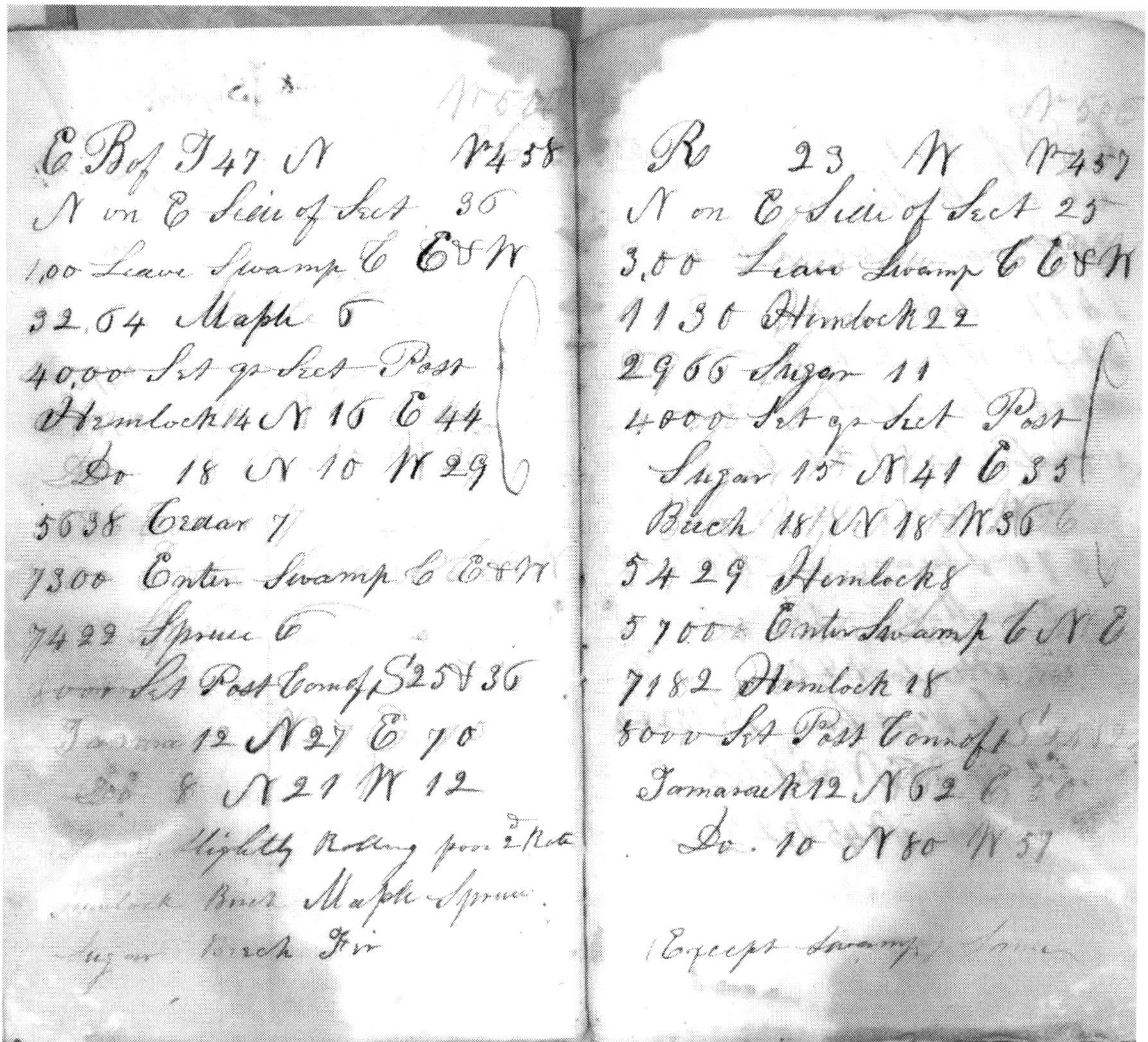

E B of T 47 N No 458
N on E Side of Sect 36
1,00 Leave Swamp E E&W
32.64 Maple 6
40,00 Set qr Sect Post
Hemlock 14 N 16 E 44
Do 18 N 10 W 29
56 38 Cedar 7
73 00 Enter Swamp E E&W
74 22 Spruce 6
80 00 Set Post Cor of S 25 & 36
Tamarac 12 N 27 E 70
Do 8 N 21 W 12
Slightly Rolling poor 2nd Rate
Hemlock Birch Maple Spruce
Sugar Birch Fir

R 23 W No 457
N on E Side of Sect 25
3,00 Leave Swamp E E&W
11 30 Hemlock 22
29 66 Sugar 11
40 00 Set qr Sect Post
Sugar 15 N 41 E 35
Beech 18 N 18 W 56
54 29 Hemlock 8
57 00 Enter Swamp E N E
71 82 Hemlock 18
80 00 Set Post Cor of S
Tamarack 12 N 62 E 30
Do 10 N 80 W 57
(Except Swamp) Same

Burt's description of trees found in this area of the east border of T47N R23W
These are similar to the trees in T47N, R22W.

APPENDIX
Chapter VIII
Item 6

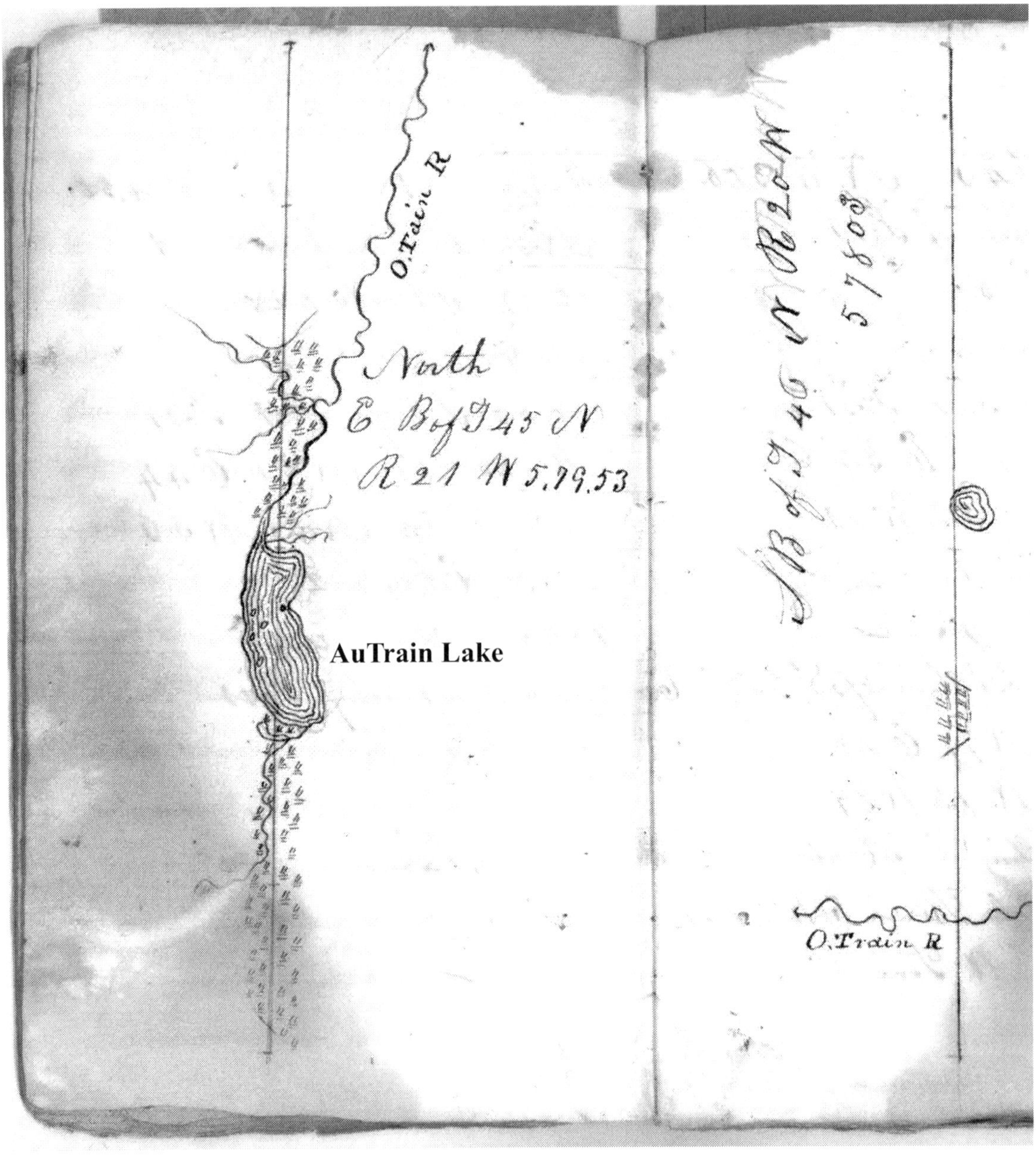

Burt's drawing of the area of Au Train River and Lake T45N, – R21W

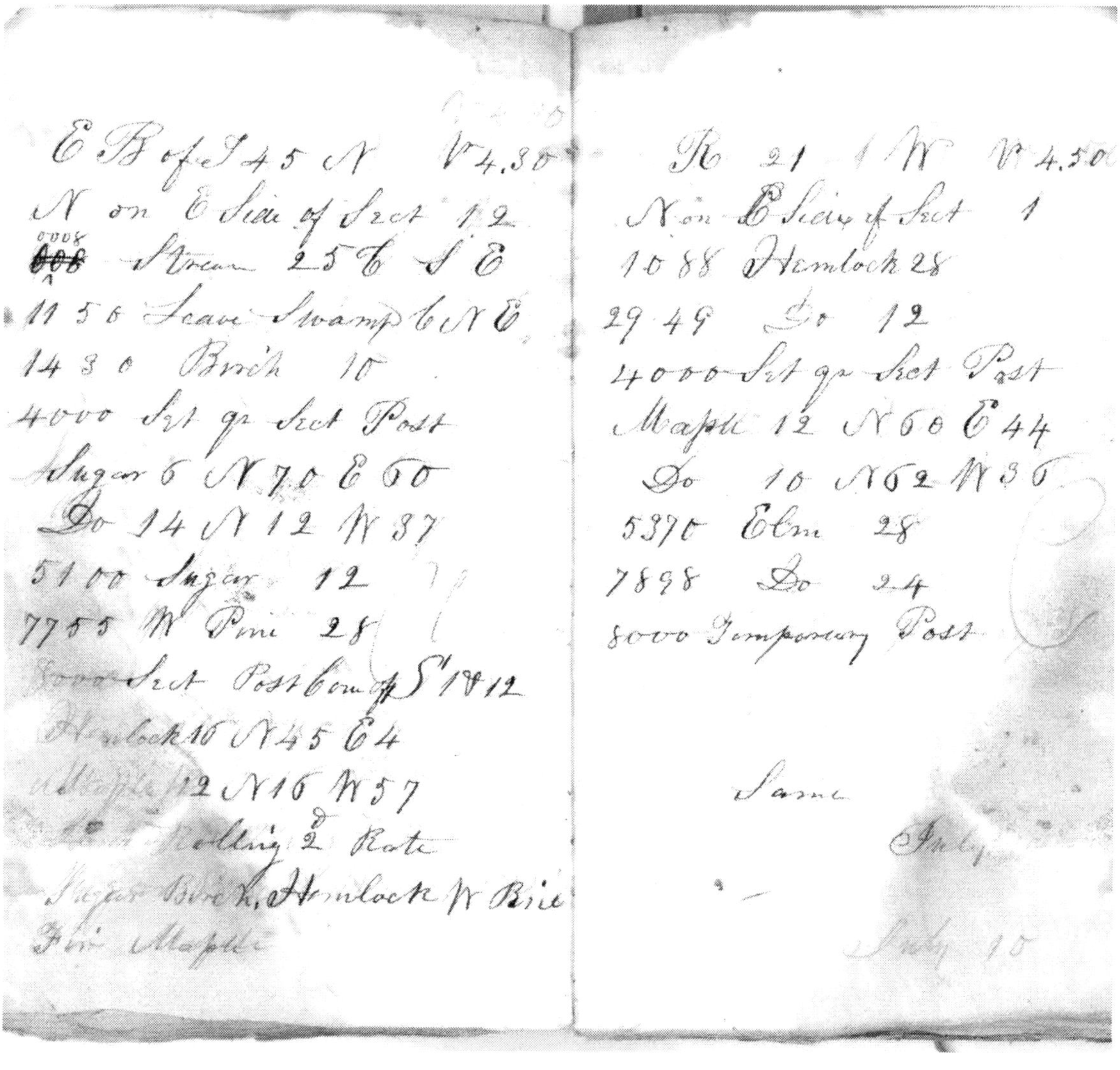

E B of T 45 N V 4.30
N on E Side of Sect 12
~~008~~ 0008 Stream 25 E S E
1150 Leave Swamp E N E
1430 Birch 10
4000 Set qr Sect Post
Sugar 6 N 70 E 60
Do 14 N 12 W 37
5100 Sugar 12
7755 W Pine 28
8000 Set Post Com of S 1 & 12
Hemlock 16 N 45 E 4
Maple 12 N 16 W 57
Land Rolling 2d Rate
Sugar Birch, Hemlock W Birch
Fir Maple

R 21 W V 4.50
N on E Side of Sect 1
1088 Hemlock 28
2949 Do 12
4000 Set qr Sect Post
Maple 12 N 60 E 44
Do 10 N 62 W 36
5370 Elm 28
7898 Do 24
8000 Temporary Post

Same

July

July 10

Burt's description of trees found in area of T45N, – R21W

APPENDIX
Chapter VIII
Item 7

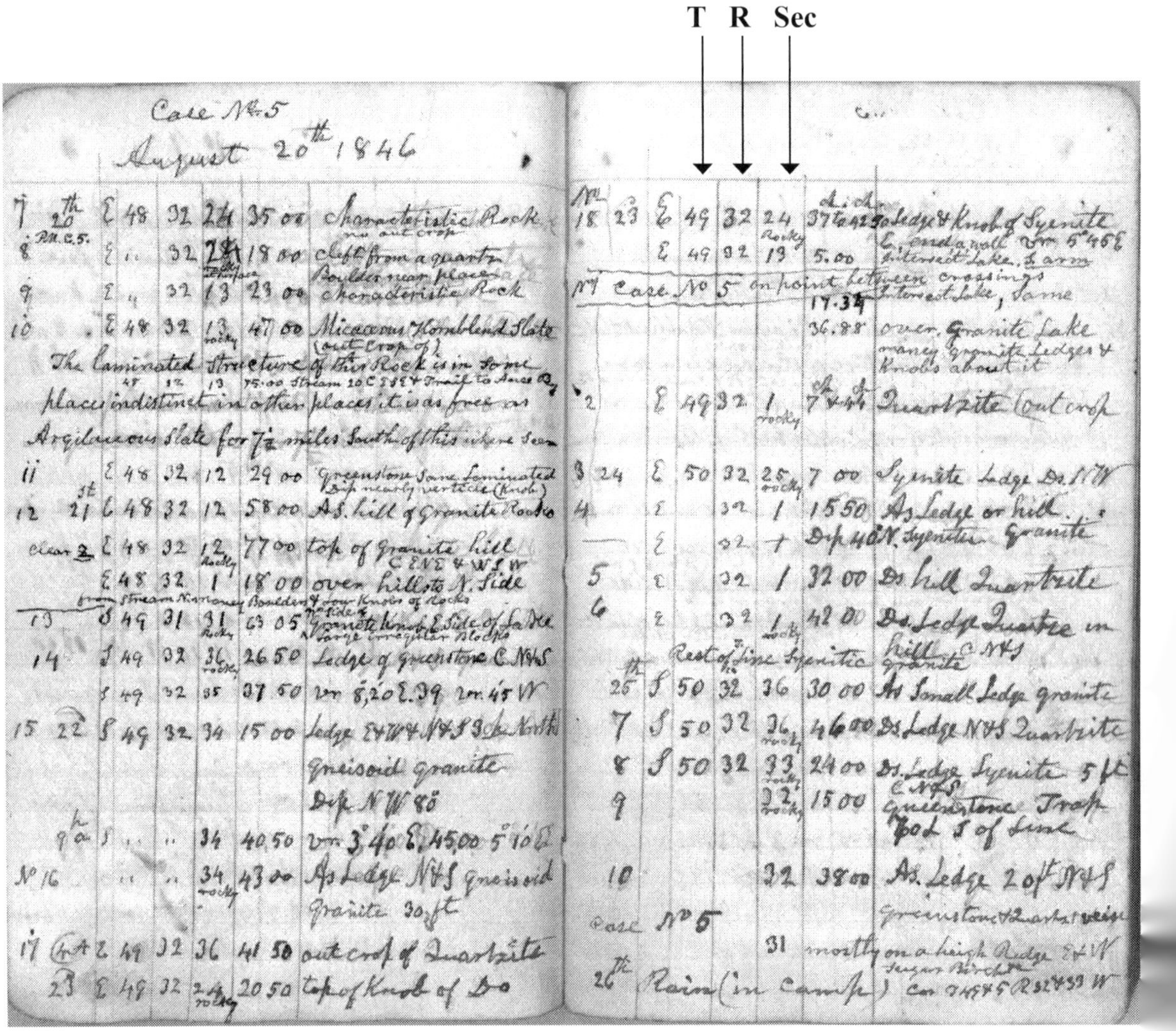

William Burt's description of geology of the area to the west of Marquette in 1846 Areas surveyed are in T44 through T50, beginning with L'anse and extending two-thirds the way between Lake Superior and Lake Michigan. The Upper Peninsula is an extremely diverse geological region. Rock formations include: Sandstone, quartz, micaceous, hornblende slate, argillaceous slate, greenstone, gneissoid granite, syenite granite, iron ore, and mica.

APPENDIX
Chapter VIII   Item 8

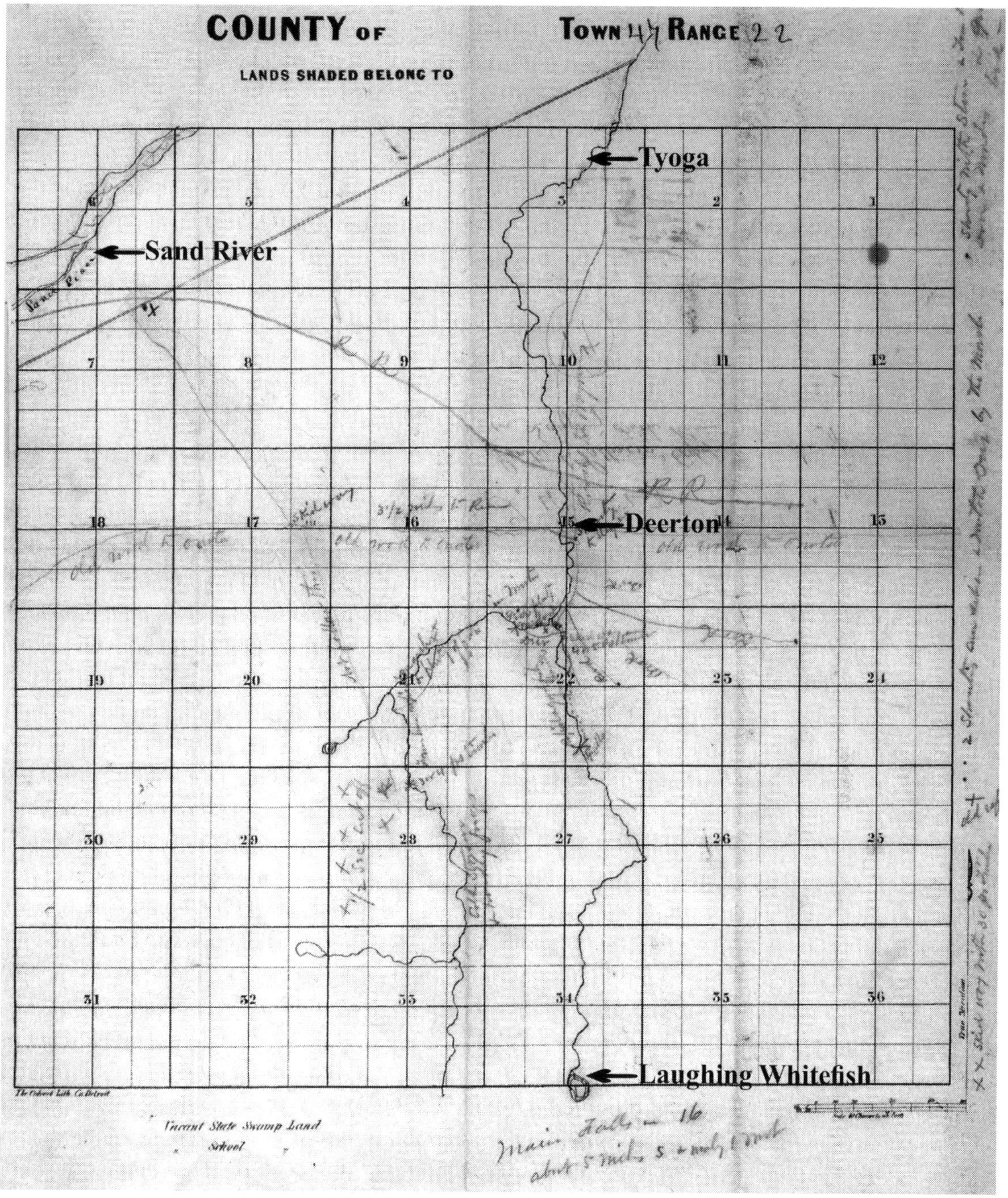

Maps of Township 47 and 46 North – Range 22 West
A.H. Matthews' law office (Marquette) – 1880's Township 47 includes Deerton
Township 46 includes Laughing Whitefish Falls and valley with Laughing Whitefish Lake on the border between T47 and T46.

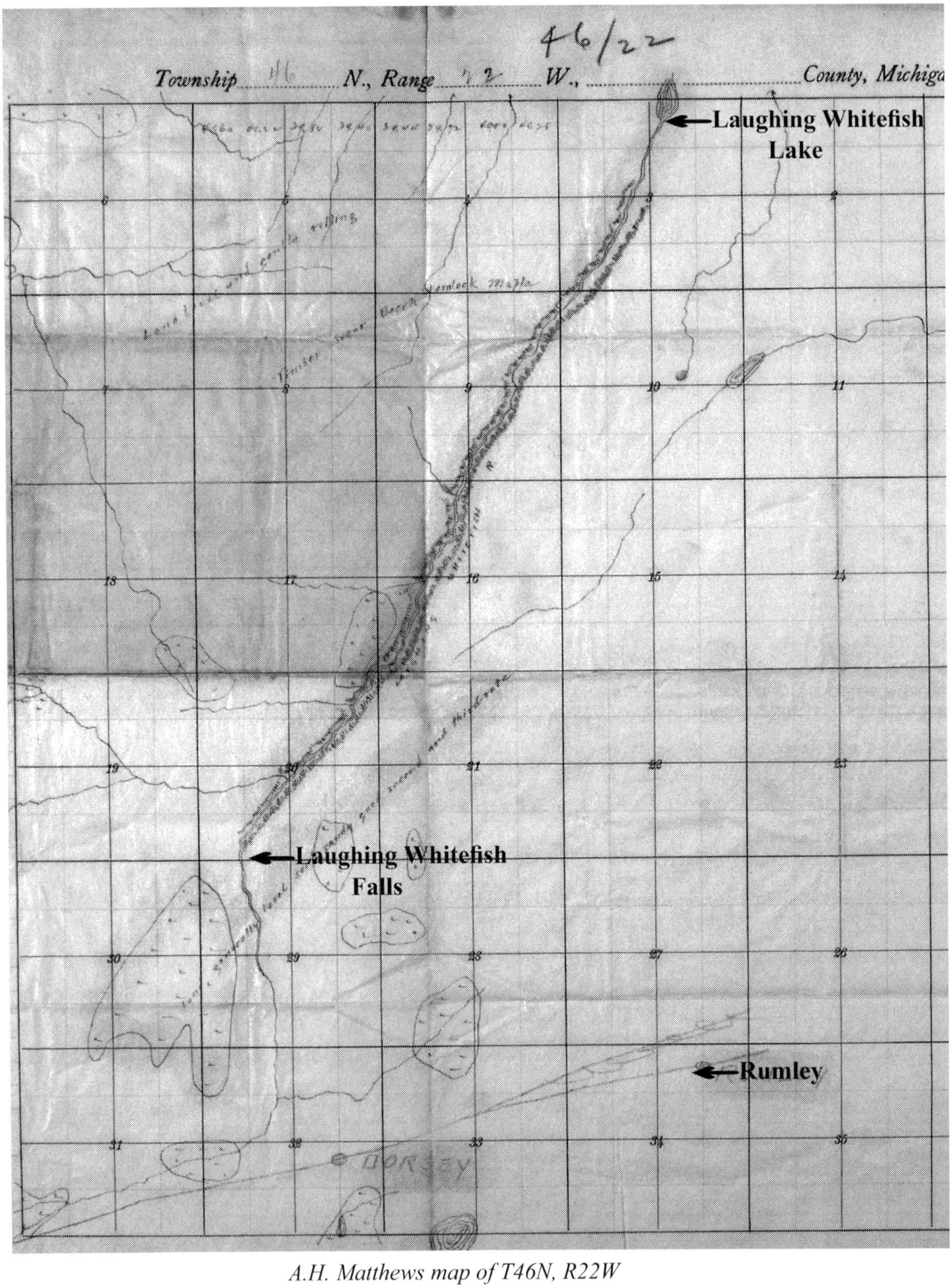

A.H. Matthews map of T46N, R22W

APPENDIX
Chapter VIII
Item 9
Map by C.A. Baldwin ca. 1880

T48, R22

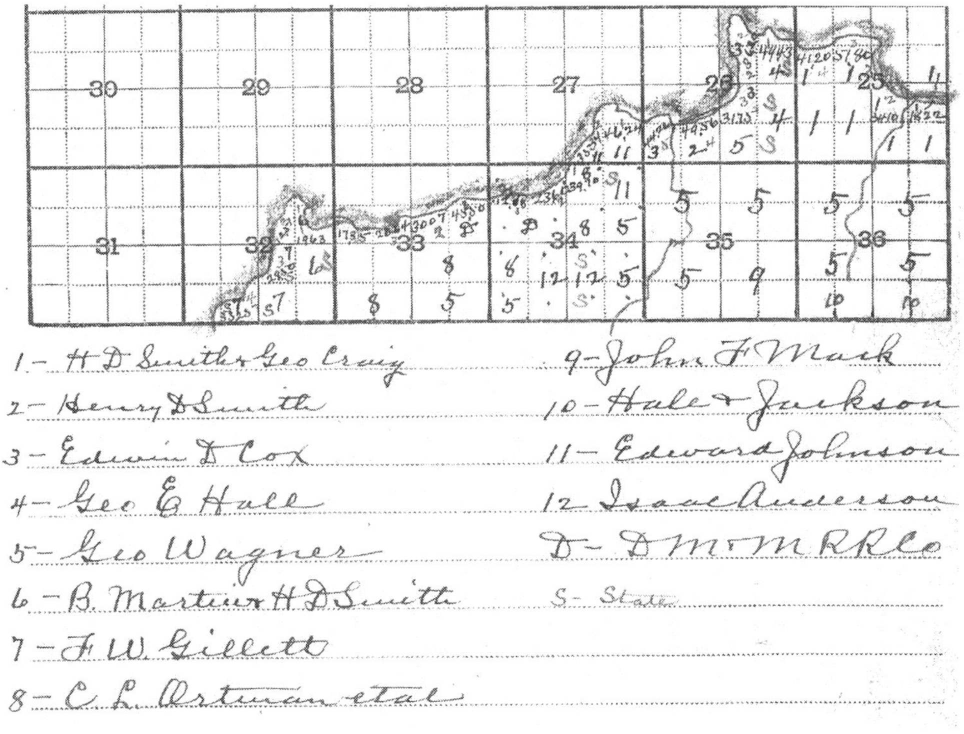

Courtesy Alger County Title Company

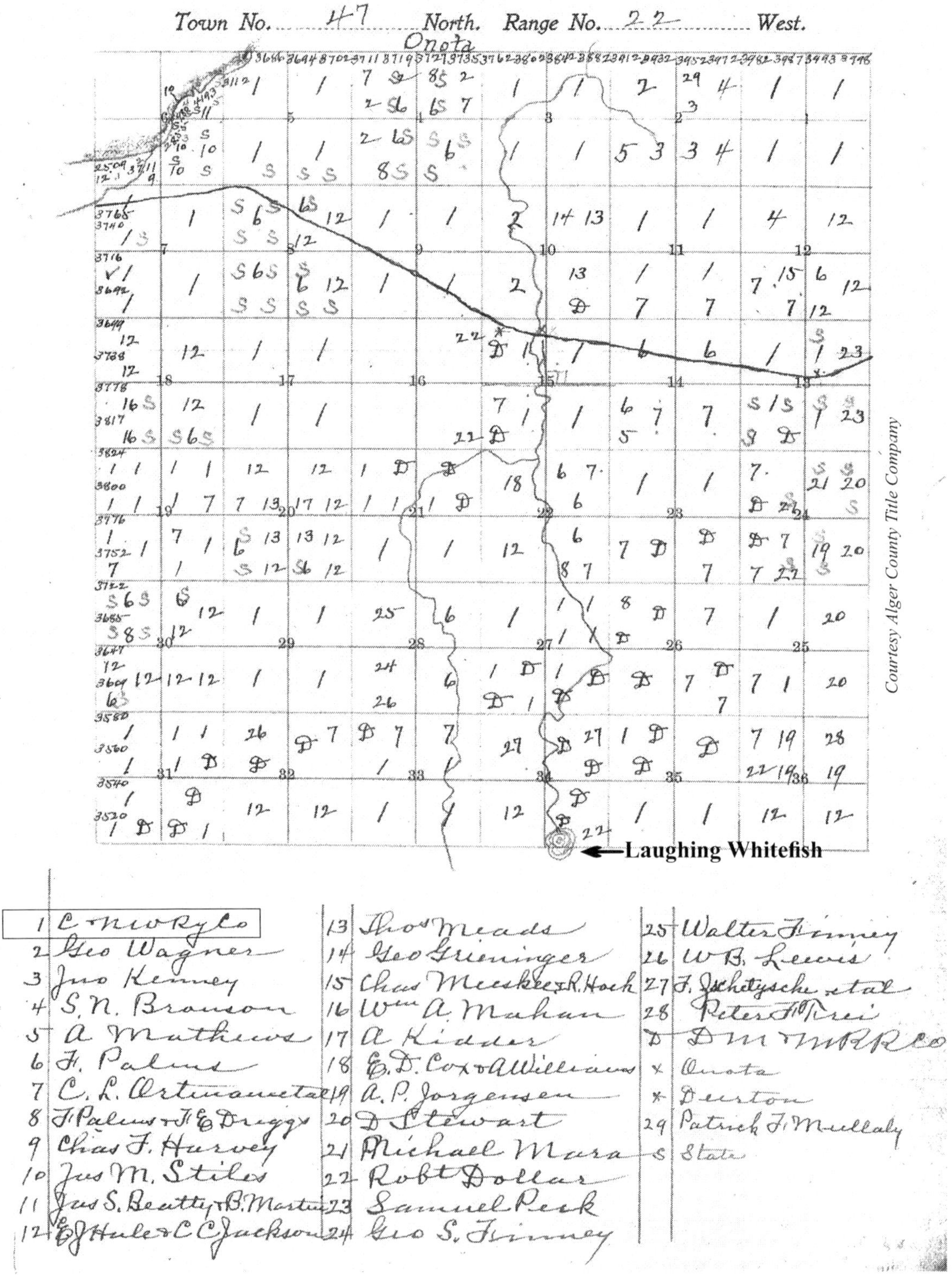

Courtesy Alger County Title Company

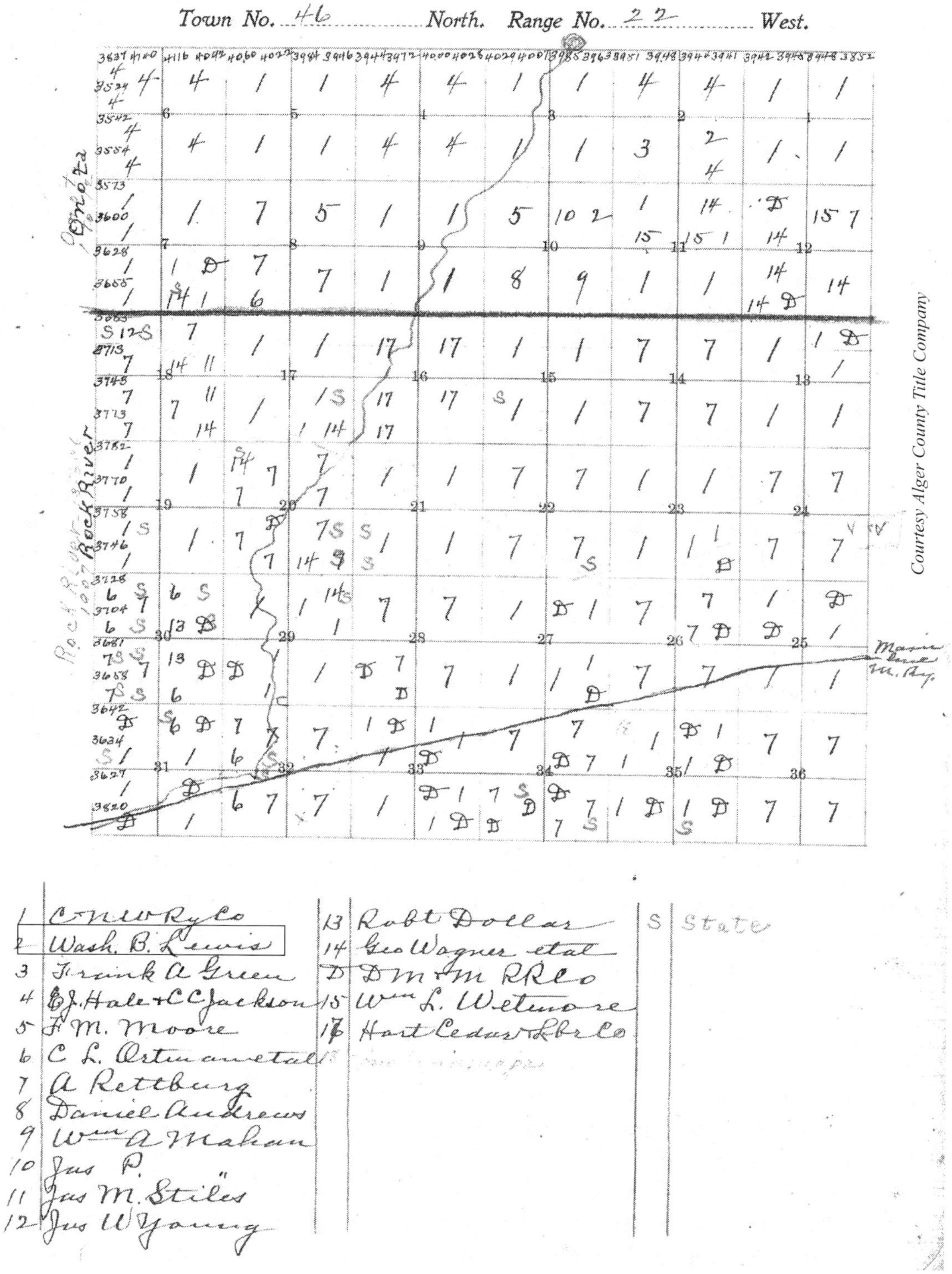

Courtesy Alger County Title Company

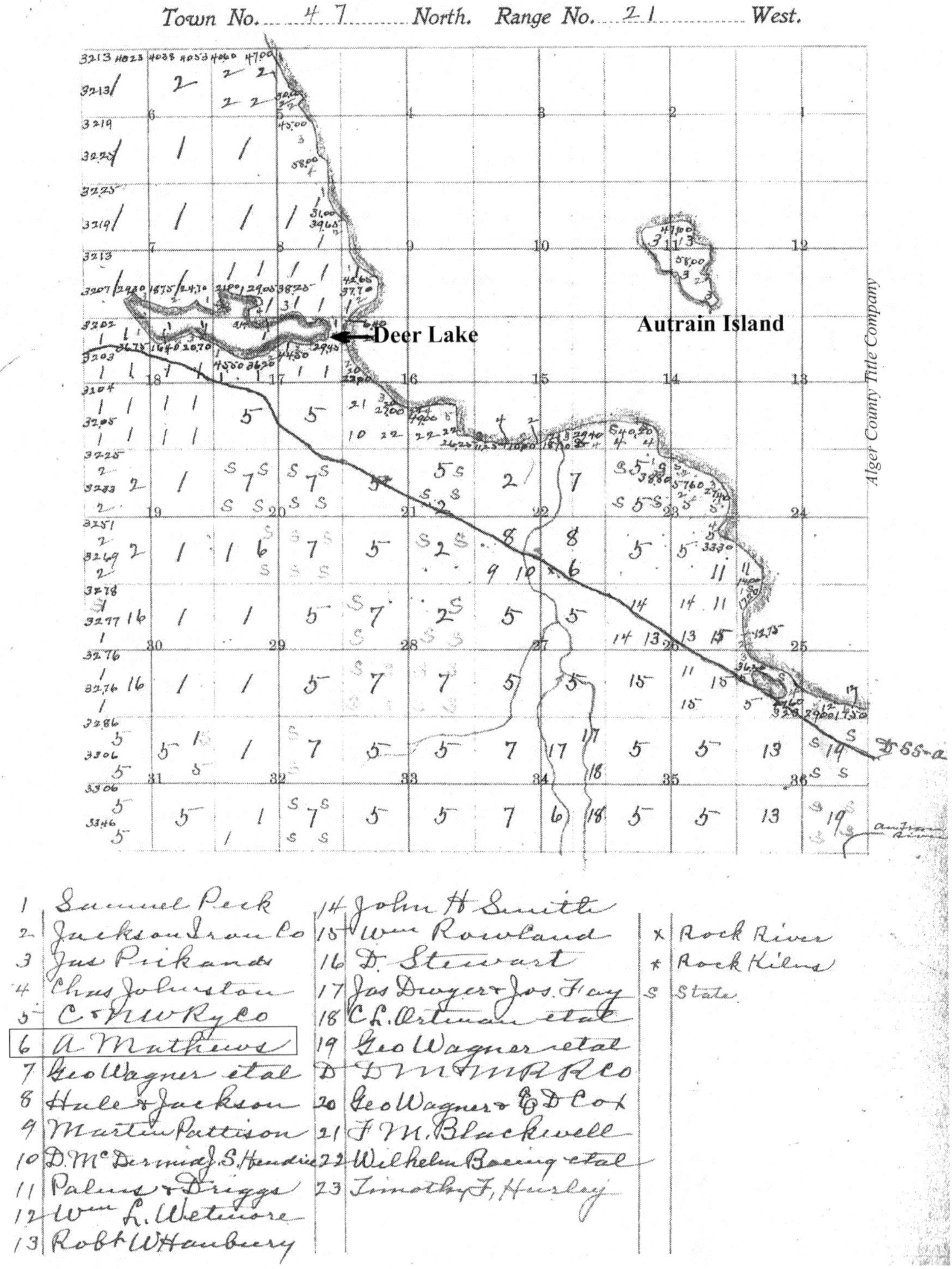
Town No. 47 North. Range No. 21 West.
Deer Lake
Autrain Island
Alger County Title Company
1 Samuel Peck
2 Jackson Iron Co
3 Jas Pickands
4 Chas Johnston
5 C & N W Ry Co
6 A Mathews
7 Geo Wagner etal
8 Hale & Jackson
9 Martin Pattison
10 D. McDermid J. S. Hendrie
11 Palms & Driggs
12 Wm L. Wetmore
13 Robt W Hanbury
14 John H Smith
15 Wm Rowland
16 D. Stewart
17 Jas Dwyer & Jos. Fay
18 C. L. Ortman etal
19 Geo Wagner etal
D D M & M R R Co
20 Geo Wagner & E D Cox
21 F. M. Blackwell
22 Wilhelm Boeing etal
23 Timothy F. Hurley
x Rock River
* Rock Kilns
S State

APPENDIX
Chapter VIII
Item 10

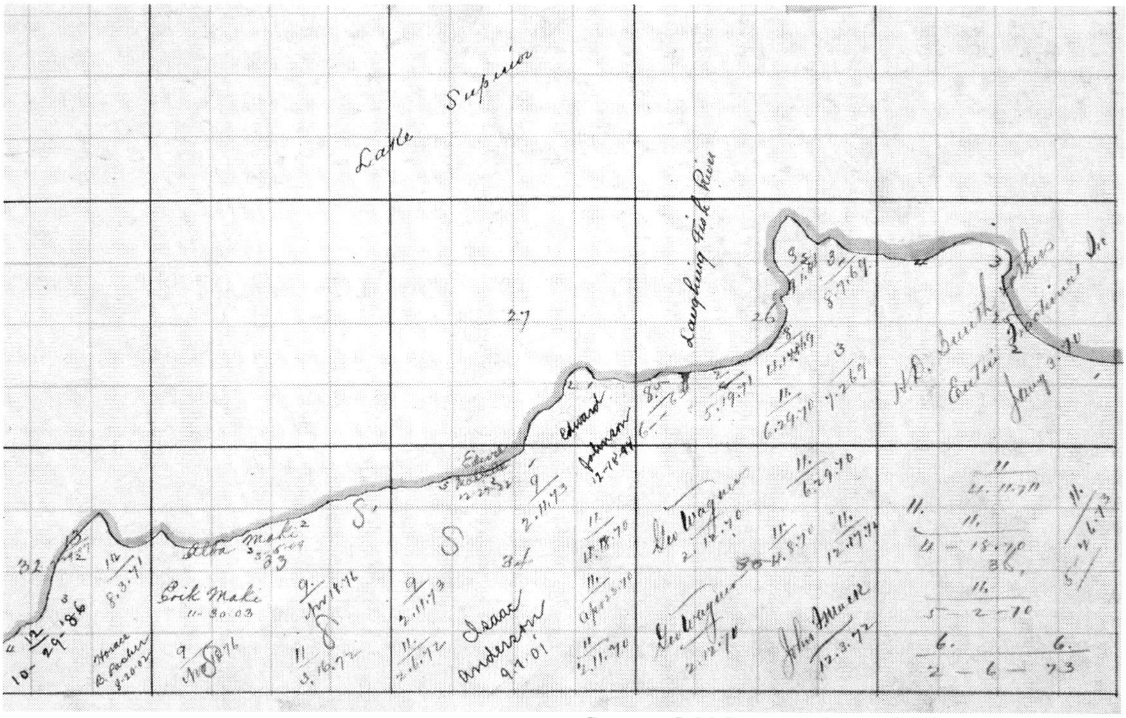

Courtesy J. M. Longyear Research Library

Map of Township 47 North – Range 22 West A. H. Matthews', 1870-1901
The owners up to 1901 are superimposed on the original 1841 Burt map. George Wagner owned the western half of section 35 which included the area of the Tyoga Lumber Company.

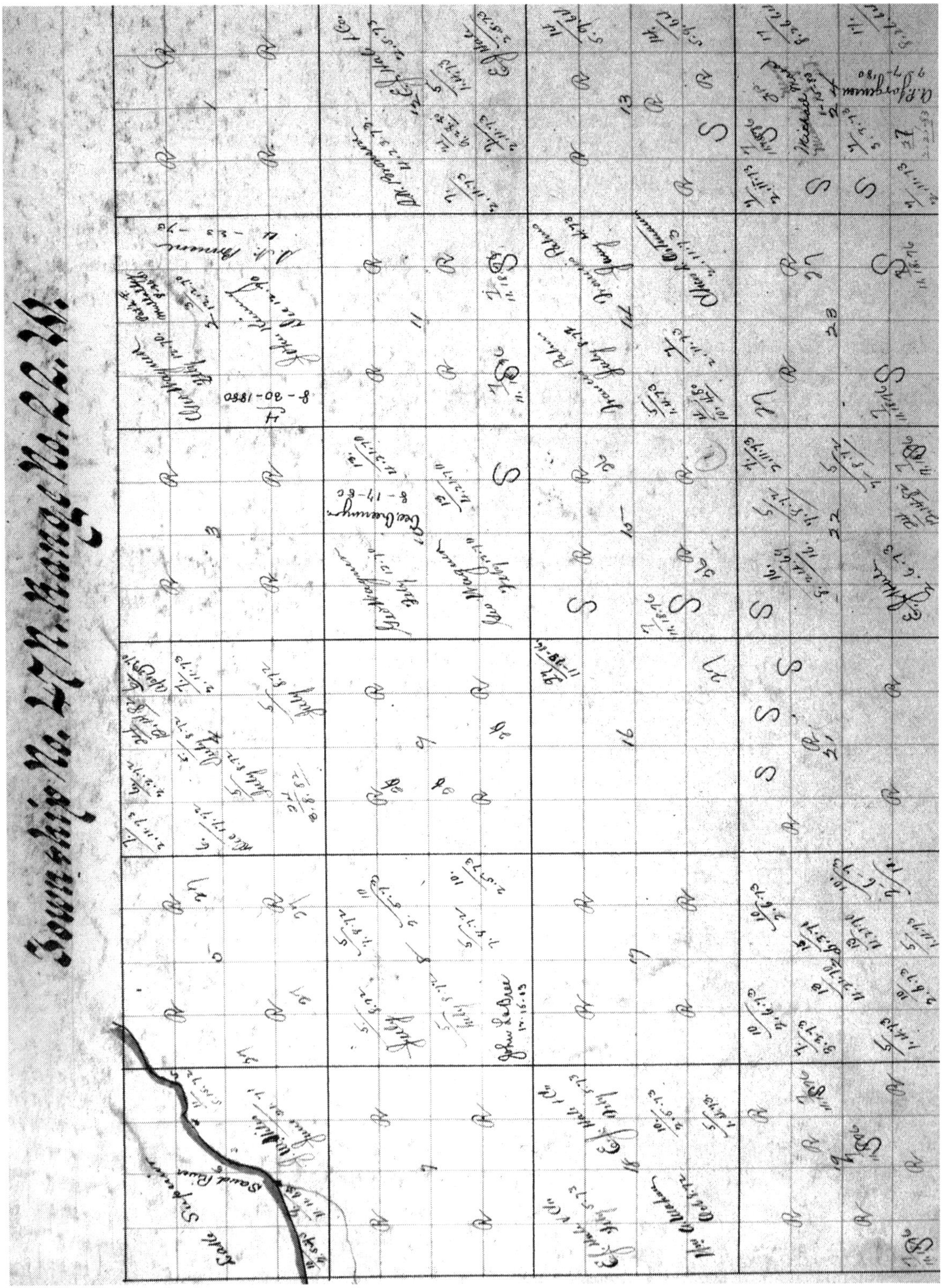

Courtesy J. M. Longyear Research Library

DSS&A labeled as "A"

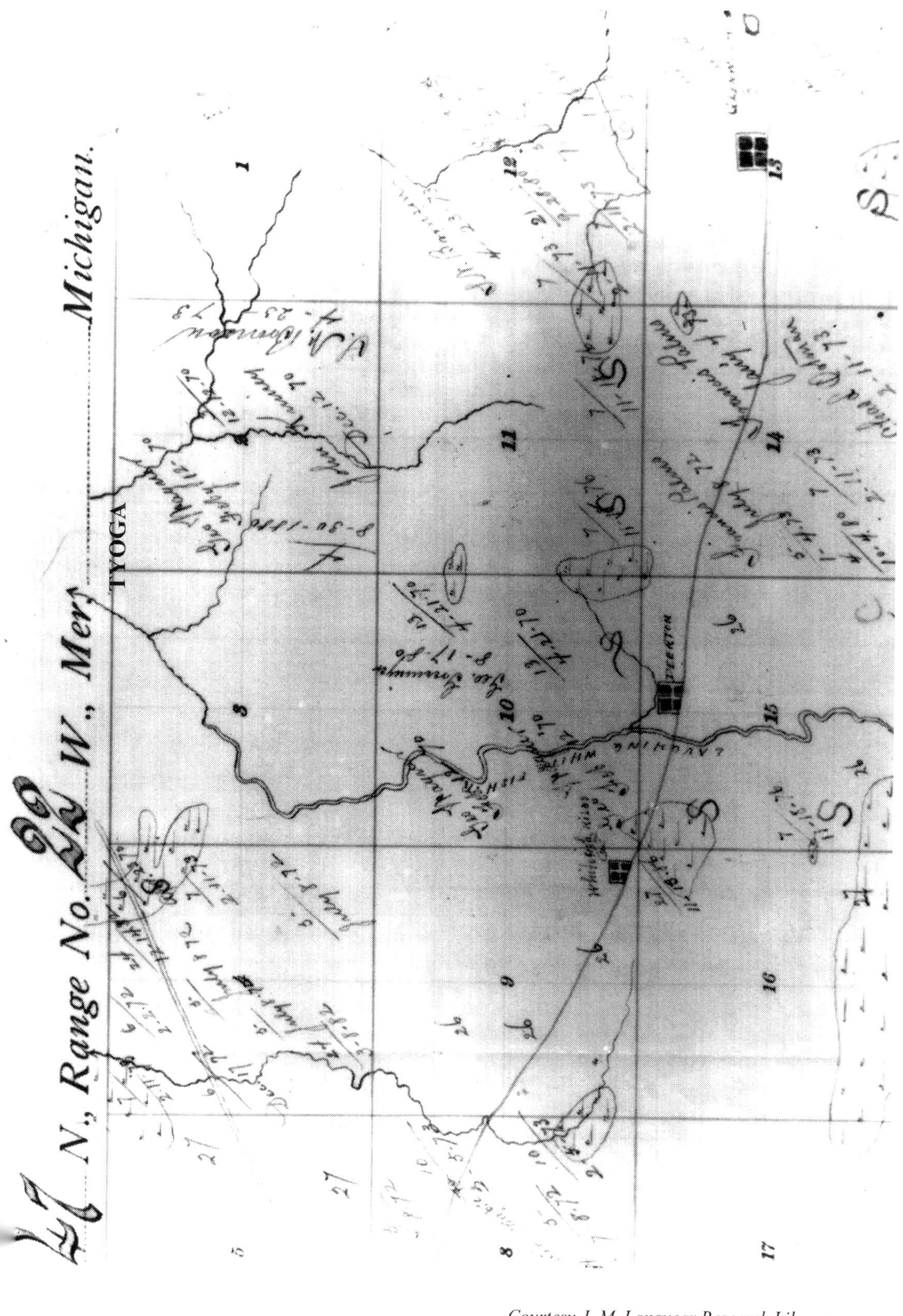

Courtesy J. M. Longyear Research Library

APPENDIX
Chapter VIII
Item 12
Logging camps in Eastern Alger and Western Schoolcraft Counties, 1910-1930

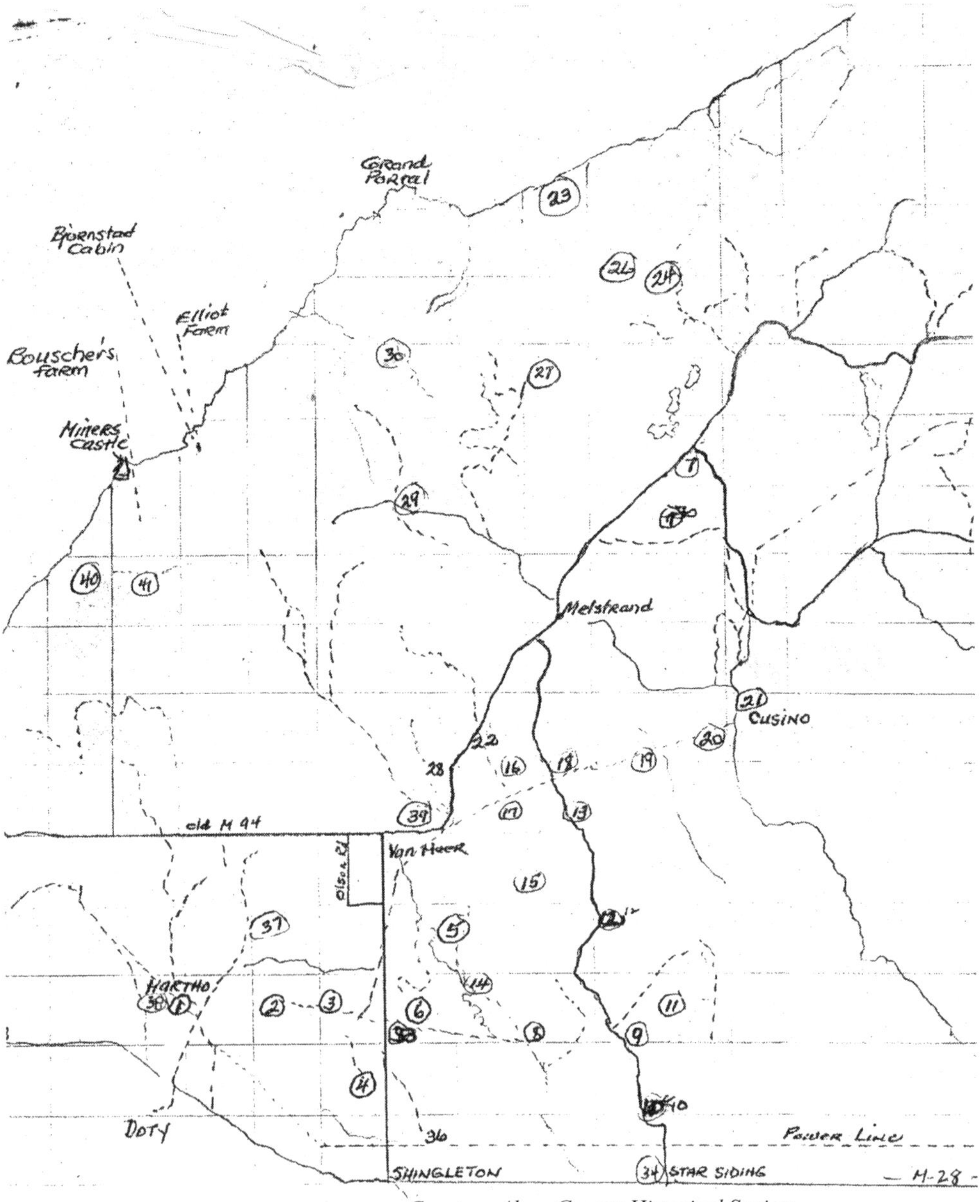

Courtesy Alger County Historical Society

NAMES OF LOGGING CAMPS
Eastern Alger and Western Schoolcraft Counties
Most after 1900-1930
Legend

1. Original Hartho
2. Chapman
3. Percy
4. Camp 19 (Niemi)
5. Sam Johnson Farm
6. Master's Lumber Camp
7. Camp I – Art Bower
8. Metzer's Camp III
9. Camp II – Tom Dolaskie
10. Petrel
11. Fred Blank Camp
12. Linne Finn Camp
13. Mummy Bros. Camp
14. Mud Cordwood Camp
15. Foot's Lumber Camp
16. St. Claire's Camp
17. Myren Branch
18. Higgins Camp
19. Bouscher's Camp
20. LRue's Camp
21. Original Cusino – Mill
22. Original Frost's Camp (later William's Crossing)
23. Jack Frost Camp
24. Webster's (No Buck) Camp
25. Harris Beaver Creek Camp
26. Earl Hall, Logging
27. Micheau's Camp
28. Harry Travis – Logging
29. Camp 15 – Tom Dolaskie (1927-1928)
30. Lung Camp – CCI
31. Guy Burrell 1926
32. Guy Burrell – 1928 (later Fournier's – S.E. of Shingleton)
33. Home Bill Groh (Traveling camp photographer)
34. Star Siding
35. Clement's Camp
36. Skinner's Camp
37. Orson Livermore
38. Kraus Jobbed
39. H. North Camp
40. Hank North Camp (CCI)
41. Winn Camp (CCI)

APPENDIX
Chapter VIII
Item 13

List of area rivers, when they were first drawn, labeled, and labeled current term designations

	Chocolay River	Sand River	Laughing Whitefish River	Rock River	Au Train River	Grand Island
1650 N. Sanson (internet)	ND	ND	ND	ND	ND	ND
1668 Allouez Map of New France 1610-1791 WHS	ND	ND	ND	ND	ND	Grand Island
1671 Allouez Jesuit Map of LS (internet)	N?	ND	N?	ND	N?	Les Grandes Isles
1673 Jacques Marquette Autograph Map of the Mississippi WHS	ND	ND	ND	ND	ND	Les Grandes Isles

Courtesy John Parlin

ABBREVIATIONS

DN	Drawn but not named
N	Named
N?	Named but unreadable
ND	Not drawn
NI	Area not included

See Bernard Peter's *Lake Superior Place Names* for origin of names of streams

	Chocolay River	Sand River	Laughing Whitefish River	Rock River	Au Train River	Grand Island
1688 Cornell WFL	ND	ND	N?	ND	N?	ND
1831 Burr MSU						
1835 Bradford Michigan and the Great Lakes (internet)	N	DN	DN	ND	DN	Grand Isle
1841 Wm. Burt Chocolay to Grand Marais DNR	N	DN	DN	DN	N	DN
1886 Wm. Burt Revised A. Matthews	NI	N	N	NI	NI	NI
1841 Bela Hubbard Lake Superior Journal PWL	Big Sandy	Little Sandy or Cypress	LFR	Flat or Slippery Rock	Dog River or Montaban	Grand Isle
1847 T.W. Bristol Chart of Lake Superior PWL	Chocolate	Fish	DN	ND	Train	N
1872 Colton Lake Superior SV	Chocolate	Fish	DN	DN	DN	Grand Island
1873 T.B. Brooks Map of the Upper Peninsula PWL	Chocolate	Fish	Au Sable River	Laughing Fish River	DN	Grand Island
1896 N. Codarette Map of the Environ of Marqutte, Nyaun & Ishpining MCHS	NI	N	LFR	NI	NI	NI
1900 Lake Superior Chart #2 Corps of Engineers PWL	N	N	Sable	Laughing Fish River	NI	NI
1910-1920 Platt Map Quote T48NR21T22W MCHS	NI	NI	DN	NI	NI	NI

Courtesy John Parlin

	Chocolay River	Sand River	Laughing Whitefish River	Rock River	Au Train River	Grand Island
1928 Dept of Land Economic Survey MCHS	NI	NI	N	NI	NI	NI
1931 Hinsdale MCHS	NI	NI	DN	N	N	NI
1940 US Forest service Ariel Photo Hiawatha Forest						
1943 Lake Superior Coast Chart #93 PWL	NI	N	DN	NI	NI	NI
1944 NOAA PWL	N	N	N	N	N	N
1983 NOAA PWL	N	NI	Sable	DN	N	NI
1985 Topo 1:24000	NI	N	N	NI	NI	NI
? Lake Superior Watershed Project	N	N	N	N	N	DN
2001 Ariel Mouth of LWR						
2006 Joe Wagner GPS Maps						
Eugene Sinervo Map of the US Drawn at Sand River 1969 ________ son Vincent	ND	N	ND	ND	ND	N

Courtesy John Parlin

BIBLIOGRAPHY

Adams, David Wallace, *Education for Extinction,* University Press of Kansas, Lawrence, 1995

Alger County, A Centennial History 1885-1985, Bayshore Press, Munising, 1986

American Lumberman, Chicago

Anderson, Oliver M., *The Shining Big-Sea-Water*, Bay Shore Press, Munising, 1989

Ashworth, William, *Late Great Lakes*, Alfred & Knopf, NY, 1986

Benton-Banai, Edward, *Mishomis Book of the Voice of the Ojibway*, Indian Country Communications, 1988

Botti, William B. and Moore, Michael D., *Michigan's State Forests: A Century of Stewardship*, Michigan State University Press, East Lansing, 2006

Bryce, Kenyon, *Old Trails & Roads*, WNNU Radio Broadcast 10-3154 PWL - Marquette, Michigan

Burt, John S., *They Left Their Mark – William Austin Burt*, Edward Brothers, Ann Arbor, 1985

Carter, James M., *A Map of Logging Railroads*, Pilot Press, Marquette, 2006

Carter, James M., *Historical Guide to Seney Stretch*, Pilot Press, Marquette, 2005

Castle, Beatrice H., *Grand Island Story*, John M Longyear Research Library, Marquette, 1974

Cleland, Charles E., *Rites of Conquest*, University of Michigan Press, Ann Arbor, 1992

Cranston, J. H., *Etienne Brüle – Immortal Scoundrel*, Ryerson Press, Toronto, 1949

Crowe, Wm. S., *Lumberjack*, North County Publishing, Skandia, 2002

Densmore, Frances, *Chippewa Customs*, Minnesota, Historical Society Press, St. Paul, 1979

Foster, J. W. and Whitney, J. D., *Report on Geology of Lake Superior*, Printed for House of Reps,1851

Graham, Loren R., *Face in the Rock*, University of California Press, Berkley, 1998

Harpers New Monthly Magazine, Pictured Rocks of Lake Superior, Vol. 34 Issue 204, Cornell Library, Ithaca, 1967

Hinsdale, W. B., *Primitive Man in Michigan*, University of Michigan Press, Ann Arbor, 1925

Hinsdale, Wilbert, *Archeological Atlas of Michigan*, University of Michigan Press, Ann Arbor, 1931

Hubbard, Bela, *Lake Superior Journal, Account of the 1840 Houghten Expedition;* Northern Michigan University Press, Marquette, 1983

Hulse, Charles A.,_*An Archaeological Evaluation of Rock Kilns and Finn Hollow*, MSA, Survey report #17, East Lansing, 1977

Hyde, Dayton O., *Thunder Down the Track*, Macmillan, New York, 1986

Karamanski, Theodore, *Deep Woods Frontier*, Wayne State University Press, Detroit, 1989

Karpinski, Louis, *Bibliography of the Printed Images of Michigan, 1804 – 1880*, Michigan Historical Commission, Lansing, 1931

Kidder, Homer H., *Ojibway Narration*, Wayne State University Press, Detroit, 1994

Kohl, J. G., *Discovery of America*, Chapman & Hill, London, 1862

Kohl, Johann Georg, *Kitchi-Gami – Life Among the Lake Superior Ojibway*, Minnesota Historical Press, St. Paul, 1885

LaFayette, Kenneth D, *The Way of the Prairie: Forest Industries of Marquette County during the White Pine era, 1878 - 1912,* printed in Marquette, 2008

Larson, Amanda, *Finnish Heritage in America*, Bicentennial pub of Delta Chapter, Delta Kappa Gamma Society, Marquette, 1976

Legislative Manual of State of Michigan 1895-6, Robert Smith & Co, Lansing

Lindquist, Fred, *They Came to Laughing Whitefish*, Saari Press, Marquette, 1978

Magnaghi, Russell, *Guide to Indians of Michigan's Upper Peninsula. 1621-1900*, Northern Michigan University, Marquette, 1984

Magnaghi, Rusell, *Native Americans of Michigan's Upper Peninsula*, Northern Michigan University, Marquette, 2009

Martin, Helen M., *Ne-Saw-Je-Won – A Tale of the Waters that Run Down from Lake Superior to the Sea*, Wm. Feather Co., Cleveland, 1939

Martin, Susan R., *Wonderful Power – The Story of Ancient Copper Working in the Lake Superior Basin*, Wayne State University Press, Detroit, 1999

Mason, Philip P., *Schoolcraft Ojibwa Lodge Stories*, Michigan State University Press, East Lansing, 1997

Mukkala, Ben, *Copper, Timber, Iron, and Heart*, Still Waters Publishing, Marquette, 2003

Monette, Clarence J., *History of Jacobsville and Other Sandstone Quarries*, University of Michigan, Ann Arbor 1976

Oliver, *"Moccasin" Bill, Diary of –1874 to 1931*, private collection

O'Rourke, Dan, *Rock Kilns*, Michigan Technical University, Houghton, 1997

Peters, Bernard C. *Lake Superior Place Names: From Bawating to the Montreal,* Northern Michigan University Press, Marquette, 1996

Paul, Helen Longyear, *Landlooking into Upper Peninsula of Michigan*, J. M. Longyear Library Press, Marquette, 1983

Penney, Charles W., *North to Lake Superior*, J. M. Longyear Library Press, Marquette, 1976

Perron, Wesley, *Railroad Collection*, Peter White Library, Marquette

Powell, J. W., *Bureau of American Ethnology 1896-7*, Washington Government Printing Office, 1899

Practical Home & School Education Vol VI, Dixon-Hanson Co., Chicago, 1910

Reimann, Lewis C., *When Pine was King*, Avery Color Studios, Au Train, 1981

Rydholm, C. Fred, *Michigan Copper – The Untold Story*, Winter Cabin Books, Marquette, 2006

Rydholm, C. Fred, *Superior Heartlands, A Backwoods History*, Brown-Brumfield, Inc, Ann Arbor, 1989

Schoolcraft, Henry R., *History of Indian Tribes of the U.S.*, Historical American Indian Press, J.P. Lippincott, Philadelphia, 1857

Sheldon, E. M., *Early History of Michigan*, A. S. Barnes, New York, 1856

Shiras III, George, *Hunting Wild Life with Camera and Flashlight*, National Geographic Society, Washington, D.C., 1898

Symon, Charles, *Who Were Those People*, Alger Co. Historical Society, Munising, 1984

Traditional Ojibway Resource in Western Great Lakes, Bureau of Applied Research in Anthropology, University of Arizona (Tuscon), 2000

Vogel, John N., *Great Lakes Lumber on the Great Plains – The Laird, Norton Lumber Company in South Dakota*, University of Iowa Press, Iowa City, 1992

Warren, George Henry, *The Pioneer Woodsman as is Related to Lumbering in the Northwest*, Press of Hahn & Harmon Company, Minneapolis, 1914

Warren, William W., *History of Ojibway People*, Minnesota Historical Society Press, St. Paul, 1984

Whittlesey, Charles, *Ancient Mining on the Shores of Lake Superior*, Smithsonian, Washington DC, 1862

Williams, Ralph, *The Honorable Peter White*, Freshwater Press, Cleveland, 1986

INDEX

H

I

J

K

L

M

N

O

ACKNOWLEDGEMENTS

Thanks to all those friends who have contributed in many ways:

Bill Botti, Martin and Delores Dolan, Tom Figgins, Robert Graves, Sr., Tina Hall, Shawn Hayden, Tim Hayden, John Hermann, Don Johnson, Tim Johnson, Clint Jones, Marvin June, Cheryl Lane, Carl Lindquist, Carl Linna, Toivo Maki, Andy Mika, Don Paajanin, Paul Petosky, Al Pierce, Terry Pihlainen, Phyllis Reynolds (dec. 2010), Terry Reynolds, June Rydholmn, Vincent Sinervo, Bill and Mary Soderna, Linda Sturvist, JoanTouranini, Bruce Ventura, Joe Wagner, Family and Friends

A special thanks to those helping with research in making this amateur effort somewhat more professional by suggestions for content and editing:

Alger Title Agency;

Bill Brondyke, Michigan DNR, Gwinn Office;

Jean Brown, Archivist of Alger County Historical Society;

Jim Carter, Editor of Chapter VII;

Jack Deo, Superior View Studios, Marquette;

Echoe Deibert, Manuscript Editing and Author of Essay;

Ginny Forman, Manuscript Editing;

Scott Hickman, Birder;

Catherine Kimar, *Trees of Tyoga Historical Pathway* brochure;

Delores LeVecque, Editor of Chapter II;

Rosemary Michelin, J.M. Longyear Library in the Marquette Regional History Center (formerly Marquette County Historical Society);

Marcus Robyns, Northern Michigan University Archives;

June Rydholm;

Bruce Sarjeant, Olsen Library, NMU;

Snowbound Books, Adrian, Diane, and Ray;

Bruce Veneberg, DNR Retired, Author of Chapter IV;

Peter White Library Reference Staff, Caroline Jordan and Diana Menhennick;

Jack O'Mally, editor, wordsmith, and author of preface.

Last but by no means least, thanks to my Jackie Wilkens, dear friend and transcriptionist for thirty-four years at Clermont Mercy Hospital, Batavia, Ohio, who typed and re-arranged this manuscript more times than I would like to admit. The quality of her work approaches perfection. Stacey Willey and her associates at Globe Printing always had the answer to questions about layout, the result being a more professional look for this manuscript. My wife, Tory, has always been kind and considerate of all my various projects with invaluable constructive criticism.

ABOUT THE AUTHORS

Bernice (Bea) Parkkila Anderson, born in Chatham, Michigan, left home at age 18 to work for Pratt and Whitney in Detroit from 1942 through 1944, our own original Rosie-the-Riveter. She then enrolled in the cadet corps of nurses for six months at the University of Michigan and at the Women's Hospital in Detroit for a six month clinical experience. She was awarded an honorary BSN from NMU in 1972. Bea lived in Marquette together with her family, including sons Bill, Dave, and John. Bea died in September 2011.

John Parlin received a BA (History) in 1965 Trinity College, Hartford, Connecticut. Following medical school and specialty training in radiology at the University of Cincinnati between 1965 and 1973, he practiced general diagnostic radiology at Clermont Mercy Hospital from 1973 through 2007. He became a homeowner in Alger County in 1990. He and his wife, Tory, spend as much time as possible in the U.P.

Mike Zuidema, DNR (Ret), was born in Muskegon and received a Bachelor's Degree in Forest Management from Michigan Tech (Houghton, Michigan) in 1964. Following college, he was employed by the South Dakota Game, Fish, and Parks from 1964 through 1966, and then took a position as U.S. Forest Service ranger in Munising (1966). The remainder of his career was with the Michigan DNR in both Ishpeming and Gwinn offices between 1967 and 1997. Mike lives with his wife, Deena, and family in Escanaba. He continues his interest in cougar sightings in Michigan. He also enjoys ice hockey, occasionally taking to the ice rink himself.

Trailhead of Tyoga Historical Pathway
Zuidema, Parlin, and Anderson
beginning a workday at Tyoga in 2009

Made in the USA
Monee, IL
08 November 2026

68007854R00177